EXS 68

Conservation Genetics

Edited by V. Loeschcke
J. Tomiuk
S. K. Jain

Birkhäuser Verlag
Basel · Boston · Berlin

Editors

Dr. V. Loeschcke
Institute of Ecology and Genetics
University of Aarhus
Ny Munkegade, Building 540
DK-8000 Aarhus C
Denmark

Dr. J. Tomiuk
Section of Clinical Genetics
University of Tübingen
Wilhelmstr. 27
D-72074 Tübingen
Germany

Dr. S. K. Jain
Department of Agronomy
and Range Sciences
University of California
Davis, CA 95616
USA

Library of Congress Cataloging-in-Publication Data
Conservation genetics
/ edited by V. Loeschcke, J. Tomiuk, S. K. Jain.
—(EXS; 68)
Includes bibliographical references and index.
ISBN 3-7643-2939-4—ISBN 0-8176-2939-4
1. Germplasm resources—Congresses. 2. Nature conservation—Congresses. 3. Population genetics—Congresses.
I. Loeschcke, V. (Volker), 1950–. II. Tomiuk, J. (Jürgen) III. Jain, Subodh K., 1934–.
IV. Series.
QH75.A1C6655 1994
639.9′01′5751—dc20

Deutsche Bibliothek Cataloging-in-Publication Data
Conservation genetics / ed. by V. Loeschcke. . .—Basel;
Boston; Berlin: Birkhäuser, 1994
(EXS; 68)
ISBN 3-7643-2939-4 (Basel . . .)
ISBN 0-8176-2939-4 (Boston)
NE: Loeschcke, Volker; GT

© 1994 Birkhäuser Verlag, PO Box 133, CH-4010 Basel, Switzerland
Printed on acid-free paper produced from chlorine-free pulp
Cover illustration: Jon Fjeldså
Printed in Germany

ISBN 3-7643-2939-4
ISBN 0-8176-2939-4

9 8 7 6 5 4 3 2 1

Contents

vi

Preface

It follows naturally from the widely accepted Darwinian dictum that failures of populations or of species to adapt and to evolve under changing environments will result in their extinction. Population geneticists have proclaimed a centerstage role in developing conservation biology theory and applications. However, we must critically reexamine what we know and how we can make rational contributions. We ask: Is genetic variation really important for the persistence of species? Has any species become extinct because it ran out of genetic variation or because of inbreeding depression? Are demographic and environmental stochasticity by far more important for the fate of a population or species than genetic stochasticity (genetic drift and inbreeding)? Is there more to genetics than being a tool for assessing reproductive units and migration rates? Does conventional wisdom on inbreeding and "magic numbers" or rules of thumb on critical effective population sizes (MVP estimators) reflect any useful guidelines in conservation biology? What messages or guidelines from genetics can we reliably provide to those that work with conservation in practice? Is empirical work on numerous threatened habitats and taxa gathering population genetic information that we can use to test these guidelines?

These and other questions were raised in the invitation to a symposium on conservation genetics held in May 1993 in pleasant surroundings at an old manor house in southern Jutland, Denmark. A group of people, now represented by contributors and editors, had felt the need of newly assessing the role of genetics in conservation biology. We asked contributors to evaluate what has happened in conservation genetics in the last decade and to identify the present and potential future role of genetics in conservation biology.

At the same time, we used the occasion to celebrate the accomplishments of our retiring colleague, coauthor, and friend, Klaus Wöhrmann, to whom this book is affectionately dedicated. It acknowledges his wide ranging interests in genetics, population biology, and nature conservation.

We are grateful to the participants, the chairman, and discussion leaders for sharing their goals and perspectives with us, and to the reviewers of the manuscripts for their prompt and helpful advice. We thank the Danish Research Council of Natural Sciences (grant No. 11-0111-1) and the Faculty of Sciences, University of Aarhus, for financial support. We wish to thank Doth Andersen and Ulla Bording from the Department of Ecology and Genetics in Aarhus for their excellent help toward the preparation of the meeting and for its success at Sandbjerg.

Aarhus, Denmark,   Volker Loeschcke, Jürgen Tomiuk
September 1993   and Subodh Jain

Part I

Genetics and conservation biology

Conservation Genetics
ed. by V. Loeschcke, J. Tomiuk & S. K. Jain
© 1994 Birkhäuser Verlag Basel/Switzerland

Introductory remarks: Genetics and conservation biology

V. Loeschcke[1], J. Tomiuk[2] and S.K. Jain[3]

[1]*Department of Ecology and Genetics, University of Aarhus, Ny Munkegade, Building 540, DK-8000 Aarhus C, Denmark*
[2]*Section of Clinical Genetics, University of Tübingen, Wilhelmstrasse 27, D-72074 Tübingen, Germany*
[3]*Department of Agronomy and Range Sciences, University of California, Davis, CA 95616, USA*

Genes are the basic material of evolutionary changes and, at the same time, hold in memory the records of events that have occurred in the past. Thus conservation of species is also conservation of the result of an evolutionary process manifested in the genetic and genomic structures of a population. Studies of evolution and genetic variation are clearly essential topics for the development of theory and practice in conservation biology.

Population and evolutionary genetics contribute to conservation biology in numerous ways: they provide, firstly, tools to describe genetic diversity within and among populations at different hierarchical levels of genetic organization, and, secondly, a basic theory to understand the mechanisms that bring about evolutionary changes and affect the pattern of genetic variation in natural and artificial populations. This theory then allows us to analyze the evolutionary fate of small populations and to develop management guidelines for the protection of endangered taxa *in situ* and *ex situ*. Most importantly, however, population and evolutionary genetics can significantly contribute to an integrated concept of biodiversity conservation that is needed to define the goals and methods as well as identify the priorities in conservation programs.

Historically, the avoidance of inbreeding and the maintenance of genetic variation were considered as the main issues in conservation genetics. Parallel to the population ecology side of conservation biology that was dominated by the rather static theory of island biogeography, during the 1960s and 1970s population genetics applied rather simplistic and basic rules to conservation biology. Since then conservation genetics has developed into a more dynamic and theoretically rich field, but since the benchmark publications of Frankel and Soulé (1981) and

Schonewald-Cox et al. (1983) no monograph has presented the whole spectrum of developments in the field of conservation genetics of the last decade. Two recent volumes (Orians et al., 1990; Falk and Holsinger, 1991) dealt with the genetic issues with greater emphasis on *ex situ* conservation.

Population genetic theory was originally developed for dealing with animal or plant breeding, for analyzing human pedigrees, and for microevolutionary studies. Using models of variation and selection, it describes conditions for maintaining variation in populations, including a variety of models for subdivided structure, multilocus systems, changing environments, etc. Genetic behavior of small populations was always of special interest in terms of potential harm due to inbreeding, as well as failures to respond to selection due to some overriding effects of genetic drift. More recently, applications in conservation biology sought to describe genetic variation (and therefrom, define population structure and phylogenetic relationships), to provide markers for measuring parameters of gene flow and mating systems, and to infer or predict evolutionary responses to habitat losses, reduced population size, novel mutations, etc. Populations can be genetically monitored for all these changes. However, conservation biology in practice actually emphasizes the following: (1) protection of habitats in reserves for which selection of sites and design often requires little information from genetics; (2) optimal reserve area based on population viability analysis in which both genetics and demographic criteria might be important (Lande, 1988); (3) the fact that species survival plans for rare and endangered taxa often involve an *ex situ* component (e.g., breeding in zoos and reintroduction); this has strong population genetic inputs for making various decisions on maintaining maximum amounts of genetic variation, maximized N_e/N_a (effective/actual population size) ratios and controlled gene flow. Many new theoretical developments call for a new research agenda with more rigorous studies on inbreeding, drift, and metapopulation properties, all with a greater use of ecological parameters (adaptive response, social behavior, population regulation, spatial scales of dispersion). Clearly, the gap between the conceptual role of population genetics and the actual practices will only be filled with greater empirical support and realism.

The present volume grew out of the view that a reassessment of the role of conservation genetics and its contribution to a synthetic theoretical basis of conservation biology is timely and useful. We have focused on genetic issues of conservation biology from an empirical and theoretical perspective and considered both plants and animals in order to address the following questions: (1) Have population genetic advances during the last decade reaffirmed the prominent role of population genetics in conservation biology? (2) What new applications are available by the recent molecular variation assays? (3) How useful do we find older models and applications vis-a-vis these new developments?, and, finally, (4) can we make good prognoses of the critical research agenda for the next 10–15 years? We believe that various chapters in this book give an unequivocal affirmative answer to the first question and we summarize the answers to the other questions in the concluding remarks.

An introductory discussion by Vida provides a global perspective on species conservation. He stresses the utmost importance of preserving genetic variation in a changing world and appropriately emphasizes the need to place genetic issues within the framework of species richness patterns and their losses under continuing environmental degradation.

Part II is concerned with the relationship between genetic variation and fitness and its implications for conservation. Savolainen presents data on pines that show a high differentiation with respect to presumably adaptive quantitative variation, but it is not certain whether heterozygosity at individual marker loci correlates with fitness. Vrijenhoek discusses the relation between genetic diversity and fitness in small populations based on long-term field and experimental studies of desert stream- and pond-dwelling fishes. Here, it is interesting to note that the origin of linkage disequilibrium during population bottlenecks allows multilocus associations to be investigated for certain selection components. Couvet and Ronfort study the potential negative consequences of increasing effective population sizes, a recommended practice in conservation biology, as this may lead to a decrease in individual fitness due to an increase in the mutational load. Gabriel and Bürger discuss the accumulation of deleterious mutations and its effect on individual and population fitness and thereby on population size. The reduction in population size again enhances the accumulation of deleterious mutations and such synergistic interaction may lead to extinction.

Part III is concerned with inbreeding, population structure, and social interactions. Templeton and Read discuss the different meanings of inbreeding and show that they must be kept separated or erroneous management recommendations and evaluations can occur. They review some of the confusion and its consequences that continue to appear in the literature despite Jacquard's (1975) seminal attempt to settle this issue. The use of the different meanings is illustrated with a worked example, the North American captive population of Speke's gazelle. Gliddon and Goudet critically review metapopulation models and their implications for conservation biology. They give an overview of population genetic elements and demographic theory in metapopulation models and clarify the different meanings of effective population size, which is a very basic term in conservation genetics. Hauser et al. analyze in a simulation study fitness effects of inbreeding due to a drop in population size. The results are used to suggest experimental designs aimed at more rigorous inbreeding studies in plants and ways to manage gene flow. Noordwijk studies the combined effects of inbreeding depression and environmental stochasticity on the extinction risk of populations. His results suggest that current population genetic and population dynamic models are inappropriate to predict the extinction risk of small populations, which in a fluctuating environment depends on the social system as well as on the maximum reproductive rate. de Jong et al. report on their work on long-tailed macaques that live in social groups. The hypothesis that the social system and migration patterns of macaques prevent inbreeding depression is tested by computer simulations. The results suggest that formerly protected deleterious alleles in natural populations can cause negative effects when social structures are disrupted, as for instance in zoos. Brakefield and Saccheri summarize a research paradigm, using butterflies as model organisms, for studying the behavior of small populations. Special emphasis is paid to the components of quantitative genetic variation under bottlenecks and subsequent population recovery.

In Part IV molecular aproaches to conservation are presented. Hedrick and Miller review the current information on the MHC (major histocompatibility complex – a well studied multiallelic system) relevant to conservation genetics and discuss the sampling of rare and advantageous alleles, considering transferrin variants in breeding groups of Przewalski's horses. Arctander and Fjeldså use molecular data to identify closely related bird species of Tapaculos which are difficult to distinguish morphologically and share similar habitats. Crozier and Kusmierski discuss procedures for the setting of conservation priorities in the sampling of endangered species on the basis of phylogenetic relationships and advocate to maximize the preservation of genetic information. Witting et al. integrate ecology and phylogeny of species in their method for optimizing the preservation of phylogenetic diversity and minimizing the risk of extinction.

In Part V case studies on natural and experimental populations of plants and animals are presented. Bijlsma et al. report on genetic erosion in natural populations of two plant species, *Scabiosa columbaria* and *Salvia pratensis*, that are threatened by extinction in the Netherlands due to habitat destruction. Hansen and Loeschcke present data and discuss the effects of the release of hatchery-reared trout on the structure of wild brown trout populations. They suggest some procedures for the release into and the management of wild populations. Jain presents a study on endemic Californian plant species of vernal pools in which genetic and demographic data over several years begin to provide some conservation guidelines. Patch dynamics of a colonizing species (rose clover) is discussed in terms of the extinction-recolonization process. Krebs and Loeschcke discuss the adaptability of populations to stress and the fitness costs of acclimation based on field observations and experimental data on geographic populations of *Drosophila buzzatii* that were exposed to high temperatures. Hindar discusses morphological and life history variation in terms of selective forces maintaining such polymorphisms. In particular, he develops some applications in conservation practice for salmonid fish populations. Altukhov provides data on long-term monitoring of natural and agricultural populations and discusses some implications for conservation-oriented breeding strategies.

Part VI gives some contributions to genetic resource conservation. Brown and Schoen address sampling strategies in core collections of crop plants. Hurka considers the role of botanic gardens in conservation biology based on many years of practical experience. Barker discusses conservation genetics in an animal breeding context which relates to questions or preservation of rare and endangered breeds or populations, and to utilization for improving viability, productivity, and efficiency of production.

Furthermore, five scenarios are presented to illustrate a few highly significant topics in conservation practice in which geneticists and ecologists will find a great many interactive research projects. These include population monitoring in social organisms, restoration of habitat, genetic resource collection and evaluation in crop plants, host-pathogen coevolution, and plant-pollinator mutualism. We emphasize some direct applications of population genetics in diverse ways. Finally, in a summary chapter we attempt an overview of the development of conservation genetics in the last decade, and attempt to give an assessment of the current and future role of genetics in a synthetic approach to conservation biology. Following our original queries about the gap between the desired inputs and the actual role of geneticists in conservation practice, we propose some research needs and issue another urgent call for the long-awaited joining of ecological and genetic approaches. They need each other in order to firm up our theses and empirical evidence.

References

Falk, D. A. and Holsinger, K. E. (1991) *Genetics and conservation of rare plants.* Oxford University Press, New York.
Frankel, O. H. and Soulé, M. E. (1981) *Conservation and evolution.* Cambridge University Press, Cambridge.

8

Jacquard, A. (1975) Inbreeding: One word, several meanings. *Theor. Popul. Biol.* 7: 338–363.
Lande, R. (1988) Genetics and demography in biological conservation. *Science* 241: 1455–1460.
Orians, G. H., Brown, Jr., G. M., Kunin, W. E. and Swierzibinski, J. E. (1990) *The preservation and valuation of biological resources.* Univ. Washington Press, Seattle.
Schonewald-Cox, C. M., Chambers, S. M., MacBryde, B. and Thomas, L. (1983) *Genetics and conservation: A reference for managing wild animal and plant populations.* Benjamin-Cummings, London.

Conservation Genetics
ed. by V. Loeschcke, J. Tomiuk & S. K. Jain
© 1994 Birkhäuser Verlag Basel/Switzerland

Global issues of genetic diversity

G. Vida

Department of Genetics, Eötvös Lóránd University, Múzeum krt. 4/a, H-1088 Budapest, Hungary

Summary. Genetic diversity within species is highly significant during their adaptation to environmental changes and, consequently, for their long-term survival. The genetic variability of species is also the basis for the evolution of higher levels of biodiversity, the evolution of species, and it might be an indispensible prerequisite for the functioning of our biosphere. Studies which promote understanding of the maintenance and the functional aspects of biodiversity at any level are therefore essential for the future welfare of mankind.

Introduction

Many ecological textbooks contain vegetational maps in which the areas of the major biomes are shown by different colors. However, the study of most regions on our planet reveals a rather similar surface formation – urban areas, croplands, cultivated forests and grasslands, etc. Only few regions have maintained their original landscape as natural biomes, which is shown on the vegetational map. The rule is that only remnants of original natural communities can be found in any part of our earth and, moreover, it is almost impossible to find an ecosystem which has not been influenced by human activities.

Four thousand years ago the environmental impact of *Homo sapiens* was hardly detectable, but during the last few centuries the environmental changes caused by humans have progressed exponentially. Four thousand years might be a very long period in the history of mankind, but it is almost negligible on the geological timescale. Even the evolution of any biological community has taken a hundred thousand times longer than that of the short history of human activity, which, however, can have significant future consequences for the biosphere. The rapid changes of our biosphere are unique in at least two aspects. First, they are caused by *Homo sapiens*, a single species, which was formerly a normal member of the natural biosphere; second, the change is characterized by a growing proportion of unusual ecosystems which have a substantially lower species diversity with lower genetic variation than most of the natural systems. In other words, there is a global loss of both species and their genetic diversity.

Genetic diversity and survival of species

It cannot be accidental that low biodiversity is so exceptional in Nature. Evolution is a fascinating process which remembers every success, but forgets all failures. A suitable amount of genetic diversity might be such a success. Uniformity can be superior in some particular environment, but it is condemned to extinction in a changing world.

Natural ecosystems are organized in a completely different manner than ecosystems which have changed through human influence. In contrast to natural ecosystems, a reduced diversity of species, each with more or less low genetic diversity, is present in most cultivated areas. Human management of nature often intends to increase the biomass of economically important species. The cultivated species are evolutionarily successful in the short-term because they are successful as long as humans keep their environment fairly constant and stable, while Nature's strategy seems to derive from a thousand million years of experience in improving the adaptability of ecosystems to environmental uncertainties. The "biodiverse" strategy which has been tested by natural conditions is now globally replaced by the human monotony. To date, the enormous amount of evolutionary memory which is engraved into the genetic information of millions of species is becoming drastically reduced. The genetic libraries of a few selected species (the species useful for human purposes) are carefully catalogued while hundreds of other species libraries are extinguished every day.

The evolutionary importance of genetic diversity for the adaptation of species has long been recognized. The simple Darwinian mechanism is capable of explaining the surprisingly intricate abilities of living creatures to cope with their environment. The relevant amount of genetic variation has always been a major issue in population genetics. Fisher's Fundamental Theorem of Natural Selection or Kimura's maximization principle, or Wright's shifting balance theory are all based on the amount and dynamics of genetic variation. This picture of Darwinian evolution was confused to some extent by the so-called neutral theory of evolution which regards the huge amount of genetic variation present in most populations as selectively neutral and, accordingly, gives a very simple solution of amino acid substitution rates in molecular evolution. But we should not forget, however, the adaptiveness of species and the obvious fact that neutral evolution *per definition* has nothing to do with adaptation.

Nevertheless, genetic variation cannot be easily qualified as neutral or adaptive. For a given allelic form a certain environment is selective while another is neutral, as is well known in experimental genetics. Biotic and abiotic environments in nature are much more complex and change with time and space. Strict neutrality is therefore unlikely, although in small populations and with the increasing dominance of stochastic processes neutral models of evolution will adequately approach the reality.

This does not mean, however, that the majority of genetic variation is not relevant to the processes of genetic adaptation. Recent studies (even in the "neutral camp") report evidence of adaptive evolution at the molecular level; moreover, it is acknowledged that present neutral genetic variation can serve as a preadaptation to an unknown environment of the future (Selander et al., 1991; Kimura and Takahata, 1991).

The question of utmost importance is whether our biosphere is prepared to face a global change or not (Solbrig et al., 1992). In other words, do species have in store the suitable gene forms, and do they have enough time to select and rearrange them to maintain adaptedness?

We are witnessing an accelerating global change. While we may appreciate technological developments and the overall improvement in human welfare in many parts of the world, it is becoming increasingly difficult to ignore negative effects on our local, regional, or global environment. Ecologists and economists have long held differing opinions on the state of this changing world (Brown, 1991). The optimistic view is often blinded by science and technology: "We shall overcome every environmental problem." I wish they were right. But to my knowledge, biodiversity is something different. We can restore the polluted Rhine, Thames, or Danube in a couple of years, but how could we bring back millions of species, even neglecting intraspecific genetic diversity? We already know that the genetic information is written into the DNA with the four-letter alphabet, but we have great difficulty in organizing a scientific mega-program to decipher a single genome, "the" human genome, a 3-billion-letter-long message. Lost species, especially the undiscovered ones, are therefore lost forever.

Evolution is capable of restoring species diversity. Catastrophically high extinction rates at the end of the Permian or at the end of the Cretaceous periods resulted in a diminished generic (and very probably also specific) diversity. It took roughly 10 million years to attain the pre-catastrophic degree of diversity (Hoffman, 1987). Thus, if species diversity appears to be important for the functioning of the biosphere, we should take much better care of it.

The future of species diversity is in the genetic diversities of the species. In general, the higher the maintained genetic diversity, the higher the adaptability and, consequently, the survival probability of species in a changing world.

Dynamics of genetic diversity

Forty years have passed since the discovery of the structure of DNA, a discovery gave us a clue to understand the simple, yet miraculous way of storing and replicating genetic information. It is really remarkable how accurately this information can be copied. Thanks to the sophisticated molecular mechanism of replication consisting of several enzymes and other proteins, the copying fidelity is of the order of 10^{-9}/base pair/replication. The amount of DNA/cell varies between 10^6 bp of bacteria to $10^9 - 10^{11}$ bp in higher plants and animals. Consequently, "mistakes" or mutations somewhere in the genome are by no means rare events. An adult higher plant or animal is a result of millions of cell divisions. In order to avoid the high mutational load of DNA replication, most of the animals and, as recently discovered, also the flowering plants (Klekowski, 1988) separate the so-called germ line early in the development. This germ cell line appears to be parsimonious with divisions, reaching the gametes in a few tens or hundreds of divisions. Even so, a species with 10^9 bp genome size (DNA/cell) is unlikely to preserve the very same genetic information from zygote to the meiotic division ($10^{-9} \cdot 10^9 \cdot 10^2 = 100$ bp changes *per* genome).

The rate of mutation in living systems is usually very difficult to estimate. It is also known that the probability of change (mutation) of a DNA base pair can vary enormously along a DNA molecule, and since the structure (and function) of the copying machinery is coded by the DNA, it creates a feed-back loop (positive or negative) on the mutation rate. In addition, a substantial fraction of mutation appears to be independent of replication. On the other hand, single base-pair change is just one kind, albeit a very important one, in the mutational arsenal. Nevertheless, for the sake of simplicity let us assume that only base pair substitutions are occurring with constant rate of 10^{-9}/base pair/replication. We have already seen that this will lead to genetically unique individuals. For dynamic studies therefore, a more suitable unit would be a piece of a genomic DNA, such as the gene or even a single base pair. An average gene is $10^3 - 10^4$ bp long, consequently, for a gene the mutation rate is expected to be between 10^{-6} and 10^{-5} *per* replication, and around 10^{-4} *per* generation for a higher multicellular organism. Since the number of possible base pair substitutions is very high (a 10^4-bp-long DNA can do so in $3 \cdot 10^4$ ways), practically every mutation creates a new gene form (allele). Neglecting the actions of selection for a while, this diversity generating process is firmly controlled by the finite size of the population. Actually, in the absence of mutation every finite population loses genetic diversity, eventually forming a uniform population. Fortunately, mutation is acting against this process and provided the rate of mutation (u) and size of the population (N) are kept constant, the genetic diversity ($\hat{H}$) approaches a steady-state equilibrium value. The simple equation describing this state,

$$\hat{H} = \frac{4Nu}{4Nu + 1},$$

gives an idea of the amount of genetic diversity (cf. Fig. 1) which could be present at each gene locus (e.g., if $N = 10^4$ and $u = 10^{-4}$ the expected heterozygosity is 4/5 with more than 5 alleles in the population). Unfortunately, there are manifold problems with this equation: (1) We cannot be sure how close a population has approached the equilibrium; (2) The size of the population (N) is meant here as the so-called effective size, which, among others, is very much dependent on the past history. A bottleneck situation can drastically reduce the effective size, as compared with the present numbers; (3) Since the effect of selection is disregarded, it applies only for neutral variation, a condition which is difficult if not impossible to be met in nature; (4) If we are concerned about adaptability (survival) of a species or population, it at first seems nonsensical to look at the dynamics of useless neutral variation.

However, we should not be too rash in condemning the neutral model, for it is useful as a null-hypothesis. If we introduce selection, we have to specify fitness values for each genotype combination (e.g., 15 estimated values in the case of five alleles and in addition, it is necessary to randomize the rest of the genome to obtain unbiased results, a difficult task indeed. Simple directional selection models predict a reduction of genetic diversity as compared with the neutral model, unless selection is very weak and the population is small ($Ns \ll 1$, where s is the selection coefficient). On the other hand, special kinds of selection, such as heterozygote superiority, frequency-dependent selection and others can act to increase genetic diversity (although not always). Several recent studies indicate that such balance of selective forces is not rare in Nature. Thus, we may find the non-neutral gene diversity not too far from the neutral one.

For the sake of simplicity, let us neglect the difficulties of defining neutral *vs.* adaptive allelic forms and the inadequacy of single locus

---

$$H = -\sum p_i \log p_i \quad \text{(Shannonian index)}$$

$$H_e = 1 - \sum p_i^2, \quad \text{(Simpsonian index or expected heterozygosity)}$$

where p_i is the frequency of the ith allele in the population

$$\pi = \sum \frac{\pi_{ij}}{n_c} \quad \text{(Nei's nucleotide diversity index)}$$

where π_{ij} is the proportion of different nucleotides between ith and jth allelic sequences in the population; n_c is the total number of comparisons available.
Simplest proposition to incorporate mutational distances:

$$D = \frac{V_T}{N},$$

where V_T is the smallest number of mutational steps to reach a uniform population; N is the number of genomes in the population.

Fig. 1. Some measures of genetic diversity.

treatment in this matter. It is sufficient to say that a certain fraction of gene diversity is necessary for adaptation. The amount of gene diversity can be very different depending on the type and strength of selection, but is ultimately limited by the effective size of the population.

The loss and its consequences

We are all well aware of the loss of species diversity on a global scale (cf. Wilson, 1988; Ehrlich and Ehrlich, 1981; Ehrlich, 1988). Is there a similar case for the within-species genetic diversity as well? My answer is a definite "yes" (Vida, 1987). But while in the case of species diversity we lose those that we dislike or that we are ignorant of, genetic diversity loss is also substantial in our pampered cultivated plants and domestic animals. In wild species we lose genetic diversity because we do not pay attention to their population size, while in domesticated ones we lose genetic diversity because we breed them.

Classical breeding is a manipulation of genetic diversity of populations. Selection aims to increase the frequency of alleles or allele combinations with favorable effects at the expense of others, eventually eliminating many of them. The more advanced a breeding program the more a uniform population results. Drastic reduction of genetic diversity is expected from various methods of genetic engineering. In my opinion, the risk of releasing manipulated agricultural species is not due to their potential environmental damage, but just the other way round. Because the target of genetic manipulation is usually a single cell or its components, the resulting genetically modified organisms (GMOs) are descendants of the same single cell. The released uniform populations could be very promising and popular at the onset, replacing the older varieties, but as soon as they become abundant they induce rapid pest adaptation. Depleted genetic diversity makes GMOs very susceptible to an epidemic disaster. Of course, this will keep the geneticists busy, as long as they can find more resistance genes every 4 or 5 years, a condition which cannot be guaranteed. Thus, it seems as if science-aided human eagerness were marching into a trap (Fig. 2).

Of course, we need to produce more food to feed the hungry, but another trap turns up here. In many parts of the world an increased food supply is rapidly transformed into more people, and more hunger. The expanding human population demands an increasing share of the resources of the biosphere. We already use 40 percent of the total terrestrial net primary production. The war against Nature goes on in most parts of the Globe. The alarming rate of species extinction is nothing but the visible tip of the iceberg. The submerged part is the massive damage to genetic diversity of the existing species.

Years bp	Selecting	From	By means of	Area of cultivated land Productivity Susceptibility Environmental problems	Diversity, cultivated and wild	Success depends on access to
10^4	wild species	local biota	hunting, gathering			
10^3	domesticated races	species	domestication			collection of lines and breeds
10^2	land varieties	populations	selection in populations	increasing →	decreasing	
10^1	modern lines	individuals	manipulation and selection			wild relatives
10^0	genes	cells, DNA	biotechnology, genetic engineering		?	global biodiversity

Fig. 2. The breeding trap: The connection between the development of breeding methods and the genetic diversity (in both cultivated and wild species). Diminishing global diversity might prevent further success in breeding. Reduction of genetic diversity in cultivated species is a consequence of the decreasing hierarchical level of the manipulation. (bp = before present)

How many alleles have been lost? It is impossible to estimate even the order of magnitude. If we do not know how many species are living with us in the biosphere, how numerous they are, how they interact with each other and with the physical environment, or what sort of genetic system they have, to mention but a few of the larger questions, we know next to nothing about the genetic richness of the biosphere. There is one thing we cannot deny, however. The individual number of most species has been reduced enormously.

Let me use my homeland, Hungary, as an example. Less than 5% of our land surface carries biotic communities believed to be close to the natural situation. Because of our incomplete knowledge on relations of the area to genetic richness, effective size of populations, migration rates, etc., it is difficult to judge how much total genetic diversity has been lost in the 95% transformed area, but it is hardly negligible. Furthermore, even if we safeguard the remaining areas, additional loss of genetic diversity is expected in many populations because of their reduced sizes.

Unfortunately, this is not the end of the possible scenario. These populations are members of a very complex biotic community. They have been evolving together for a long period, adjusting their gene pool in a very special way. Genetic diversity in this coevolutionary process could find remote adaptive peaks on the mutational landscape (Fig. 3). If the size of a population is kept small for many generations, it has a double risk of extinction. Extinction is either due to stochastic fluctuation of the number of individuals, and/or inbreeding or lack of adaptability. In a small reserve of natural community adaptability is needed in many populations if a new species enters the community, if a former member of the community disappears, or if the physical environment changes drastically. All three have actually occurred. I do not know if Hungary will definitely lose this tiny 5% natural state area because the reduced genetic diversity will result in collapse of the communities, but this is a danger which should be studied carefully, since genetic diversity is the basis of the higher rank biodiversity; species diversity, ecosystem diversity, etc.

The situation is better or worse in other countries, but there are drastic changes everywhere. Mankind's global-scale experiment is proceeding with unpredictable results. Some rather straightforward effects, such as changes of the Earth's radiation balance, are gradually being accepted by the scientific community. Yet, we cannot immediately suspend the dangerous experiment. Social inertia prevents rapid action. On the other hand, the global biosphere-atmosphere-hydrosphere-geosphere system can have a much longer time lag in its reactions. Consequently, by the time we recognize an effect, it might be too late to control it. It may be the eco-equivalent of the tragic Chernobyl nuclear disaster.

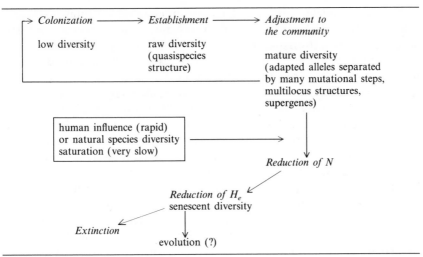

Fig. 3. Succession of genetic diversity of a population. Colonization usually starts with low gene diversity (a single or a few individuals only). In the next step (hundreds of generations) mutation begins to accumulate in the neighborhood of the colonizer's nucleotide sequences. As the genetic diversity increases and the interactions with other populations perform selection upon the raw diversity, mature adapted diversity evolves (thousands of generations). By entering additional populations into the community, finally (millions of generations?) it will be saturated with species, many of them necessarily being present in small population size (N) because of the finite space. This, in turn, inevitably leads to reduction of genetic diversity (H_e) and increases the probability of extinction. Artificial reduction of population sizes (as occurs in small reserves) can lead to the same result.

Vistas and conclusions

Current science is not in very promising shape to tackle the problems of the supraindividual organization (cf. Solbrig, 1991). If we look at the contents of leading scientific journals with broad coverage such as *Nature*, *Science*, *PNAS* and others, the overwhelming dominance of molecular-level studies with very narrow scope is surprising indeed. A reasonable proportion is devoted to the other extremity, the celestial bodies, and very little remains in between.

Diversity is often avoided by many experimental biologists for the simple reason of desiring less "noise" in the system. In general, scientists love diversity in nature but hate it in the laboratory. However, for a number of very important issues diversity is the solution, and consequently, the presently dominating "Procrustean" biology cannot understand it. (I call it Procrustean after the name of the legendary bandit who fit victims to a bed by stretching them or cutting off bodily parts.)

As a convincing support for the functional importance of diversity, the immune system can be mentioned. Survival would be very difficult without the immense variability of immunoglobulins. Various defects of

18

the immune system make an otherwise mild infection fatal, as in the case of AIDS. HIV viruses apply the same strategy to conquer the immune system's diversity. They create even more kinds of themselves, thanks to their rather inaccurate RNA replication, and in this way finally exhaust the diversity of the immune system (Weiss, 1993). It is tempting to apply this lesson to supraindividual levels up to the biosphere's biodiversity. We should play a "Gedankenexperiment" with our biosphere, replacing all the natural terrestrial communities with a few presently favored agrarian systems, such as fields of wheat, rice, corn, sweet potato, etc., and see if the so-called "ecosystem services" of the biosphere, so crucial for us (see Keeling, 1993), are provided as before or not. If not, can we assist in restoring or substituting it by means of some technology, or do we need Mother Nature's help?

Such important points should have the highest priority. In this respect, the recent Rio Convention on Biodiversity is rather disappointing. It has fallen into another trap, where property right (be it material or intellectual) is in the center and "our common future" remains only a phrase (see Fig. 4).

Sir Winston Churchill said "The longer you can look back the further you can look forward." He was not referring to biodiversity, but the statement is applicable. Rich libraries of genetic history still surround us. We can read them, although the language is not fully understandable. The most important message, however, is becoming more appar-

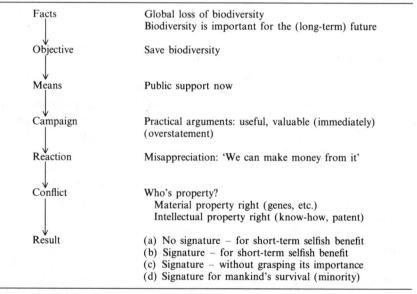

Fig. 4. The UNCED Trap. United Nations Conference on Environment and Development, Rio de Janeiro, June 1992.

ent: without genetic diversity there is no future, no hope for longer survival. There is no shortage of genetic diversity in our species, *Homo sapiens*, although estimated values are lower than those of *Drosophila* (Li and Sadler, 1991). Our species, however, cannot survive without a properly functioning biosphere. Furthermore, this process cannot be expected without a suitable amount of genetic diversity and higher level diversities, especially in a rapidly changing environment. Contributors to this volume therefore focus on the important question: How much genetic diversity is "the suitable amount," and how can we maintain it under the present constraints of space, time, and money for the sake of higher level biodiversities?

References

Brown, L. R. (1991) The new world order. *In:* Brown, L. R., Durning, A., Flavin, C., French, H., Jacobson, J., Lenssen, N., Lowe, M., Postel, S., Renner, M., Ryan, J., Starke, L. and Young, J. (eds), *The state of the world.* W. W. Norton and Co., New York, London, pp. 3–20.

Ehrlich, P. R. (1988) The loss of diversity: causes and consequences. *In:* Wilson, E. O. (ed.) (1988) *Biodiversity.* National Academy Press, Washington, D.C. pp. 21–27.

Ehrlich, P. R. and Ehrlich, A. H. (1981) *Extinction. The causes and consequences of disappearance of species.* Random House, New York.

Hoffman, A. (1987) Neutral model of taxonomic diversification in the Phanerozoic: A methodological discussion. *In:* Nitecki, M. H. and Hoffman, A. (eds), *Neutral models in biology.* Oxford Univ. Press, pp. 133–146.

Keeling, R. F. (1993) Heavy carbon dioxide. *Nature* 363: 399–400.

Kimura, M. and Takahata, N. (1991) *New aspects of the genetics of molecular evolution.* Japan Sci. Soc. Press and Springer Verlag, Tokyo-New York.

Klekowski, E. J., Jr. (1988) *Mutation, developmental selection, and plant evolution.* Columbia Univ. Press, New York.

Li, W. H. and Sadler, L. A. (1991) Low nucleotide diversity in man. *Genetics* 129: 513–523.

Selander, R. K., Clark, A. G. and Whittam, T. S. (1991) *Evolution at the molecular level.* Sinauer, Sunderland, MA.

Solbrig, O. T. (1991) *From genes to ecosystems: A research agenda for biodiversity.* IUBS Press, Paris.

Solbrig, O. T., van Emden, H. M. and van Oordt, P. G. W. J. (1992) *Biodiversity and global change.* IUBS Press, Paris.

Vida, G. (1987) Genetic diversity and environmental future. *Environ. Conserv.* 5: 127–132.

Weiss, R. A. (1993) How does HIV cause AIDS? *Science* 260: 1273–1279.

Wilson, E. O. (1988) *Biodiversity.* National Academy Press, Washington, D.C.

Part II

Genetic variation and fitness

Conservation Genetics
ed. by V. Loeschcke, J. Tomiuk & S. K. Jain
© 1994 Birkhäuser Verlag Basel/Switzerland

Introductory remarks

In population genetics one major focus is to study the correlation between genetic variation and fitness. The fitness concept used may be based either on the attributes of individuals or of populations. Individual or population fitness varies in dependence on the environment experienced, which means fitness is a relative value. Knowledge of both genetic and environmental factors that influence fitness are necessary in order to succeed in conservation biology practice. Furthermore, we need to understand the genetic basis of traits affecting variation in fitness.

Newly arising mutations are often deleterious recessives, but remain in the population in low frequency (mutation-selection balance), as heterozygotes are not affected. They contribute to the mutational load of a population which is created by the elimination of recurrent harmful mutations (Haldane, 1937). Crow (1958) gave a mathematical definition of the genetic load as the fraction by which the population fitness is reduced in comparison to the most fit genotype. Early theoretical work in this area demonstrated that the genetic load in a population would be enormous if genetic variation at most loci is maintained by balancing selection as proposed by Dobzhansky (1965). Kimura (1983) discusses in detail various selection models which have been proposed to demonstrate how selection maintains most genetic variation, even at the molecular level, without increasing the load to unacceptable size in natural populations. However, the population size influences the genetic load. In small populations the mutant frequency drifts away from its equilibrium value, which may cause a considerable increase in the genetic load of small populations (Kimura et al., 1963). In small populations also codominant alleles that are disadvantageous may become fixed by random genetic drift as drift may dominate selection where selective differences are small. With time the number of loci where (slightly) deleterious mutations become fixed increases (Muller's ratchet; Muller, 1964) and the mean population fitness decreases. The reduction in population fitness in turn may lead to a reduction in population size which again increases the rate of fixation of deleterious mutations. Population fragmentation and habitat loss often is followed by a loss of genetic variation that is associated with increased levels of genetic drift and inbreeding. This result links processes that usually are described separately in ecology and genetics.

Other than low frequency deleterious alleles, much of the genetic variation at the molecular level is usually considered neutral, but some presumably is adaptive. Some researchers have found a positive correlation between allozyme variation and fitness traits (e.g., Zouros and Foltz, 1987), but others have not. Even with a positive correlation, we do not know whether it is the variation at particular loci *per se* that is adaptively significant, or whether it involves linked gene systems. Variation at the quantitative level is more commonly considered ecologically relevant, but variability at this level is not always concordant with the variation at allozyme or molecular marker loci. Further, we need more direct assessment of the role of genetic variation in the avoidance of extinction, in establishing new colonies, or in some regulation of population number (e.g., Solbrig, 1991).

Ledig (1986) reviewed genetic variation patterns in temperate forest species, with the objective of testing the dominance *vs.* overdominance models of heterosis. Assuming juvenile growth and survivorship to be good fitness measures for these long lived species, heterosis in many interpopulation hybrids, as well as inter-neighborhoods (within-stand crosses), was explained using a dominance model, but some overdominant loci were not entirely ruled out. In general, high rates of cumulative mutational load (due to slightly deleterious alleles at many loci; see Lande and Barrowclough, 1987, for theory) were found in various inbreeding response experiments. Ledig (1986) also noted that the outcome of such studies might make little change in conservation practices, in which maximization of genetic variation both *in situ* and *ex situ* is the primary goal, albeit with no *a priori* information on the ecological properties of any such variation. In contrast, in many fish and insect studies, recent emphasis on life history related variation (e.g., dispersibility polymorphism, uni- or multivoltine races, annual/perennial ecotypes) is valuable for redirecting efforts in ecological genetic research.

In a recent book, *Genes in Ecology* (Berry et al., 1993), several authors review attempts to ascribe the adaptive role for specific polymorphic loci. They also deal with the heterogeneity of biotic and spatial environmental factors in relation to genetic variation patterns. For conservation genetics we need specific (perhaps case-by-case) answers to the following: does inbreeding depression matter in terms of lifetime loss of reproductive value and thereby failure of a population to persist without human intervention? Alternatively, can purging of genetic load under selection and inbreeding lead to more fit gene pools, at least until new mutational load accumulates? Are sexual species more prone to extinction than asexuals? How do interfamily and intrafamily selection methods influence viability in a population? Some answers to these questions are the topic of chapters in Part II.

References

Berry, R. J., Crawford, T. J. and Hewitt, G. M. (1993) *Genes in ecology.* Blackwell, London.

Crow, J. F. (1958) Some possibilities for measuring selection intensities in man. *Hum. Biol.* 30: 1–13.

Dobzhansky, Th. (1965) Genetic diversity and fitness. *In:* Geerts, S. J. (ed.), *Genetics today* (Proc. XI Int. Cong. Genet.). Pergamon Press, Oxford, pp. 541–552.

Haldane, J. B. S. (1937) The effect of variation on fitness. *Am. Nat.* 71: 337–349.

Kimura, M. (1993) *The neutral theory of molecular evolution.* Cambridge University Press, Cambridge.

Lande, R. and Barrowclough, G. F. (1987) Effective population size, genetic variation, and their use in population management. *In:* Soulé, M. E. (ed.), *Viable populations for conservation.* Cambridge University Press, Cambridge, pp. 87–123.

Ledig, F. T. (1986) Heterozygosity, heterosis, and outbreeding plants. *In:* Soulé, M. E. (ed.), *Conservation biology: The science of scarcity and diversity.* Sinauer, Sunderland, MA, pp. 77–104.

Muller, H. J. (1964) The relation of recombination to mutational advance. *Mutat. Res.* 1: 2–9.

Solbrig, O. T. (1991) *From genes to ecosystems: a research agenda for biodiversity.* IUBS, Cambridge.

Zouros, D. and Folz, E. W. (1987) The use of allelic isozyme variation for the study of heterosis. *In:* Rattazi, M. C., Scandalios, J. G. and Whitt, G. S. (eds), *Isozymes: Current topics in biological and medical research,* Vol. 13. Alan R. Liss, New York, pp. 1–59.

Conservation Genetics
ed. by V. Loeschcke, J. Tomiuk & S. K. Jain

Genetic variation and fitness: Conservation lessons from pines

O. Savolainen

Department of Genetics, University of Oulu, SF-90570 Oulu, Finland

Summary. Pines are an important component of the boreal ecosystem, and many species are important crop plants. Information on patterns of variation and on the nature of inbreeding depression are needed both for conservation and tree breeding purposes. Populations of *Pinus sylvestris* are highly differentiated with respect to some adaptive quantitative variation. However, marker loci (enzyme loci, DNA) have so far not provided similar patterns of variability. It is one goal of molecular marker studies to locate clinal variation in the DNA itself. The genetic basis of inbreeding depression in *Pinus sylvestris* has been studied. Results on correlation between individual heterozygosity and fitness provide little evidence for the overdominance model. Instead, deleterious recessive genes are well documented in this and other species of conifers. Most species carry large numbers of embryonic lethal equivalents. Further, in later life stages may more deleterious recessives influence growth, survival, and reproduction. The implications of the genetic load for conservation and breeding are discussed.

Introduction

The maintenance of genetic variation and avoidance of inbreeding depression are two major goals in conservation genetics (e.g., Frankel and Soulé, 1981). While any genetic variation is informative as a reflection of population structure and history, it is also important to conserve genetic variation that is directly related to adaptation. Studies on nearly 700 plant species have reported on the variation at enzyme loci in higher plants (Hamrick et al., 1982). However, the relationship of such variation to survival or fertility in different environments is largely unknown. Numerous studies in conifers allow comparisons of patterns of variation between adaptive traits and marker genes.

Inbreeding depression and its counterpart, heterosis, were already documented in many plant species by Darwin (1876). The two main hypotheses to account for the phenomenon are overdominance and partial to full dominance (e.g., Charlesworth and Charlesworth, 1987; Charlesworth, 1991). Most evidence to date suggests that deleterious recessive alleles are the main cause (Frankel, 1983). The genetic basis is of significance both for conservation biology and for plant breeding. The nature of the deleterious recessive genetic load is also important; it is not irrelevant whether the load is due to a few vary harmful mutations or to many slightly deleterious alleles (Barrett and Charlesworth, 1991; Hedrick, 1994).

The purpose of this contribution is to address these issues in the light of the information that is available from conifers, especially the genus *Pinus*, and in particular *Pinus sylvestris*.

Of what interest are pines to conservation genetics?

Pinus sylvestris ranges from eastern Siberia to western Scotland and has the widest distribution area of all pines (Mirov, 1967). Even if Scots pine populations are not threatened, in many respects their genetic variation is a relevant topic in a conservation context.

First, even if numbers of pine trees are not declining, the genetic composition of the populations may be changing due to domestication. In central Europe, hardly any original natural forests are left. In other areas, more natural forests remain. In Finland, about half of the annual regeneration is through planting and sowing (Savolainen and Kärkkäinen, 1992). At an early stage of domestication it is still possible, and important, to conserve the original genetic variation.

Second, future pine breeding programs include schemes where small populations are allowed to inbreed, or are intentionally inbred (e.g., Kang and Nienstaedt, 1987). The consequences of this may resemble the population fragmentation that is the fate of many species. Several other conifer species do have limited and fragmented distributions, e.g., *Pinus radiata* in California and Mexico, or *Picea omorika* in Serbia (Mirov, 1967). We thus need to consider what consequences a change in the level of inbreeding may have.

Third, few wild plant populations have been studied as extensively as trees (e.g., Langlet, 1971; Adams et al., 1992). Patterns of variation of other species with large outbreeding populations may resemble those found in these populations.

Genetic variation and its relevance to fitness

The distribution of genetic variability at marker loci in trees have been recently reviewed by Hamrick et al. (1992). This review was based on a total of 198 woody species. *Pinus sylvestris* serves as a good example of the patterns of variation (see also Muona, 1989).

Pinus sylvestris has large and continuous populations. Pollen flow is extensive (Koski, 1970). Regeneration often takes place in an opening in a forest, where an initial seed crop of a few million seed *per* hectare give rise to less than a thousand adult trees (100 years) *per* hectare (see Savolainen and Kärkkäinen, 1992, for further references). While there are some selfs at the seed stage, the adult population has no evidence of inbreeding (Muona, 1989).

The distribution of variability has been documented for a large variety of traits in *Pinus sylvestris* (Tab. 1). Of secondary metabolic products, e.g., monoterpenes, one component (3-carene) has been found to vary between populations, but others are homogeneous over latitudes (Tigerstedt et al., 1979; Muona et al., 1986). At enzyme loci, the populations are quite homogeneous over long distances (Gullberg et al., 1985; Muona and Harju, 1989). The G_{ST} values are less than 0.02 within Scandinavia, and even when populations are included from Eastern China, the estimate is about 0.03 (Wang et al., 1991). Similar patterns of variability seem to hold in initial studies of variation at the DNA level. Ribosomal DNA has proved highly variable, but no clinal patterns were detected (Karvonen and Savolainen, 1993).

This pattern of variation suggests genetically homogeneous populations, but many transfer experiments show that *Pinus sylvestris* is closely adapted to the local climatic conditions (e.g., Eiche, 1966). If seedlings are planted farther to the north or at higher altitudes than their origin, mortality increases significantly even with small latitudinal (less than a 100 km) or altitudinal transfer differences. Experimental studies of growth rhythm show that the populations are highly differentiated with respect to timing of growth cessation. Mikola (1982) grew samples of Finnish populations from different latitudes in a common garden experiment. More than 80% of the variation in date of budset was between populations, with a very regular clinal pattern. This variability is important for adaptation to current environmental conditions, and relevant to adaptability to future environmental change.

This distribution of variability is, of course, not unique to *Pinus sylvestris*. Similar data are available for many other species, e.g., for *Pseudotsuga menziesii* in North America. There is highly significant variation between populations in growth characteristics (Campbell, 1979), but much less in allozymes (Li and Adams, 1989).

When genetic diversity is sampled for breeding or conservation purposes, the variability that is directly related to survival or fertility is of importance. In the case of *Pinus sylvestris*, marker studies have so far

Table 1. Distribution of variation in various genetic characters in *Pinus sylvestris*, Finnish populations

Character	% of variation between populations	Ref.
Allozymes, 13 loci	2%	(1)
rDNA	14%	(2)
Bud set date	80–90%	(3)

References: (1) Muona and Harju (1989); (2) Karvonen and Savolainen (1993); (3) Mikola (1982).

not been a substitute for direct measurements of quantitative variation. In general, patterns of variation generated by mutation and migration in large outcrossing populations can be expected to be different from those generated mainly by selection-migration balance. However, population history can also be reflected for many generations even in the distribution of quantitative variability, as, for example, in *Picea abies* (Lagercranz and Ryman, 1990). In species with other kinds of population structure, where marker genes are differentiated, such as many selfing plants, patterns due to drift may not accurately reflect patterns in variation due to selection. Nevertheless, marker gene variability can be a good guide in many parts of a gene conservation program.

Molecular population genetics should also aid in finding methods for studying variation at the quantitative trait loci. This will initially only be possible in some model species. One approach is to find linked markers using saturated linkage maps and quantitative trait loci (QTL) mapping (e.g., Paterson et al., 1988). Another would be to try to assess variation directly at these loci. McKay and Langley (1991) showed that molecular variation at one locus accounted for a proportion of bristle number variation in *Drosophila melanogaster*. Similar approaches should soon be possible in a model plant species such as *Arabidopsis thaliana*.

Is there evidence for overdominance in trees?

Conifers are known to express severe inbreeding depression in many life stages (e.g., Koski 1971; Eriksson et al., 1973; Wilcox 1983; Sorensen and Miles, 1984; and review by Williams and Savolainen, 1993). The genetic basis of the inbreeding depression is still being discussed. Correlations between number of heterozygous loci *per* individual and fitness traits have been used for studying whether heterozygotes have an inherent fitness advantage. It has been suggested that positive correlations are evidence for overdominance at the individual loci (for forest trees, e.g., Bush and Smouse, 1992). However, even under the partial dominance model, correlations could arise in partially inbreeding populations, if selfs with a low number of heterozygous loci are included in the sample (Haldane, 1949; Charlesworth, 1991). Similarly, correlations could come about in substructured or small populations due to gametic disequilibrium. Nevertheless, in large random mating populations, correlations should only arise through overdominance, because other causes are absent (Houle, 1989). Empirical results are mixed: there are many negative findings but positive correlations between number of heterozygous loci and fitness have been often reported in two groups of organisms, marine invertebrates (e.g., Gaffney et al., 1990) and coniferous trees (Bush and Smouse, 1992).

Savolainen and Hedrick (1993) studied the correlations between individual heterozygosity and fitness related traits (growth, pollen production, seed production) in three populations of *Pinus sylvestris* in Finland. The populations were large, and the adult trees showed no evidence of inbreeding. Two of the three populations were even aged seed orchards, where we were able to measure accurately both fertility and growth variables due to vegetatively replicated genotypes. The third population was an old natural population in the north at timberline. The number of heterozygous loci *per* individual ranged from zero to seven. The number of heterozygous loci *per* individual, heterozygosity at single loci, or the loci jointly did not account for more of the fitness variation than can be expected by chance. In previous studies, where single locus effects in conifers have been found, they have varied between populations (see Bush and Smouse, 1992, for a review). From our study and a review of other conifer data, we concluded that there is little evidence for overdominance at enzyme loci.

Deleterious alleles, lethals and inbreeding depression

Tab. 2 summarizes some evidence on the level of early inbreeding depression in different species of conifers. The proportion of full seed is much reduced after selfing. This is due to embryonic mortality, because conifers have no self-incompatibility system.

The observed inbreeding depression can be transformed into an estimate of the number of lethal equivalents (Morton et al., 1956). Several methods have been used to estimate the numbers of lethals in conifers (e.g., Sorensen, 1969; Koski, 1971; Bramlett and Popham,

Table 2. Estimates of empty seed frequencies in selfs (R_s) and controlled crosses (R_c), and inbreeding depression ($\delta = 1 - R_s/R_c$) in different species of conifers

Species	Empty seeds			
	Self	Cross	δ	Ref
Pinus resinosa	29.0	28.0	0.02	(1)
Picea omorika	67.6	47.7*	0.38	(2)
			0.31	(3)
Pinus radiata	65.8	21.1	0.56	(4)
Pinus sylvestris	84.4	25.6*	0.80	(5)
North	75.4	23.7*	0.68	(5)
South	86.5	24.8*	0.82	(5)
Picea abies	86.2	15.0*	0.84	(2, 6)
Pseudotsuga menziesii	92.1	32.8	0.88	(7)

*Open pollination. References: (1) Fowler (1965); (2) Koski (1973); (3) Kuittinen and Savolainen (1992); (4) Griffin and Lindgren (1985); (5) Kärkkäinen et al. (1993); (6) Koski (1971); (7) Sorensen (1971).

1971). The methods provide slightly different results (Savolainen et al., 1992). Tab. 3 lists some of these estimates for conifers (and other organisms). The estimates for conifers were obtained by using published data on average proportion of full seed after selfing (R_s) and outcrossing (R_c). The relative self-fertility ($R_f = R_s/R_c$) can be used for estimating the numbers of lethal equivalents *per* zygote $n = -4 \ln R_f$ (Morton et al., 1956; Sorensen, 1969). *Pinus sylvestris* and most other conifer species have high numbers of embryonic lethal equivalents, although Tab. 3 also lists some exceptional species with a very low number of lethals (e.g., *Pinus resinosa*).

Inbreeding effects in later life stages have not been characterized as thoroughly. The growth data in Tab. 4 are an underestimate of later total inbreeding depression, because the data are only for those trees that survived from planting to measurement. Furthermore, the little data that are available show that inbreeding depression is severe for the fertility component of fitness as well. Orr-Ewing (1969) reported that only 14% of planted first generation selfs of *Pseudotsuga menziesii* produced considerable numbers of male and female strobili. It seems that inbreeding depression in conifers is much higher than in many other organisms, such as predominantly outcrossing shorter-lived plants

Table 3. Estimates of number of lethal equivalents per zygote in different species

Species	Number of lethals	Ref.
Pinus sylvestris	6.2	(1)
Pinus radiata	3.3	(2)
Pinus resinosa	0.1	(3)
Phlox drummondii	0.8	(4)
Homo sapiens	3–5	(5)
Drosophila melanogaster	0.3–3.0	(6)
D. persimilis	1.5	(6)
D. pseudoobscura	1.6	(6)

References: (1) Kärkkäinen et al. (1993); (2) Griffin and Lindgren (1985); (3) Fowler (1965); (4) Levin (1989); (5) Morton et al. (1956); (6) Lewontin (1974).

Table 4. Estimates of post-embryonic inbreeding depression in conifers after selfing

Species	Trait	Age	δ	Reference
Picea omorika	height	15	0.27	(1)
Pinus radiata	diameter		0.12	(2)
Pinus sylvestris	height	6–7	0.32	(3)
	survival	6–7	0.21	(3)
Picea abies	height	61	0.22	(4)

References: (1) Geburek (1986); (2) Wilcox (1983); (3) Dengler (1939); (4) Eriksson et al. (1973).

(e.g., *Phlox*), *Drosophila* (review in Lewontin, 1974), or man (Morton et al., 1956) (Tab. 3).

The expected equilibrium frequency of recessive lethals depends on the mutation rate and the level of inbreeding (Haldane, 1940). Conifer populations at the adult stage have no evidence of selfing. However, in most species there is a low proportion of selfs among mature seed (Muona, 1989). Given the high mortality of selfs, this shows that there must be considerable self-fertilization. This is also shown by direct measurements of labeled self-pollen around the crown (Koski, 1970). Partial inbreeding should result in a low equilibrium frequency of lethal alleles (Haldane, 1940). Thus, mutation rates must be high to maintain these levels of lethals, as has been suggested by Ledig (1986) and Sedgley and Griffin (1989). Evidence for high mutation rates in *Pinus sylvestris* is available for chlorophyll mutants, summarized by Kärk-käinen et al. (1993).

Haldane's (1940) result related to single locus mutation rates. However, we do not know how many loci mutate to lethals in conifers. Namkoong and Bishir (1987) used conventional mutation rates (10^{-6} to 10^{-5}) to infer that there must be more than 10^4 loci. We have assumed that the genes are completely recessive, but in fact their heterozygous effects have not been studied. Further, we know even less about the number of loci, or the distribution of the fitness effects, or mutation rates for those loci influencing inbreeding depression in later life stages. Such important information is mainly available for model organisms like *Drosophila melanogaster* (Simmons and Crow, 1977).

Can deleterious recessives and lethals be purged?

Whatever the cause for the high numbers of lethal genes, they do present a problem for the tree breeder, or for a conservation geneticist. In tree breeding, inbreeding has been presented as an alternative (Kang and Nienstaedt, 1987). These authors suggest that the deleterious recessives can be purged. However, unless populations are very large, some of the deleterious recessives may become fixed. The probable fate of the deleterious genes depends on the nature of the genetic load (Barrett and Charlesworth, 1991). If the inbreeding depression is due to few highly deleterious alleles, then purging is likely. These alleles will be eliminated rapidly, before drift has a chance to fix them. This will lead to a rapid recovery of fitness in subsequent generations. On the other hand, when there is a large number of less deleterious genes, they can probably become fixed by drift. In such cases, there may be little recovery of fitness in the inbred populations. Population fitness can be recovered by crossing these populations to each other. The possibility of maintaining large numbers of offspring in conifer breeding programs suggests that it is possible to purge high numbers of deleterious genes, at least far more easily than, for example, in populations of captive animals with low offspring number (Hedrick, 1994).

Some pine species listed in Tab. 3 have very low levels of lethals, e.g., *Pinus resinosa*. The variation in the number of lethal equivalents among pines is in contrast to *Drosophila* species which have fairly uniform levels of lethal equivalents. These conifer species have probably lost the lethals (and much other genetic variability) in a bottleneck followed by inbreeding. *P. resinosa* now has a wide distribution in eastern North America (Mirov, 1967), but also very little genetic variability in other traits (Fowler and Morris, 1977). This example, and others like it, show that the lethal equivalents can be purged.

For the use of conservation genetics and tree breeding, we would need quantitative predictions about the rate of purging. However, more information will be needed on the nature of the genetic load, e.g., number of loci and the distribution of selection coefficients.

Acknowledgements
This research has been financed mostly by the National Research Council for Agriculture and Forestry of Finland. Numerous discussions with Katri Kärkkäinen have helped in clarifying the issues. I thank Christian Damgaard for his insightful comments on the manuscript.

References

Adams, W. T., Strauss, S. H., Copes, D. L. and Griffin, A. R. (1992) *Population genetics of forest trees.* Kluwer, Dordrecht, The Netherlands.

Barrett, S. and Charlesworth, D. (1991) Effects of a change in the level of inbreeding on the genetic load. *Nature* 352: 522–524.

Bramlett, D. L. and Popham, T. W. (1971) Model relating unsound seed and embryonic lethal alleles in self-pollinated pines. *Silvae Genetica* 20: 192–193.

Bush, R. M. and Smouse, P. E. (1992) Evidence for the adaptive significance of allozymes in forest trees. *New Forests* 6: 179–196.

Campbell, R. K. (1979) Genecology of Douglas-fir in a watershed in the Oregon Cacades. *Ecology* 60: 1036–1050.

Charlesworth, D. (1991) The apparent selection on neutral marker loci in partially inbreeding populations. *Genet. Res.* 57: 159–175.

Charlesworth, D. and Charlesworth, B. (1987) Inbreeding depression and its evolutionary consequences. *Annu. Rev. Ecol. Syst.* 18: 237–268.

Darwin, C. (1876) *The effects of self- and cross fertilization in the vegetable kingdom.* J. Murray, London.

Dengler, A. (1939) Über die Entwicklung künstlicher Kiefernkreuzungen. *Zeit. Forst. Jagdwesen* 71: 457–485.

Eiche, V. (1966) Cold damage and plant mortality in experimental provenance plantations with Scots pine in northern Sweden. *Studia Forestalia Suecica* 36: 1–218.

Eriksson, G., Schelander, B. and Åkerbrand, V. (1973) Inbreeding depression in an old experimental plantation of *Picea abies. Hereditas* 73: 185–194.

Fowler, D. P. (1965) Effects of inbreeding in red pine, *Pinus resinosa* Ait. *Silvae Genetica* 12: 12–23.

Fowler, D. P. and Morris, R. W. (1977) Genetic diversity in red pine: Evidence for low genetic heterogeneity. *Can. J. For. Res.* 7: 343–347.

Frankel, R. (1983) *Heterosis. Reappraisal of theory and practice.* Springer Verlag, N.Y.

Frankel, O. H. and Soulé, M. (1981) *Conservation and evolution*. Cambridge University Press, Cambridge, UK.

Gaffney, P. M., Scott, T. M., Koehn, R. K. and Diehl, W. J. (1990) Interrelationships of heterozygosity, growth rate and heterozygote deficiencies in the coot clam, *Mulinia lateralis*. *Genetics* 124: 687–699.

Geburek, T. (1986) Some results of inbreeding depression in Serbian spruce (*Picea omorika* (Panc.) Purk.). *Silvae Genetica* 35: 169–172.

Griffin, A. R. and Lindgren, D. (1985) Effect of inbreeding on production of filled seed in *Pinus radiata* – experimental results and a model of gene action. *Theor. Appl. Genet.* 71: 334–343.

Gullberg, U., Yazdani, R., Rudin, D. and Ryman, N. (1985) Allozyme variation in Scots pine (*Pinus sylvestris*) in Sweden. *Silvae Genetica* 34: 193–201.

Haldane, J. B. S. (1940) The conflict between selection and mutation of harmful recessive genes. *Ann. Eugenics* 10: 417–421.

Haldane, J. B. S. (1949) The association of characters as a result of inbreeding and linkage. *Ann. Eugenics* 15: 15–23.

Hamrick, J. L., Godt, M. J. W. and Sherman-Broyles, S. (1992) Factors influencing levels of genetic diversity in woody plant species. *New Forests* 6: 95–124.

Hedrick, P. W. (1994) Purging inbreeding depression. Manuscript.

Houle, D. (1989) Allozyme-associated heterosis in *Drosophila melanogaster*. *Genetics* 123: 789–801.

Kang, H. and Nienstaedt, H. (1987) Managing long-term tree breeding stock. *Silvae Genetica* 36: 30–39.

Karvonen, P. and Savolainen, O. (1993) Variation and inheritance of ribosomal DNA variation in Scots pine (*Pinus sylvestris* L.). *Heredity*, (in press).

Koski, V. (1970) A study of pollen dispersal as a mechanism of gene flow. *Commun. Inst. For. Fenn.* 70: 1–78.

Koski, V. (1971) Embryonic lethals of *Picea abies* and *Pinus sylvestris*. *Commun. Inst. For. Fenn.* 75: 1–30.

Koski, V. (1973) On self-pollination, genetic load and subsequent inbreeding in some conifers. *Commun. Inst. For. Fenn.* 78: 1–42.

Kuittinen, H. and Savolainen, O. (1992) *Picea omorika* is a fully self-fertile but outcrossing conifer. *Heredity* 68: 183–187.

Kärkkäinen, K., Koski, V. and Savolainen, O. (1993) Geographical variation in the early inbreeding depression of Scots pine. Manuscript.

Lagercranz, U. and Ryman, N. (1990) Genetic structure of Norway spruce (*Picea abies*): concordance of morphological and allozymic variation. *Evolution* 44: 38–53.

Langlet, O. (1971) Two hundred years of genecology. *Taxon* 20: 653–722.

Ledig, F. T. (1986) Heterozygosity, heterosis, and outbreeding plants. *In:* Soulé, M. (ed.), *Conservation biology: The science of scarcity and diversity*. Sinauer Assoc., Sunderland, MA, pp. 77–104.

Levin, D. A. (1989) Inbreeding depression in partially self-fertilizing *Phlox*. *Evolution* 43: 1417–1423.

Lewontin, R. C. (1974) *The genetic basis of evolutionary change*. Columbia University Press.

Li, P. and Adams, W. T. (1989) Range-wide patterns in allozyme differentiation in Douglas-fir (*Pseudotsuga menziesii*). *Can. J. For. Res.* 19: 149–161.

McKay, T. F. C. and Langley, C. H. (1991) Molecular and phenotypic variation in the achaete-scute region of *Drosophila melanogaster*. *Nature* 348: 64–66.

Mikola, J. (1982) Bud-set phenology as an indicator of climatic adaptation in Scots pine in Finland. *Silva Fennica* 16: 178–184.

Mirov, N. T. (1967) *The genus Pinus*. The Ronald Press Company, New York.

Morton, N. E., Crow, J. G. and Muller, H. J. (1956) An estimate of the mutational damage in man from data on consanguineous marriages. *Proc. Natl. Acad. Sci. USA* 42: 855–863.

Muona, O. (1989). Population genetics in tree improvement. *In:* Brown, A. H. D., Clegg, M. T., Kahler, A. L. and Weir, B. S. (eds), *Plant population genetics, breeding, and genetic resources*. Sinauer, Sunderland, MA, pp. 282–298.

Muona, O. and Harju, A. (1989) Effective population size, genetic variability, and mating system in natural stands and seed orchards of *Pinus sylvestris*. *Silvae Genetica* 38: 221–228.

Muona, O., Hiltunen, R., Shaw, D. and Morén, E. (1986) Analysis of monoterpene variation in natural stands and plustrees of *Pinus silvestris*. *Silva Fennica* 20: 1–8.

36

Namkoong, G. and Bishir, J. (1987) The frequency of lethal alleles in forest tree populations. *Evolution* 41: 1123–1127.

Orr-Ewing, A. L. (1969) Inbreeding to the S2 generation in Douglas-fir. *Second World Consultation on forest tree breeding.* Washington, 7–16 August 1969. FAO, Rome.

Paterson, A. H., Lander, E. S., Hewitt, J. D., Peterson, S., Lincoln, S. E. and Tanksley, S. D. (1988) Resolution of quantitative traits into Mendelian factors by using a complete linkage map of restriction fragment polymorphisms. *Nature* 335: 721–726.

Savolainen, O. and Hedrick, P. W. (1993) Heterozygosity and fitness: no association in Scots pine. Manuscript.

Savolainen, O. and Kärkkäinen, K. (1992) Effect of forest management on gene pools. *New Forests* 6: 329–345.

Savolainen, O., Kärkkäinen, K. and Kuittinen, H. (1992) Estimating numbers of embryonic lethals in conifers. *Heredity* 69: 308–314.

Sedgley, M. and Griffin, A. R. (1989) *Sexual reproduction of tree crops.* Academic Press, New York.

Simmons, M. J. and Crow, J. F. (1977) Mutations affecting fitness in *Drosophila* populations. *Annu. Rev. Genet.* 11: 49–78.

Sorensen, F. C. (1969) Embryonic genetic load in coastal Douglas-Fir. *Am. Nat.* 103: 389–398.

Sorensen, F. C. (1971) Estimate of self-fertility of Douglas-Fir from inbreeding studies. *Silvae Genetica* 20: 115–120.

Sorensen, F. C. and Miles, R. S. (1984) Inbreeding depression in height, growth, and survival of Douglas-fir, ponderosa pine, and noble fir to 10 years of age. *Forest Sci.* 28: 283–292.

Tigerstedt, P. M. A., Hiltunen, R., Chung, M. S. and Morén, E. (1979) Inheritance and genetic variation of monterpenes in Scots pine (*Pinus silvestris* L.) *In:* Rudin, D. (ed.), *Proc. Conf. Biochem. Genet. Forest Trees.* Umeå, Sweden. Gotabb. Stockholm, pp. 29–38.

Wang, X. R., Szmidt, A. E. and Lindgren, D. (1991) Allozyme differentiation among populations of *Pinus sylvestris* (L.) from Sweden and China. *Hereditas* 114: 219–226.

Wilcox, M. (1983) Inbreeding depression and genetic variances estimated from self- and cross-pollinated families of *Pinus radiata. Silvae Genetica* 32: 89–95.

Williams, C. G. and Savolainen, O. (1993) Inbreeding depression in conifers. Manuscript.

Conservation Genetics
ed. by V. Loeschcke, J. Tomiuk & S. K. Jain
© 1994 Birkhäuser Verlag Basel/Switzerland

Genetic diversity and fitness in small populations

R. C. Vrijenhoek

Center for Theoretical and Applied Genetics, and Institute of Marine and Coastal Sciences, Rutgers University, New Brunswick, NJ 08903-0231, USA

Summary. Geographical fragmentation and small population size can have manifold effects on the distribution and content of genetic diversity in endangered populations. In very small populations remnant polymorphism is likely to become correlated with variation at unobserved linked loci that affect fitness. To support these arguments, I report studies of two topminnow species in the genus *Poeciliopsis*. Laboratory studies of the endangered Sonoran topminnow, *P. occidentalis*, revealed that remnant variation marked by allozyme loci is associated with greater survival, growth, fecundity, and developmental stability. Studies of a second species, *P. monacha*, revealed that a rapid loss of variation during an extinction/recolonization event was associated with concomitant losses of developmental stability, tolerance to physical extremes, competitive ability, and an increase in parasite load. The relative fitness of this population was restored following the introduction of genetic variation from a nearby population. Heterozygosity as marked by allozyme loci is essential for the immediate fitness and survival of these *Poeciliopsis* populations. Even though most easily identified genetic polymorphisms may themselves be adaptively neutral, because of the likelihood of genic correlations arising in small populations, all remnant variation should be considered precious and worth protecting.

Introduction

Loss of genetic variability associated with inbreeding and genetic drift is believed to increase the probability of extinction of small populations (Gilpin and Soulé, 1986). Despite the potential significance of this hypothesis for conservation management, we lack evidence that inbreeding and genetic drift have caused an "extinction vortex" in nature. Consequently, a debate exists over the relative importance of management practices aimed at preserving genetic variability *versus* those aimed at controlling extinction probability due to demographic or environmental stochasticity (Lande, 1988). While I do not dispute the real problem in conservation biology – vanishing natural habitats as a consequence of anthropogenic disturbances – I argue that remnant genetic variability in endangered small populations is likely to be correlated with fitness and thereby can affect population viability, especially if the population must deal with competitors and parasites.

I base these arguments on 25 years of field and experimental studies with small populations of stream-dwelling fishes of the genus *Poeciliopsis* (Poeciliidae). Herein, I review studies of two species that occur in the Sonoran Desert region of Arizona (US) and northwestern Mexico.

38

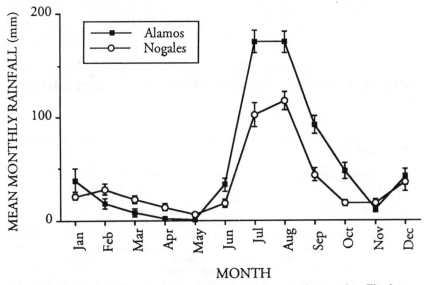

Fig. 1. The annual cycle of rainy and dry seasons in the Sonoran desert region. The data are for Nogales, Sonora, which is near the Arizona *Poeciliopsis occidentalis* sites; and Alamos, Sonora, which is near the *P. monacha* sites. Data from Hastings and Humphrey (1969).

Although still abundant in Sonora (MX), *Poeciliopsis occidentalis* is federally listed as endangered in the Gila River drainage of Arizona (U.S. Dept. Interior, 1980). *Poeciliopsis monacha* has a restricted and highly fragmented distribution in southern Sonora (MX) and northern Sinaloa (MX). During the long dry season in this arid region (Fig. 1), stream habitats typically dry up and the fish are restricted to residual pools and permanent springs. In years of severe drought local extinctions are common. The rainy season lasts from mid-summer to November, restoring rivers and allowing recolonization of temporary habitats. Theoretical considerations (below) lead us to believe that recurrent extinction and colonization events should result in losses of selectively neutral genetic polymorphism. Notwithstanding, allozyme polymorphism is maintained in many of these fish populations (Vrijenhoek et al., 1992).

Genetic markers, genomic diversity and fitness

Hughes (1991) recently asserted that "loss of diversity should not be a cause for concern because the vast majority of genetic polymorphisms are selectively neutral." Such biased statements appear to ignore much of what we have learned about genetic variation over the past 25 years!

Nevertheless, it is legitimate to ask – what kind of variation should conservationists worry about? Allozyme surveys have provided most of the information we have about genetic variation in natural populations. Yet, allozymes are generally assumed to be selectively neutral. What can allozymes and other presumably neutral markers like mitochondrial DNA polymorphism tell us about variation in ecologically relevant traits (i.e., variation in competitive abilities, niche requirements, physiological tolerances, life history characters, immune response, etc.)?

Despite the common assumption of selective neutrality, we cannot afford to dismiss allozymes and most other polymorphisms as irrelevant to the concerns of conservation biology (Vrijenhoek and Leberg, 1991). The debate over selective neutrality of allozymes has not yet reached a consensus (Kimura, 1983; Zouros and Foltz, 1987). The persistence of this problem is evident in a recent report by Karl and Avise (1992). Allozyme gene frequencies in the American oyster, *Crassostrea virginica*, were known to be relatively homogeneous across this species' range, but a historical boundary separating Atlantic and Gulf of Mexico populations was revealed by mtDNA and anonymous single-copy nuclear (scnDNA) patterns. Karl and Avise assumed that the DNA polymorphisms are more likely to reveal historical patterns of isolation because variation at synonymous nucleotide sites or in non-coding regions is likely to be selectively neutral. They concluded that the homogeneity of allozyme frequencies across this boundary was probably due to some kind of balancing selection on the allozymes. Many of the loci used in routine electrophoretic surveys encode energy-producing enzymes, and adaptively relevant phenotypes have been associated with a number of well-studied protein polymorphisms (Koehn and Hilbish, 1987; Watt, 1985). Furthermore, numerous investigators have reported positive associations between allozyme-based estimates of genomic heterozygosity and fitness-related traits like survival, fecundity, growth rate, and developmental stability (Zouros and Foltz, 1987). Even if we assume allozymes are only neutral markers for linked "effector" loci, the burden of evidence suggests that we cannot just dismiss these markers as insignificant.

Why should allozyme-based estimates of individual heterozygosity be correlated with phenotypic values controlled by other loci? In a large panmictic population, correlations between allozyme markers and effector loci should be very weak (Chakraborty, 1981; Mitton and Pierce, 1980; Turelli and Ginzburg, 1983). However, "associative overdominance" of neutral markers can exist because each time a new deleterious mutation occurs it is in linkage disequilibrium with the genes around it (Ohta, 1981). Although such associations decay rapidly in a large population, they decay slowly in small populations and in populations that inbreed. It is especially these features (i.e., geographically isolated, highly inbred, small populations) that pertain to most endangered species.

In permanently small populations of effective size N, the correlation between alleles at two loci will reach an equilibrium, a function of the recombination rate c between genes (Hill and Robertson, 1968):

$$r = \sqrt{\frac{1}{1 + 4Nc}}.$$

In a fragmented population with limited dispersal, the development of such correlations within small colonies coupled with random drift among colonies contributes to relatively large linkage disequilibria, if one considers genic correlations within colonies relative to those of the entire population (Ohta, 1982). Fusion among genetically divergent colonies can generate significant linkage disequilibria that will take many generations to decay for closely linked genes (Nei and Li, 1973). Small population size and fusion occur commonly in human tribal isolates and lead to considerable linkage disequilibrium for blood group and protein loci (Smouse et al., 1983). Stable linkage disequilibria also are evident among allozyme loci in some *Poeciliopsis* populations (Vrijenhoek et al., 1992).

Inbreeding also can contribute to the origin and maintenance of correlations among loci (Weir et al., 1972). Furthermore, individuals that are multi-locus homozygotes in an allozyme survey are more likely to be the products of inbred matings than are heterozygous individuals (Ledig, 1986). Deviations from panmixia in a structured population will increase correlations between neutral markers and other loci. Since small fragmented populations are a principle concern of conservation biologists, we cannot discount the potential significance of genetic correlation.

The endangered Sonoran topminnow

Associations between allozyme diversity and fitness-related traits were evident in the Sonoran topminnow, *Poeciliopsis occidentalis*. In recent history, this species was the most abundant fish in lowland streams, springs, and marshes of the Gila River in southern Arizona (Hubbs and Miller, 1941). As a consequence of stream modifications, ground-water pumping, and the widespread introduction of exotic species, it presently is restricted to fewer than a dozen remnant habitats (Minckley, 1973). Most Arizona populations contain very low levels of allozyme diversity compared with *P. occidentalis* populations across the border in Sonora, Mexico (Vrijenhoek et al., 1985). In an effort to restore this species to reclaimed habitats, the U.S. Fish and Wildlife

Service used a hatchery stock derived from one of these remnant populations from Monkey Spring, a thermally stable spring-fed pool in southern Arizona. Based on genetic, ecological, and physiological criteria, Vrijenhoek et al (1985) suggested that the Monkey Spring stock was not the best choice for restoration efforts. We suggested that a genetically variable stock of fish from a thermally fluctuating stream, Sharp Spring, be used instead.

To test the wisdom of this advice, Quattro and Vrijenhoek (1989) examined fitness-related traits in three *P. occidentalis* stocks, chosen simply because of their differences in allozyme variation based on a sample of 25 loci. Of the Arizona populations, mean heterozygosity was lowest in the Monkey Spring fish (0.0%), highest in the Sharp Spring fish (3.7%), and intermediate in Tule Spring fish (1.5%). We transported gravid females from each of the three sites to a thermally controlled culture facility at Rutgers University. Laboratory-born progenies from 10 wild females of each stock were tested for survival and growth rate during a 12-week grow-out period. Fish were sacrificed at the end of this period and measured; fecundity was determined and adjusted with regression techniques to account for differences in size of the females. We also assayed fluctuating asymmetry of seven bilateral meristic characters. More heterozygous individuals typically exhibit reduced fluctuating asymmetry, which is a sign of increased developmental stability (Palmer and Strobeck, 1986).

The results were striking! The homozygous Monkey Spring fish exhibited the highest mortality, the slowest growth rate, the poorest fecundity, and the weakest developmental stability. The relative poor quality of the Monkey Spring stock strongly suggested inbreeding depression. In contrast, the genetically variable Sharp Spring fish had the best growth rate, fecundity, and developmental stability. They were not different from the Tule Spring topminnows for survival. It is unlikely that these differences in fitness-related traits were environmentally induced. All progeny grew from eggs that were produced 1 to 2 months after their mothers were placed in the culture facility. The eggs were produced by mothers that had been fed identical foods, and the progenies were gestated under identical conditions of temperature and light. We raised the newborns individually in randomly assigned containers within a flow-through culture system that exposed all fish to the same water, light, temperature, and food. Although the Monkey Spring fish performed poorly in our laboratory, it has been argued that they are well adapted to their thermally stable springhead (Constantz, 1979). Regardless of the underlying causes of their poor performance in the laboratory, the Monkey Spring fish clearly were not the best choice for re-introduction into novel environments in Arizona. The U.S. Fish and Wildlife Service has since adopted aspects of our proposal to use genetically more variable fish for its current recovery plan (Simons et al., 1989).

It is possible that the fitness differences among the three strains were only coincidentally correlated with levels of allozyme heterozygosity. Different levels of heterozygosity may reflect no more than their respective histories of population bottlenecks. Differences in performance in the laboratory may reflect their respective histories of dealing with natural environmental fluctuations *versus* stability. Although we cannot determine causal relationships in a study like this, we need to consider the possibility that such correlations between remnant genetic diversity and fitness might be typical for small endangered populations.

Relative fitness of *Poeciliopsis monacha* populations

Long-term field studies of *P. monacha* provide more insight into possible mechanisms behind relationships between heterozygosity and fitness. This species coexists with a gynogenetic biotype, *P. 2 monacha-lucida*, that we have used as a genotypically uniform control to gauge changes in the relative fitness of *P. monacha*. The gynogen requires sperm from males of *P. monacha* to activate embryogenesis, but inheritance is strictly maternal and clonal. Two clones *MML/I* and *MML/II* occur at our study sites in the Arroyo de los Platanos and neighboring streams (Fig. 2). The Platanos descends a steep gradient with many little waterfalls, and a steep barrier separates this stream from the mainstream of the Arroyo de Jaguari. My students and I have followed the population dynamics and genetics of fish at these sites since 1975.

Genetic diversity and a founder event

The uppermost pools in the Platanos (Deep Pool through Heart Pool) were completely desiccated during a severe drought in 1976. By the spring of 1978, these pools had been recolonized by *P. monacha* and *MML/I*. Casual inspection of the new colony revealed that many *P. monacha* individuals exhibited spinal curvature and other deformities that were not apparent in co-occurring *MML/I* individuals. Electrophoretic examination revealed that the upstream population of *P. monacha* was essentially homozygous for polymorphic loci that segregated at more permanent downstream sites (the Log Pool area and the mainstream Jaguari) and used to occur at the upstream site before the extinction/recolonization episode. From the drastic loss of heterozygosity in the founder population, we estimated that the genetic equivalent of one gravid female, had recolonized these uppermost pools, probably

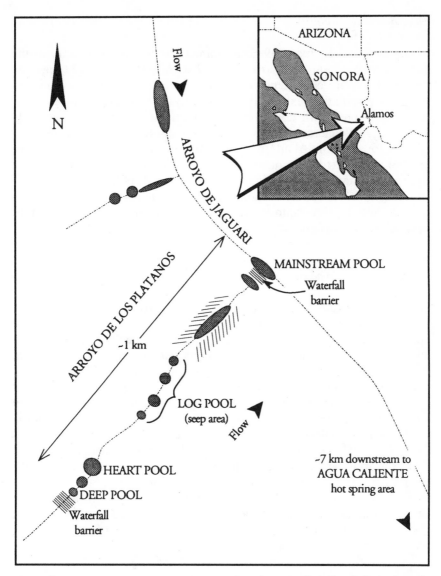

Fig. 2. The Arroyo de los Platanos and surrounding sites. Modified after Vrijenhoek et al., 1992.

a consequence of stepwise founder events during recolonization (Vrijen-hoek and Lerman, 1982). Although *MML/I* must have recolonized in the same way, it retained high heterozygosity. Clonal reproduction permanently fixes its heterozygosity. The other clone *MML/II* recolonized the upper Platanos several years later.

Developmental stability

Vrijenhoek and Lerman (1982) examined fluctuating asymmetry in the founder population of *P. monacha*. The results from eight meristic characters were consistent and unequivocal. The homozygous *P. monacha* population exhibited significantly elevated fluctuating asymmetry compared with downstream populations that retained relatively high levels of heterozygosity. The clone exhibited no comparable up-stream-downstream trend in fluctuating asymmetry, and its average level of asymmetry was not different from that of the genetically diverse sexual populations occurring downstream. Rapidly increased homozygosity of the founder population of *P. monacha* was associated with a loss of developmental stability, a classical manifestation of inbreeding depression.

Survival during hypoxic stress

A subsequent study of survival under hypoxic stress confirmed the poor fitness of the founder population of *P. monacha*. Experimental and statistical techniques are described elsewhere (Vrijenhoek et al., 1992). Fish were crowded into small containers with a limited surface area. Within 2 hours (h) when dissolved oxygen levels dropped below 2 ppm, the fish began to die. We monitored mortality during the 4-h experiment (Fig. 3). Electrophoretic examination of the dead and surviving fish from the downstream population revealed that survival of *P. monacha* (males and females) was slightly less than that of the two clones ($\chi^2 = 17.7$, 8 df, $P = 0.025$). However, the homozygous *P. monacha* from the founder population exhibited much poorer survival than the clones ($\chi^2 = 66.7$, 8 df, $P < 0.001$). I do not know why survival of the two clones was so much better at the upstream site than at the downstream site. Perhaps local conditioning played a role in these differences. If true, conditioning did not help survival of homozygous *P. monacha* from the upstream site.

Balancing selection and fluctuating stress

To better understand the relationship between heterozygosity and survival in *P. monacha*, Vrijenhoek et al. (1992) examined survival of individual genotypes at four polymorphic loci (*Idh-2*, *Ldh-1*, *Pgd*, and *Ck-A*) in the mainstream population. Each locus segregated for a + allele that is common in this species and a variant allele (*v*) that is restricted to the Platanos/Jaguari system. The results were remarkably consistent. At each of the four loci, +/+ homozygotes survived better

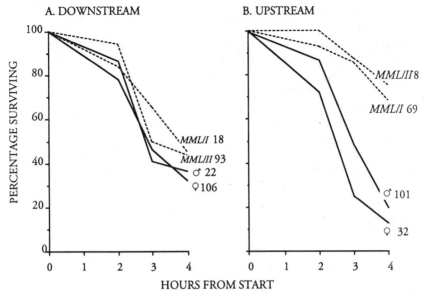

Fig. 3. Survival during hypoxic stress in *P. monacha* males and females and two clones of *P. 2 monacha-lucida* from the mainstream Jaguari (A) and the upper Platanos (B). Numbers following the labels indicate the beginning sample sizes. For the upstream Platanos sample, the numbers of *P. monacha* were enhanced beyond their normal frequency. The two clones could not be discriminated in the field, so their numbers reflect their frequencies at the two sites.

than *v/v* homozygotes. Heterozygotes were either intermediate (*Pgd*), partially dominant with respect to the + allele (*Idh-2* and *Ldh-1*), or overdominant (*Ck-A*).

These results raised a significant paradox with respect to three of the loci (*Pgd*, *Idh-1*, and *Ldh-1*). If the *v* alleles are so poor, why do the polymorphisms persist at these loci? The *v* alleles are not rare and allelic frequencies at most localities (except the upper Platanos) are relatively stable over time. Thus, Vrijenhoek et al. (1992) searched for factors that could balance the negative fitness effects of the variant alleles. We examined survival under heat stress (exposure to 40° C for 30 min) and cold stress (exposure to ~2° C for 8 min) in independent trials with *P. monacha* from a neighboring mainstream locality. The heat stress trials produced results that were nearly identical to the hypoxic stress results, even though we maintained non-stressful dissolved oxygen levels during the heat stress experiment. Again for each of the four loci, +/+ homozygotes had higher survival than *v/v* homozygotes. The parallel results from heat and hypoxic stress were not surprising, however, because dissolved oxygen ordinarily decreases with elevated temperature in nature. Adaptive responses to these stresses are probably coupled physiologically.

In contrast, we obtained opposite results during the cold stress experiment. Now v/v homozygotes had higher survival than $+/+$ homozygotes; the heterozygotes were intermediate. Alleles at each of the polymorphic allozyme loci marked a balancing process in which heterozygotes showed the lowest variance in survival across opposing kinds of stress. Although heterozygotes were generally not the best genotype (except for Ck-A under heat and hypoxic stress), they were never the worst. The elevated variance in fitnesses of the homozygotes would confer a lower geometrical mean fitness as opposing stresses vary during the year. Geometrical mean overdominance should be sufficient to maintain such polymorphism in a heterogeneous environment (Hedrick, 1976; Hedrick, 1986). Averaged across time and space, the loss of such variation would diminish the survival of *P. monacha* populations relative to that of clones in which heterozygosity is fixed.

Whether the allozymes were the effectors in these experiments or just associated markers is unknown. A high degree of linkage disequilibrium existed in the populations used for these experiments. These associations, even between unlinked loci, were a consequence of admixture between differentiated gene pools that mixed during the rainy season (Vrijenhoek et al., 1992). However, correlations also exist between chromosomally linked loci (e.g., *Ldh-1* and *Idh-2*) in partially isolated, small populations of *P. monacha* that are sheltered from admixture (Vrijenhoek, unpublished). Thus, the allozymes examined in these fish mark significant regions of their respective chromosomes. The allozymes themselves could be under selection or they might be neutral and "hitchhiking" (*sensu* Maynard Smith and Haigh, 1974). In any event, loss of these polymorphisms and associated variability at other loci can debilitate a population that must live and compete with ecologically similar species (in this case, clones).

Genetic diversity and population dynamics

Vrijenhoek (1989) reported that the loss of genetic diversity in the upstream Platanos *P. monacha* drastically altered its competitive interactions with *MML/I* and *MML/II*. The clones and *P. monacha* have relatively similar diets, but differential use of space on micro- and macro-spatial scales reduces competition between them (Schenck and Vrijenhoek, 1989). *Poeciliopsis monacha* dominates headwater areas like the Platanos, and *MML/II* is more frequent in downstream areas that are more productive. *MML/I* comprises roughly 10% of the fish in most localities and is broadly insensitive to changes in frequencies of the other two forms. Prior to the extinction/recolonization event in the upper Platanos, *P. monacha* was polymorphic at the four loci discussed above, and it constituted about 76% of the fish. After recolonization, heterozygosity was zero, and *P. monacha*'s frequency was only 5%; the rest were *MML/I*. This shift in heterozygosity and abundance remained relatively stable through the spring of 1983, through roughly 10–15 generations. No corresponding shifts in abundance occurred downstream where *P. monacha* remained numerically dominant and its gene diversity remained stable.

In 1983, I performed a simple experiment to determine whether loss of genetic diversity in the founder population of *P. monacha* had compromised its competitive ability with respect to the clone. I took 30 genetically variable *P. monacha* females from the mainstream pool (Fig. 2), and used them to replace 30 *P. monacha* females in Heart Pool at the upper end of the Platanos. By the spring of 1985, the mean gene diversity for the four polymorphic loci climbed to a level consistent with that of the source population in the mainstream Jaguari. Simultaneously, *P. monacha* regained its ecological dominance, about 80% of the fish. Both abundance and gene diversity of *P. monacha* have remained stable ever since. Strong aggressive interactions occur among *P. monacha* females and with the clonal females (Schenck and Vrijenhoek, 1989). The poor fitness manifested as diminished developmental stability and increased mortality during stress also compromised the competitive ability of the homozygous *P. monacha* population.

Parasites and the Red Queen

Genetic diversity in *P. monacha* also appears to be essential for its struggles with parasites. According to the Red Queen model, rapidly evolving micro-parasites should focus their attack on the most common host genotype (Bell, 1982; Hamilton, 1982). This leads to a frequency-dependent process that maintains genetic diversity in the host and favors the maintenance of sex. To test this hypothesis, Lively et al. (1990) examined parasite loads in fish from the Arroyo de los Platanos. Larvae of the monogenean trematode *Uvulifer* sp. infected the fish and produced black spot disease. Quantifying parasite loads was simply a matter of counting black spots and adjusting for the fish's body size. Our results were consistent with predictions of the Red Queen model – the most common clone in each habitat had the highest parasite load and *P. monacha* typically had the lowest. The sexual species was more abundant than the clones in most habitats, but each sexual individual has a unique genotype. Sexual individuals not only had fewer parasites but they exhibited about twice the variance in numbers of parasites *per* fish when compared with the co-occurring clones. Increased variance in parasite load is expected if individuals vary in susceptibility.

A highly informative exception to this pattern occurred in the upper Platanos. The homozygous founder population of *P. monacha* in Heart

Pool had a significantly higher parasite load than *MML/I*, and the variances in parasite loads were nearly identical. In 1983, we introduced genetic diversity into the founder population by translocating 30 *P. monacha* females from the variable mainstream population. By 1985, the genetically variable *P. monacha* had a lower parasite load and twice the variance in parasite load of *MML/I*, which represented return to the typical situation for these fish. Genetic variation among sexual individuals facilitates dealing with this parasite. Again, loss of allozyme variation was correlated with a general decline in ability to cope with biotic or physical challenges, this time the ecological insults of parasites. Restoration of genetic variability with a small transplant of individuals restored the robustness of the *P. monacha* population.

Genetic diversity, local extinctions and persistence

The ability of *P. monacha* to live and to compete with the clonal forms depends on genetic diversity. An extinction/recolonization event rapidly eliminated genetic diversity and produced a population that exhibited poor developmental stability, poor survival, poor competitive ability, and an increased parasite load. Its population size in the upper Platanos declined to levels where demographic stochasticity could easily result in local extinctions. Between 1978 and 1983, local extinctions of *P. monacha* occurred in several pools neighboring Heart Pool. Restoration of genetic variability in the Heart Pool population also restored fitness to *P. monacha* in adjacent upstream pools, and populations have remained stable in this area since 1983. The purpose of sexual reproduction, as far as we can tell, is to generate genotypic diversity among individuals. If you eliminate genetic diversity in *P. monacha*, the clonal form dominates the ecological contest between these forms of fish. Genetically variable *P. monacha*, on the other hand, tend to predominate in most populations.

Setting aside the special characteristics of *Poeciliopsis* for a moment, I can imagine similar interactions between a rare endangered species and its native and exotic competitors, predators, and parasites. Certainly loss of heterozygosity and inbreeding depression can have manifold effects on the general fitness of individuals. Such losses can debilitate a population that must live and interact with other native or exotic species. Genotypic diversity among individuals helps to reduce intraspecific competition for resources and it is critically needed for dealing with parasites. If you take variability away from a sexual species by fragmenting it into tiny isolated and inbred colonies, you risk destroying its very essence.

Model organisms and management implications

Although the mechanisms that cause associations between allozyme diversity and fitness-related traits remain elusive in studies such as these, we do not need to understand their causes to use these relationships in species restoration plans. If given a choice of natural stocks for a restoration program, I would recommend using the genetically most variable population, as we did for *P. occidentalis* in Arizona (Vrijenhoek et al., 1985). If a natural population has maintained "presumably neutral" variation like allozymes, it also is likely to carry adaptively relevant variation maintained by balancing forces. On the other hand, it is difficult to guess about "other" variation in a population that is monomorphic for allozymes. It might be variable at other loci; we just cannot tell, and we often do not have enough time to test for quantitative variation in ecologically relevant traits. A robust, genetically variable, natural stock would have the best chances of adapting quickly to novel conditions. Restoration programs should focus on using natural stocks rather than captively bred stocks, because natural stocks retain their history of dealing with parasites, predators, and competitors. Captive stocks typically have reduced variation, and they tend to become entrained and adapted to a protected life under domestication (Lacy, 1987). One can design a breeding program that would minimize the loss of ecologically relevant variance, and genetic markers can be useful as tools for monitoring remnant variation. However, it would be risky to breed specifically to maximize allozyme or mitochondrial DNA polymorphism. Programs aimed at maximizing variation at a small number of marker loci might accelerate loss of variation at other unlinked loci (Hedrick et al., 1986). Numerous genetic polymorphisms, dispersed throughout the chromosomes, would be desirable tools for captive breeding programs. Perhaps RAPDs (randomly amplified polymorphic DNAs) can serve this purpose.

It is occasionally suggested that inbreeding might be used as a management tool. Templeton and Read (1983) devised an inbreeding program that is intended to purge deleterious recessive mutations from a captive population of Speke's gazelle (*Gazella spekei*). Following many caveats, they justified this plan *only* as a method of last resort for a genetically debilitated stock. If the captive population survives the selective purging and manifestations of inbreeding depression are abated, the stock will very likely be reduced to little more than a domesticated breed. Although the Speke's gazelle program might save the breed, we should not view this inbreeding plan as broadly applicable. The goals of a captive breeding program should be to avoid further losses of variation and to alter a stock as little as possible.

Recent experiments with houseflies (Bryant et al., 1986) and fruitflies (Lopez-Fanjul and Villaverde, 1989) revealed that inbreeding can increase the expression of genetic variance that is available for natural selection. Apparently, inbreeding can convert variation ordinarily masked by dominance and epistasis into additive variance, thereby increasing the variance within and among inbred lines (Robertson, 1952; Goodnight, 1988). Selection across a large number of inbred strains with some migration between them should lead to more rapid change than selection in an undivided population (Wright, 1931). Although increasing the scope for natural selection might be desirable, rapid evolutionary change provides no guarantee of long-term persistence. Rapid evolution provides little help when the major cause of extinction is habitat loss. The inbreeding studies with houseflies and fruitflies revealed an additional problem associated with the increase in variance within and among lines. The average fitness of the inbred lines decreased significantly (Hedrick and Miller, 1992). An inbreeding program would appear to be a risky strategy for most endangered species.

The ultimate goal of captive breeding and *in situ* propagation plans must be restoration of the endangered species to its former range. We should choose restoration stocks based on a sound knowledge of genetic variability, population structure, and alpha-level systematics. In some circumstances, a mixed stock, containing genes from all remnant colonies, might be the best choice. Local selection pressures would preserve the best regional gene combinations. However, one would not want to mix stocks with highly divergent evolutionary histories as this might disrupt locally beneficial variation and lead to outbreeding depression (Templeton, 1986). Sometimes translocation of a small number of individuals can restore variability and viability of an endangered population, as exemplified by our *P. monacha* studies. Managed admixture among remnant populations is warranted if one is certain that recent anthropogenic disturbances were the cause of fragmentation (Allendorf, 1983; Meffe and Vrijenhoek, 1988).

Spielman and Frankham (1993) elegantly demonstrated the benefits of managed gene flow in experimental studies with *Drosophila*. They created a series of small partially inbred populations. They noted significant declines of fitness in all lines relative to an outbred control population. Introduction of just one immigrant into each of the inbred line resulted in substantial increases in the relative fitness of their respective progenies. On average, these introductions restored nearly 50% of the fitness that was lost under inbreeding. Clearly, we can learn much from experimental studies like this. Because we cannot perform similar studies with the white rhinoceros or California condor, we have little choice but to generalize from studies of model organisms. Field studies with organisms like *Poeciliopsis* provide a helpful bridge across what must seem like a leap of faith for some conservation managers. Unlike *Drosophila*, these fish have a large number of chromosomes ($2N = 48$), they exhibit natural population structures typical of fragmented species, they are endangered in Arizona, and seriously threatened in many areas of Sonora.

Field and experimental studies of model organisms like *Poeciliopsis* and *Drosophila* have contributed much to our understanding of associations between genetic diversity and fitness. Most empirical research clearly refutes cavalier statements that most genetic polymorphism is irrelevant to conservation (e.g., Hughes, 1991). It may be convenient for some theoreticians to ignore such empirical studies, or to dismiss these associations as little more than byproducts of non-random mating systems or structured populations. Yet, small fragmented populations are precisely the case for most endangered species. The remnant variation that occurs within and the differences between these population fragments should be considered precious, because of the correlations that can develop among genes. Even if most allozymes and other convenient genetic markers prove to be adaptively neutral, I suspect that conservationists cannot take for granted any remnant variation.

Acknowledgements
I thank the students and colleagues who have helped me in this research and accompanied me in the field, and particularly S. Karl, C. Craddock, and O. Folmer for their helpful comments on the manuscript. This is contribution No. D-67175-3-94 of the New Jersey Agricultural Experiment Station and No. 94-09 of the Institute of Marine and Coastal Sciences, supported by state funds and NSF grants (INT84-16427; BSR86-00661; BSR88-05361).

References

Allendorf, F. W. (1983) Gene flow, and genetic differentiation among populations. *In:* Schonewald-Cox, C. M., Chambers, S. M., McBryde, F. and Thomas, L. (eds), *Genetics and conservation: a reference for managing wild animal and plant populations.* Benjamin/Cummings, Menlo Park, CA, pp. 241–261.

Bell, G. (1982) *The masterpiece of nature. The evolution and genetics of sexuality.* Univ. of California Press, Berkeley, CA.

Bryant, E. H., Combs, L. C. and McCommas, S. A. (1986) The effect of an experimental bottleneck upon quantitative variation in the housefly. *Genetics* 114: 1191–1211.

Chakraborty, R. (1981) The distribution of the number of heterozygous loci in an individual in natural populations. *Genetics* 98: 461–466.

Constantz, G. D. (1979) Life history patterns of a livebearing fish in contrasting environments. *Oecologia* 40: 189–201.

Gilpin, M. E. and Soulé, M. E. (1986) Minimum viable populations: processes of species extinction. *In:* Soulé, M. E. (eds), *Conservation biology: the science of scarcity and diversity.* Sinauer Associates, Sunderland, MA, pp. 19–34.

Goodnight, C. J. (1988) Epistasis and the effect of founder events on the additive genetic variance. *Evolution* 42: 441–454.

Hamilton, W. D. (1982) Pathogens as causes of genetic diversity in their host populations. *In:* Anderson, R. M. and May, R. M. (eds), *Population biology of infectious diseases.* Life Sci. Res. Rep. 25, Springer, Berlin, pp. 269–296.

Hastings, J. R. and Humphrey, R. R. (1969) *Climatological data and statistics for Sonora and northern Sinaloa.* Technical Reports on the Meteorology and Climatology of Arid Regions, Report No. 19. The University of Arizona Institute of Atmospheric Physics.

Hedrick, P. W. (1976) Genetic variation in a heterogeneous environment. II. Temporal heterogeneity and directional selection. *Genetics* 84: 145–150.

52

Hedrick, P. W. (1986) Genetic polymorphism in heterogeneous environments: a decade later. *Annu. Rev. Ecol. Syst.* 17: 535–566.

Hedrick, P. W., Brussard, P. R., Allendorf, F. W., Beardmore, J. A. and Orzack, S. (1986) Protein variation, fitness, and captive propagation. *Zoo Biol.* 5: 91–99.

Hedrick, P. W. and Miller, P. S. (1992) Conservation genetics: techniques and fundamentals. *Ecol. Appl.* 2: 30–46.

Hill, W. G. and Robertson, A. (1968) Linkage disequilibrium in finite populations. *Theor. Appl. Genet.* 38: 226–231.

Hubbs, C. L. and Miller, R. R. (1941) Studies of the order Cyprinodontes, XVII. Genera and species of the Colorado River system. *Occas. Pap. Mus. Zool. Mich.* 433: 1–9.

Hughes, A. L. (1991) MHC polymorphism and the design of captive breeding programs. *Cons. Biol.* 5: 249–251.

Karl, S. A. and Avise, J. C. (1992) Balancing selection at allozyme loci in oysters: Implications from nuclear RFLPs. *Science* 256: 100–102.

Kimura, M. (1983) The neutral theory of molecular evolution. *In:* Nei, M. and Koehn, R. K. (eds), *Evolution of genes and proteins.* Sinauer, Sunderland, MA, pp. 208–233.

Koehn, R. K. and Hilbish, T. J. (1987) The adaptive importance of genetic variation. *Amer. Sci.* 75: 134–141.

Lacy, R. C. (1987) Loss of genetic diversity from managed populations: interacting effects of drift, mutation, immigration, selection, and population subdivision. *Cons. Biol.* 1: 143–158.

Lande, R. (1988) Genetics and demography in biological conservation. *Science* 241: 1455–1460.

Ledig, F. T. (1986) Heterozygosity, heterosis, and fitness in outbreeding plants. *In:* Soulé, M. E. (eds), *Conservation biology: the science of scarcity and diversity.* Sinauer Associates, Sunderland, MA, pp. 77–104.

Lively, C. M., Craddock, C. and Vrijenhoek, R. C. (1990) The Red Queen hypothesis supported by parasitism in sexual and clonal fish. *Nature* 344: 864–866.

Lopez-Fanjul, C. and Villaverde, A. (1989) Inbreeding increases genetic variance for viability in *Drosophila melanogaster. Evolution* 43: 1800–1804.

Maynard Smith, J. and Haigh, J. (1974) The hitch-hiking effect of a favorable gene. *Genet. Res. (Cambridge)* 23: 23–35.

Meffe, G. K. and Vrijenhoek, R. C. (1988) Conservation genetics in the management of desert fishes. *Cons. Biol.* 2: 157–169.

Minckley, W. L. (1973) *Fishes of Arizona.* Ariz. Game & Fish Dept., Phoenix, AZ, USA.

Mitton, J. B. and Pierce, B. A. (1980) The distribution of individual heterozygosity in natural populations. *Genetics* 95: 1043–1054.

Nei, M. and Li, W. H. (1973) Linkage disequilibrium in subdivided populations. *Genetics* 75: 213–219.

Ohta, T. (1981) Associative overdominance caused by linked detrimental mutations. *Genet. Res.* 18: 277–286.

Ohta, T. (1982) Linkage disequilibrium due to random genetic drift in finite subdivided populations. *Proc. Natl. Acad. Sci. USA* 79: 1940–1944.

Palmer, A. R. and Strobeck, C. (1986) Fluctuating asymmetry: Measurement, analysis, patterns. *Annu. Rev. Ecol. Syst.* 17: 291–421.

Quattro, J. M. and Vrijenhoek, R. C. (1989) Fitness differences among remnant populations of the Sonoran topminnow, *Poeciliopsis occidentalis. Science* 245: 976–978.

Robertson, A. (1952) The effect of inbreeding on the variation due to recessive genes. *Genetics* 37: 189–207.

Schenck, R. A. and Vrijenhoek, R. C. (1989) Coexistence among sexual and asexual forms of *Poeciliopsis:* foraging behavior and microhabitat selection. *In:* Dawley, R. and Bogard, J. (eds), *Evolution and ecology of unisexual vertebrates,* Vol. Bulletin 466. New York State Museum, Albany, NY, pp. 39–48.

Simons, L. H., Hendrickson, D. A. and Papoulias, D. (1989) Recovery of the Gila topminnow: A success story? *Cons. Biol.* 3: 11–15.

Smouse, P. E., Neel, J. V. and Liu, W. (1983) Multiple-locus departures from panmictic equilibrium within and between village gene pools of Amerindian tribes at different stages of acculturation. *Genetics* 104: 133–153.

Spielman, D. and Frankham, R. (1993) Modeling problems in conservation genetics using captive *Drosophila* populations: improvement of reproductive fitness due to immigration of one individual into small partially inbred populations. *Zoo Biol.* 12: 343–348.

Templeton, A. (1986) Coadaptation and outbreeding depression. *In:* Soulé, M. E. (ed.), *Conservation biology: the science of scarcity and diversity.* Sinauer Associates, Sunderland, MA, pp. 105–116.

Templeton, A. R. and Read, B. (1993) The elimination of inbreeding depression in a captive herd of Speke's gazelle. *In:* Schonewald-Cox, C. M., Chambers, S. M., McBryde, F. and Thomas, L. (eds), *Genetics and conservation: a reference for managing wild animal and plant populations.* Benjamin/Cummings, Menlo Park, CA, pp. 241–261.

Turelli, M. and Ginzburg, L. R. (1983) Should individual fitness increase with heterozygosity? *Genetics* 104: 191–209.

U.S. Dept. Interior (1980) Endangered and threatened wildlife and plants. *Fed. Reg.* 45: 99.

Vrijenhoek, R. C. (1989) Genotypic diversity and coexistence among sexual and clonal forms of *Poeciliopsis. In:* Otte, D. and Endler, J. (eds), *Speciation and its consequences.* Sinauer Associates, Sunderland, MA, pp. 386–400.

Vrijenhoek, R. C., Douglas, M. E. and Meffe, G. K. (1985) Conservation genetics of endangered fish populations in Arizona. *Science* 229: 400–402.

Vrijenhoek, R. C. and Leberg, P. L. (1991) Let's not throw the baby out with the bathwater: a comment on management for MHC diversity in captive populations. *Cons. Biol.* 5: 252–253.

Vrijenhoek, R. C. and Lerman, S. (1982) Heterozygosity and developmental stability under sexual and asexual breeding systems. *Evolution* 36: 768–776.

Vrijenhoek, R. C., Pfeiler, E. and Wetherington, J. (1992) Balancing selection in a desert stream-dwelling fish, *Poeciliopsis monacha. Evolution* 46: 1642–1657.

Watt, W. B. (1985) Allelic isozymes and the mechanistic study of evolution. *In:* Ratazzi, M. C., Scandalios, L. G. and Whitt, G. S. (eds), *Isozymes: current topics in biological and medical research,* Vol. 12. Alan R. Liss, New York, NY, pp. 89–132.

Weir, B., Allard, R. and Kahler, A. (1972) Analysis of complex allozyme polymorphisms in a barley population. *Genetics* 72: 505–523.

Wright, S. (1931) Evolution in Mendelian populations. *Genetics* 16: 97–159.

Zouros, E. and Foltz, D. W. (1987) The use of allelic isozyme variation for the study of heterosis. *In:* Rattazi, M. C., Scandalios, J. G. and Whitt, G. S. (eds), *Isozymes: current topics in biological and medical research,* Vol. 13. Alan R. Liss, New York, pp. 1–59.

Conservation Genetics
ed. by V. Loeschcke, J. Tomiuk & S. K. Jain
© 1994 Birkhäuser Verlag Basel/Switzerland

Mutation load depending on variance in reproductive success and mating system

D. Couvet[1] and J. Ronfort[2]

[1]Laboratoire d'Ecologie, ENS, 46 rue d'Ulm, F-75005 Paris, Cédex 05, France
[2]Center Emberger, CNRS, Route de Mende, BP 5051, F-34033 Montpellier, Cédex, France

Summary. Equalization of reproductive success of individuals, although it results in an increase of effective population size, leads also to an increase of the mutation load. The magnitude of this increase depends highly on the mode of fitness interactions between deleterious mutations, and is higher in the case of inbreeding. Recommended practices in conservation genetics must be evaluated in regards to these differing consequences of an increase of effective population size. To keep a balance between retaining genetic variability and minimizing the increase of the mutation load, equalization of reproductive success of a set of individuals rather than of every individual might be more advantageous.

Introduction

Population genetics can be a tool for investigating the history of populations of endangered species; measuring heterozygosity, number of alleles, and the distribution of the coalescence time of alleles; and using the theoretical interpretation of such indices (for review, see Hedrick and Miller, 1992). Furthermore, population genetics might be a matter of interest in itself if variations of genetic properties of populations of endangered species increase their probability of extinction. Two types of variations taking place in small populations might be worth noting: the increase of the mutation load and the restriction of genetic variability.

The factors of variation of the mutation load

Mutation load names the reduction of performances of individuals due to repeated occurrence of deleterious mutations (Haldane, 1937). This load significantly affects survival and fertility in *Drosophila* (Simmons and Crow, 1977). In other species, spontaneous abortion rates that can be as high as 50% in higher plants as well as in higher animals (Wiens et al., 1989), are supposed to reflect, in part, the existence of the mutation load; abortion rates are higher in outcrossers than in selfers (Wiens et al., 1987), which is to be expected if abortions are due to the load of deleterious mutations (Crow, 1970).

The load *per* locus will vary with the mutation rate, but not with the deleteriousness of these mutations (Haldane, 1937). The load for the whole genome will decrease when synergistic effects on fitness of deleterious mutations occur (Kimura and Maruyama, 1966). There is experimental evidence of such synergistic effects in *Drosophila* (Kimura and Maruyama, 1966; Crow, 1970) and in plants (Charlesworth et al., 1991), as well as a physiological rationale whereby the accumulation of deleterious effects should lead to an increasing break-down of developmental homoeostasis. In presence of these synergistic effects, the magnitude of the load depends on the genetic variance among progenies and is, for example, higher in the case of parthenogenesis (Kimura and Maruyama, 1966; Kondrashov, 1988). The mutation load is expected to increase in small populations, due to the fixation of deleterious alleles in such populations (Kimura et al., 1963), and also due to the accumulation of loci where such alleles are present but not fixed, a process known as Muller's ratchet (Charlesworth et al., 1993).

The restriction of genetic variability in small populations

Such restriction is expected to alter the response of populations to variations of selection pressures and thus prevent future adaptation (e.g., Franklin, 1980). The rate of loss of genetic variability increases with variance in reproductive success of individuals (Wright, 1969, pp. 215–217), a theoretical expectation supported by experiments on *Drosophila* (Borlase et al., 1992). Besides avoiding variance of reproductive success of individuals to increase retention of genetic variability, mates might be chosen according to their pedigrees. Two mating systems have been extensively studied: circular mating and maximum avoidance of inbreeding (Wright, 1969, pp. 199–204). Circular mating, where individuals in a circular arrangement are crossed with a neighboring individual, and so on, results in mates that are half-siblings. On the other hand, in the case of maximum avoidance of inbreeding, mates have the most distant genetic relationship as possible. Circular mating greatly decreases the rate of loss of genetic variability in a finite population, whereas maximum avoidance of inbreeding does not affect that rate relative to the comparable case of panmixia, that is, without variance in reproductive success (Kimura and Crow, 1963; Wright, 1969; Nagylaki, 1992). Other systems with unequal reproductive success of individuals further decrease the rate of loss of genetic variability compared to circular mating, at least in the case of a small number of individuals (Boucher and Cotterman, 1990), but as no rule exists at this point regarding the design of such systems for an arbitrary number of individuals, they will not be presently considered.

The interaction between the variance of reproductive success of individuals and the mutation load

Avoiding variance in reproductive success of individuals implies a change of the selection regime, in particular for deleterious mutations. Equalizing reproductive success of individuals implies that *familial* selection *sensu* Haldane (1924), where selection takes place among offspring produced within each family (or by each individual in the case of polygamous matings), and replaces *mass* selection, where selection takes place among the whole set of offspring produced. Different terms are used for these two selection regimes; they are called, respectively, *classical* and *global reproductive compensation* (Campbell, 1988). *Familial* selection has also been termed *intrabrood* selection (Wright, 1969, p. 143), *intra-family* selection (Nagylaki, 1992, pp. 71–73), or *equalization of family size* (Borlase et al., 1992).

Familial selection leads to a slower response of the character selected (Haldane, 1924; King, 1965); this result can in fact be deduced from Fisher's Fundamental Theorem, since the efficiency of selection depends on the available variability for fitness, and selection within families obviously decreases the variability on which selection acts. A slower response to selection is advantageous when adaptation to captivity is the trait concerned, that being an undesirable property in the wild (Allendorf, 1993). In regards to deleterious mutations, the number of individual deaths necessary to eliminate lethal genes that are completely recessive or semi-dominant was shown to be larger in the case of the equalization of individual reproductive success (Campbell, 1988), suggesting a higher mutation load in this case.

The idea of trade-offs between different kinds of genetic variability has already been investigated, but usually considers the consequences, beneficial and disadvantageous, of a decrease in genetic variability, that is, the fact that selection for a certain character leads to a higher loss of genetic variability for other characters (e.g., Haig et al., 1990). We will instead focus in this paper on the manifold consequences of maximizing genetic variability, that is, the fact that avoiding selection to minimize the loss of genetic variability will lead to a higher frequency of deleterious mutations, and hence a loss of fitness. For the sake of analytical tractability, we will first consider the case of an absence of interactions between the deleterious alleles, in terms of fitness and linkage disequilibrium (i.e., when their interaction for fitness is multiplicative and the map length of the genome is infinite so that they recombine freely), and with random mating. The consequences of these interactions will be then considered with the help of computer simulations on populations of finite size, but of sizes large enough to avoid fixation of deleterious alleles or the occurrence of Muller's ratchet in a finite number of generations. The effect of different mating systems will be compared in this latter case.

The mutation load depending on selection regime: Analytical results

We will consider the classical model with one locus and two alleles A and a, where a is recessive and deleterious. Mutation towards a happens at a rate u; the rate of backward mutation (a to A) is neglected. Fitnesses are the following:

AA	Aa	aa
1	$1 - hs$	$1 - s$,

s being the deleterious effect, and h its dominance. The mutation load for this locus is then the decrease of average selective value due to the existence of a. If y and z are frequencies of Aa and aa, respectively, then the load is $s(hy + z)$ (Haldane, 1937).

The case of mass selection

These are classical results that we will recall for comparison. The frequency of a at equilibrium, q, is when h differs from 0.5:

$$q \approx (-hs(1 + u) + \sqrt{[(hs(1 + u))^2 + 4us(1 - 2h)]})/2s(1 - 2h)$$

(Crow, 1970), which is approximated as:

$$q \approx \sqrt{(u/s)} \qquad \text{when h} = 0, \tag{1}$$

$$q \approx u/hs \qquad \text{when h} \gg \sqrt{(u/s)} \tag{2}$$

(Wright, 1969, p. 34). The load l is:

$$l = q(2h(1 - q) + q) \approx u(1 - H + \sqrt{(H(2 + H))}),$$

with $H = sh^2/2u(1 - 2h)$ (Kimura et al., 1963), which is approximated as:

$$l \approx u \qquad \text{when h} = 0 \text{ (Haldane, 1937)},$$

$$l \approx 2u \qquad \text{when h} \gg \sqrt{(u/s)}. \tag{3}$$

The increase of the load with h is rapid (see Kimura and Ohta, 1971, Fig. p. 51).

The case of familial selection

Frequencies of the different crosses, and of the progenies within each cross are given in Tab. 1. Thus, frequencies at the next generation, x', y', z' are the following (neglecting interaction terms between mutation and selection):

$$x' = x^2 + 2xy/(2 - hs) + y^2/(4 - 2hs - s) - 2ux, \tag{4}$$

$$y' = 2xy(1 - hs)/(2 - hs) + 2xz + 2y^2(1 - hs)/(4 - 2hs - s)$$
$$+ 2yz(1 - hs)/(2 - hs - s) + 2ux - uy, \tag{5}$$

$$z' = y^2(1 - s)/(4 - 2hs - s) + 2yz(1 - s)/(2 - hs - s) + z^2 + uy. \tag{6}$$

The load l is

$$l = s[y(hx + ((2h + 1)/4)y + z(h + 1)) + z(z + hx)].$$

At equilibrium:

$$z \approx y[u + y(1 - s)/(4 - 2hs - s)].$$

$$y \approx [H - \sqrt{(H^2 - 16uG)}]/4G,$$

with $H = u + hs/(2 - hs)$ and $G = (4 - 2hs - 2s)/(4 - 2hs - s) - (1 - hs)/(2 - hs)$. Frequency of a, q, equal to $(y/2 + z)$, is:

$$q \approx \sqrt{(u/s(2 - s/2))}, \quad \text{when } h = 0,$$

$$q \approx u/hs(2 - hs), \quad \text{when } h \gg \sqrt{(u/s)}.$$

The frequency of a is approximately double as compared to the case of mass selection. The load is:

$$l \approx 2u(1 - s/4) \quad \text{when } h = 0,$$

$$l \approx 4u(1 - hs/2) \quad \text{when } h \gg \sqrt{(u/s)}. \tag{7}$$

The load is approximately twice compared to the case of mass selection when hs is close to 0, that is in the case of slightly deleterious alleles. In

Table 1. Frequencies of the different offspring depending on frequencies of parents, in the case of familial selection

Genotype of parents	Frequency of crosses	Frequency of offspring within each family		
		AA	Aa	aa
$AA \times AA$	x^2	1	0	0
$AA \times Aa$	$2xy$	$1/(2 - hs)$	$(1 - hs)/(2 - hs)$	0
$AA \times aa$	$2xz$	0	1	0
$Aa \times Aa$	y^2	$1/(4 - 2h - s)$	$(2 - 2hs)/(4 - 2hs - s)$	$(1 - s)/(4 - 2hs - s)$
$Aa \times aa$	$2yz$	0	$(1 - hs)/(2 - hs - s)$	$(1 - s)/(2 - hs - s)$
$aa \times aa$	z^2	0	0	1

the case of complete recessivity, the load is halved. The increase of the load with dominance of deleterious alleles is rapid as in the case of mass selection (see above).

Dynamics of the variation of the load when the selection regime changes

Numerical iterations of the previous equations (4, 5, and 6) were performed to investigate at which time scale the increase of the load in the case of familial selection takes place, starting from the state of equilibrium defined in the case of mass selection (Eqs 1 or 2). Dominance has a significant effect on these dynamics. In the case of complete recessivity, the change of load is hardly noticeable after 50 generations, whereas there is a significant change of fitness of individuals with a moderate level of dominance (Fig. 1). This difference in the dynamics of change of the load reflects the difference for the rate of increase of frequency of a, which is slower in the case of complete recessivity.

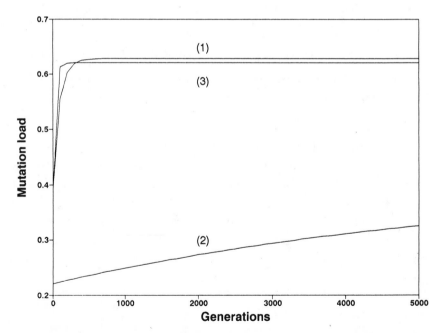

Fig. 1. Dynamics of change of load when familial selection replaces mass selection, depending on dominance of deleterious effect. Results were obtained with numerical iterations of Eqs (4), (5) and (6). Interactions for fitness are multiplicative, and recombination is free. (1): s = 0.1, h = 0.2; (2): s = 0.1, h = 0; (3): s = 0.3, h = 0.2.

The effect of mating system and interactions between deleterious mutations on the mutation load: Results from a simulation model

Methods of simulation

The model used was similar to the one used by Hauser et al. (this volume). This is a multi-locus model, with two alleles *A* and *a* (see above); the number of loci where there is polymorphism, i.e., *A* and *a* are present, reaches a stable equilibrium which depends on, *per* generation, the number of (i) loci fixed for the normal allele *A* where mutations towards deleterious alleles occur, and (ii) polymorphic loci where fixation of one of the alleles, *A* or *a*, occurs. This number can be quite variable, depending on mutation rates, population size, and selection regime (Tab. 2).

The action of selection is simulated the following way: the selective value of an offspring produced is compared to a random number generated from a uniform law between 0 and 1. The offspring is retained if its selective value is higher than this number. When the offspring is rejected, in the case of mass selection, two parents are again picked from among the population (which means that an individual can produce offspring with several mates); in the case of familial selection, offspring from the same parents are produced until one is not rejected.

Different mating systems were compared. Precisely defined mating systems like circular mating or maximum avoidance of inbreeding necessarily imply that familial selection will take place (but see Campbell, 1993). Thus, only panmixia was examined in the case of mass selection. There is obviously an interaction between the selection regime and the possibility of variance in reproductive success. Mass selection with absence of variance in reproductive success implies growing con-

Table 2. Characteristics of the mutation load over the whole genome depending on selection regime and mating system, in a population of 200 dioecious individuals. The case shown is for multiplicative interactions between deleterious mutations for fitness and an infinite map length (i.e., free recombination), $V = 0.25$ (case ■ of Fig. 2). Selection regime: F: familial, M: mass selection. Mating system: Ma: Maximum avoidance of inbreeding, Ci: Circular mating, Pa: panmixia, PN: panmixia with no variance in reproductive success

Selection regime	Mating structure	Number of polymorphic loci	Number of deleterious mutations/genome	Proportion of mutations in a homozygous state
F	Ci	1400	47.5	0.63
F	Ma	1144	40.8	0.03
F	PN	1145	41.1	0.04
F	Pa	673	40.0	0.08
M	PN	1116	41.5	0.04
M	Pa	574	21.7	0.05

62

straints on the choice of mates as offspring are produced, and there will necessarily be familial selection for the last offspring for which only one father and one mother remain to be chosen.

Values used in simulations for mutation rates over the whole genome and deleteriousness of mutations, their dominance and their fitness interactions, were based, in the absence of other data, on what is known in *Drosophila* (Simmons and Crow, 1977). To look at linkage constraints, a map length of 200 cM was chosen to maximize the constraints due to linkage that should exist in a genome of *Drosophila*. The partitioning of the genome in chromosomes, which are supposed to recombine freely, and the variation of recombination rates between sexes (e.g., absence of recombination in males in most *Drosophila* species) were not considered.

We simulated a population of 100 dioecious individuals with equal sex-ratio. For such a population size, a load *per* locus close to the load of an infinite population is expected when deleterious effects are not too

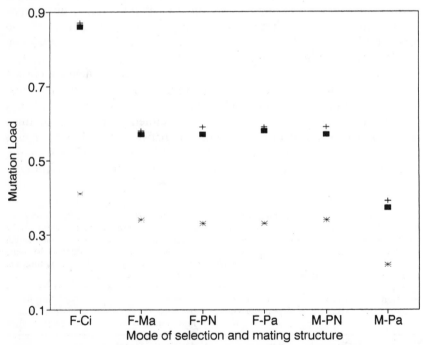

Fig. 2. Mutation load depending on interactions between deleterious mutations, mating system, and selection regime (see legend of Tab. 2 for abbreviations for selection regime and mating system). Results from a simulation model of 100 dioecious individuals. Standard-errors between runs for loads were of the order of 0.01. h = 0.2. Selective values of individuals: same parameters as for Fig. 4. *: Synergistic interactions, V = 0.25, map length: 200 cM. +: Multiplicative interactions, V = 0.25, map length: 200 cM. ■: Multiplicative interactions, V = 0.25, map length: infinite.

small, as in this case neither deleterious mutations get fixed (see Fig. 1 in Kimura et al., 1963) nor does Muller's ratchet operate (Charlesworth et al., 1993). In the absence of interactions between deleterious mutations (i.e., the so-called multiplicative interactions for fitness and free recombination), the overall fitness of the individual expected is $w = (1 - l)^n \approx e^{-nl}$, where l is the load *per* locus and n is the number of loci; $nu = V$ is the rate of appearance of deleterious mutations *per* haploid genome *per* generation. When deleterious mutations are not completely recessive, then from Eqs (3) and (7), the load expected is $1 - e^{-4V(1 - hs/2)}$ in the case of familial selection, and $1 - e^{-2V}$ in the case of mass selection, for a population of infinite size. The simulation model was checked with the help of these expectations. A slightly lower load was observed, compared to what is expected, due to inbreeding as population size is finite and due to the assumption of dioecy, with the consequence of an absence of selfing.

For each set of parameters, results in Tab. 2, and Figs 2 and 3 were an average of 5 runs of the simulation model. The value of a run was the average over 150 generations, after 100 initial generations; starting with an absence of deleterious mutations, a stable state was reached

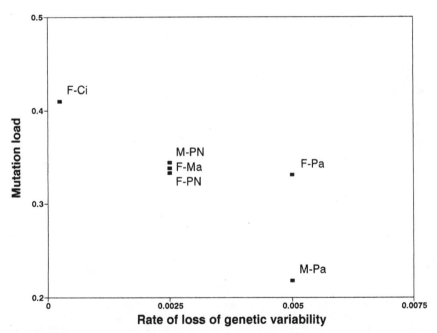

Fig. 3. Expected trade-off between load and rate of loss of genetic variability, depending on mating system and selection regime (see the legend of Tab. 2 for the abbreviations). Assumptions were the same as in case * of Fig. 2. Rate of loss of genetic variability depending on mating system: 1/4N for Ma and PN, 1/2N for Pa, and $\pi^2/4N^2$ for Ci, with N being the population size (Nagylaki, 1992, p. 233).

after 100 generations, except in the case of circular mating and multiplicative interactions, where there was an ever increasing number of deleterious mutations *per* genome, suggesting that Muller's ratchet was operating, in the absence of any fixation of deleterious mutation.

Results

Linkage constraints had a slight but consistent effect on average fitness (Fig. 2) and were a negligible factor of variation of the mutation load.

The relative increase of the mutation load due to familial selection was comparable whatever the fitness interactions (synergistic *versus* multiplicative) in terms of dynamics (Fig. 4), and of relative values at equilibrium. The load was slightly higher in the case of linkage constraints (Fig. 2).

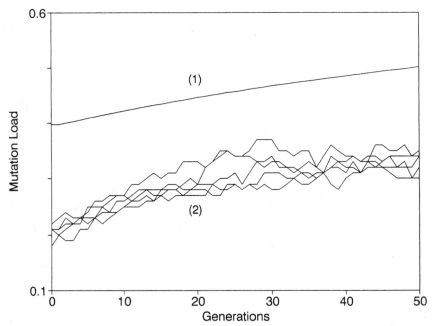

Fig. 4. Dynamics of change of fitness of individuals depending on fitness interactions and map length of the genome ($V = 0.25$, $h = 0.2$). Selective value of individuals: multiplicative interactions, $w = (1 - hs)^c(1 - s)^d$ ($s = 0.1$); synergistic interactions, $w = 1 - 0.01n - 0.01n^2$, with $n = hc + d$ (c and d being the number of deleterious recessive in the heterozygous and homozygous states, respectively). The load is 1 minus the average selective value of individuals. (1): Multiplicative interactions and free recombination; result of numerical iterations of Eqs (4), (5) and (6). (2): Synergistic fitness interactions and map length of 200 cM; results of the simulation model described in the test; the result of five randomly chosen runs is shown.

In the case of familial selection, maximum avoidance of inbreeding does not affect the load compared to the case of random mating; conversely, the load is higher in the case of circular mating. Considering both selection regime and mating system shows that the determinant factor of variation of the load is selection on reproductive success, that is, some individuals reproduce more because their offspring are more fit, and not because they mate more often; the absence of variance in reproductive success in the case of mass selection (M-PN) gives nearly the same result as familial selection without inbreeding, with variance in reproductive success (F-Pa) or without such variance (F-PN and F-Ma, see Fig. 2).

Homozygosity of deleterious mutations was much higher in the case of circular mating, and was lowest in the case of maximum avoidance of inbreeding, as expected. In regards to panmixia, in the case of variance of reproductive success, mass selection had the effect of decreasing homozygosity (M-Pa compared to F-Pa), suggesting that selection against matings between relatives took place; homozygosity was nevertheless higher than in the case of maximum avoidance of inbreeding (Tab. 2).

Discussion

Although equalization of individual reproductive success presents the advantage of minimizing both the loss of genetic variability of small populations (e.g., Frankham et al., 1992; Ralls and Meadows, 1993) and possibly adaptation to captivity (Allendorf, 1993), this selection regime might present the disadvantage to increase the mutation load. This is due to the fact that in a population of N monogamous pairs that each produce n offspring, selection of 2N individuals among 2Nn will lead on average to a higher fitness of the individuals retained than it will in choosing N times two individuals among the n produced by each family. The difference will increase with variance of fitness among families. The load *per* locus is approximately double compared to the classical case of mass selection. When the whole genome is considered, the relative increase of the load is of the same order, although the magnitude of the difference depends on parameters whose values are scarcely known (Fig. 2).

A commonly used parameter to compare expectations of genetic variability in populations of finite size is the asymptotic rate of increase of the average of the *coefficient of kinship* (Malécot, 1948). Using theoretical expectations of this parameter for the different mating systems, Fig. 3 illustrates a possible dilemma for conservation genetics: minimizing the loss of genetic variability, i.e., increasing effective population size will usually lead to a higher load. The further decrease of effective population size that will occur in the case of mass selection, due to heritability of fitness (Nei and Murata, 1966) was not considered (such heritability is expected, due to the variability for the mutation load between individuals); it would then lead to a wider trade-off between the different management strategies.

Inbreeding, due to circular mating, has the effect of increasing the load in the case of familial selection, contrary to the expectation in the case of mass selection (Crow, 1970; although the load will be initially higher during the initial phase of increase of the inbreeding coefficient of individuals due to the change of the mating system; Lande and Schemske, 1985). This discrepancy underlines the effect which selection on reproductive success of individuals has, in relation with their load of mutations. Although the load may initially appear to increase due to increase of homozygosity of deleterious mutations, selection is also more efficient due to the increased variance of the mutation load between individuals; thus, restricting selection has a more drastic effect in the case of inbreeding.

As analytical results were obtained only for theoretical cases (no linkage constraints, no synergistic fitness interactions and a population of infinite size), simulations were used to obtain expectations for cases closer to reality. Based on estimates of parameters in *Drosophila*, results of simulations show that the increase of the load might be a matter of concern in real situations (Fig. 2); the increase of the mutation load takes place rapidly for such mutations and there is a 50% increase of the load after 20 generations of familial selection in the case of synergistic fitness interactions (Fig. 4). There was a 33% decrease of fitness in the case of multiplicative interactions, and a 20% decrease in the case of synergistic interactions (Fig. 2). Besides the estimation of the different parameters, another uncertainty is the impact of this decrease on the performances of individuals, depending on their environment. The load as revealed by inbreeding depression has, for example, more negative consequences in the wild than in laboratory conditions (Hedrick and Miller, 1992). Knowledge of the magnitude of the decrease of fitness, when introduced in a Minimum Viable Population size analysis, would allow an estimation of the genetic consequences of the selection regime and mating system adopted.

In contradiction with the present theoretical results, an experimental study showed a higher fitness of individuals in the case of familial selection compared to the case of mass selection (Borlase et al., 1992). Obviously, differences of fitness observed might be due to causes other than variations of the mutation load. Monogamy was imposed in this experiment, while polygamy occurred in the present simulations; however, this difference of mating system does not seem to be a cause of the discrepancy of results, which was confirmed by additional simulations that we performed. A higher rate of increase of the inbreeding coefficient of individuals expected and observed, in the case of mass selection (Borlase et al., 1992), could account for a temporary higher load; in that case, the lower fitness in the case of random choice will last until the mutation load is at equilibrium with the presently higher inbreeding coefficients of individuals. Further studies are needed to look at the influence on the mutation load of both small population size and the number of generations since that reduction of population size occurred. In the case of a reduction of population size, the expected increase of the load appears first for slightly deleterious recessive alleles (Fig. 1 in Kimura et al., 1963; Charlesworth et al., 1993). Further theoretical investigations are needed to know how reduction of population size will interfere with absence of selection on reproductive success of individuals. It depends on, in regards to deleterious mutations, how their fixation probability and Muller's ratchet will be affected in the case of familial selection.

Conclusion

We focused in this paper on the mutation load. Several kinds of genetic loads exist which are supposed to increase in small populations (Crow and Kimura, 1965). These other loads, like the segregation load, are also expected to increase as a result of the equalization of reproductive success of individuals.

Both selection regime and mating system can be modified to improve the balance between retention of genetic variability and minimization of the mutation load. Alternative strategies might be to equalize reproductive success among units larger than a mating pair. Taking into account larger units will have the effect of lowering the increase of the mutation load, at the expense of a lower increase of effective population size; restricting migration between these units might be a way to further increase effective population size, although some migration should occur to avoid the increase of the load which occurs in small populations (e.g., Spielman and Frankham, 1992).

Finally, the goal of equalizing reproductive success of individuals seems easier to achieve, using pedigrees and/or genetic markers, than selection against an increase of genetic loads, a goal that requires reliable and economical methods to estimate the mutation load of individuals.

Acknowledgements
We thank Amy Freeman and Isabelle Olivieri for helpful linguistic and scientific comments.

References

Allendorf, F. W. (1993) Delay of Adaptation to captive breeding by equalizing family size. *Cons. Biol.* 7: 416–419.

68

Borlase, S. C., Loebel, D. A., Frankham, R., Nurthen, R. K., Briscoe, D. A. and Daggard, G. E. (1992) Modeling problems in conservation genetics using captive *Drosophila* populations. Consequences of equalization of family sizes. *Cons. Biol.* 7: 122–131.

Boucher, W. and Cotterman, C. W. (1990) On the classification of regular systems of inbreeding. *J. Math. Biol.* 28: 293–305.

Campbell, R. B. (1988) Mating structure and the cost of deleterious mutation: postponing inbreeding. *J. Heredity* 79: 179–183.

Campbell, R. B. (1993) The importance of mating structure versus progeny distribution for genetic identity under mutation. *Theor. Pop. Biol.* 43: 129–140.

Charlesworth, B., Morgan, M. T. and Charlesworth, D. (1991) Multilocus models of inbreeding depression with synergistic selection and partial self-fertilization. *Genet. Res.* 57: 177–194.

Charlesworth, D., Morgan, M. T. and Charlesworth, B. (1993) Mutation accumulation in finite outbreeding and inbreeding populations. *Genet. Res.* 61: 39–56.

Crow, J. F. (1970) Genetic loads and the cost of natural selection. *In:* Kojima, K. (ed.), *Mathematical models in population genetics.* Springer-Verlag, Berlin–New York, pp. 128–177.

Crow, J. F. and Kimura, M. (1965) The theory of genetic loads. *In: Genetics today*, Proc. of XI Int. Cong. of Genetics, Pergamon Press, Frankfurt, pp. 495–506.

Frankham, R., Loebel, D. A., Borlase, S. C., Britton, J., Woodworth, L., Nurthen, R. K., Briscoe, D. A., Spielman, D. and Craven, D. (1992) Modeling problems in conservation genetics using *Drosophila*. *CBSG News* 3: 11–13.

Haig, S. M., Ballou, J. D. and Derrickson, S. R. (1990) Management options for preserving genetic diversity: reintroduction of Guam rails to the wild. *Cons. Biol.* 4: 290–299.

Haldane, J. B. S. (1924) A mathematical theory of natural and artificial selection. *Trans. Camb. Phil. Soc.* 23: 19–41.

Haldane, J. B. S. (1937) The effect of variation on fitness. *Am. Nat.* 71: 337–349.

Hedrick, P. W. and Miller, P. S. (1992) Conservation genetics: techniques and fundamentals. *Ecol. Appl.* 21: 30–46.

Kimura, M. and Crow, J. F. (1963) On the maximum avoidance of inbreeding. *Genet. Res.* 4: 399–415.

Kimura, M. and Maruyama, T. (1966) The mutation load with epistatic gene interactions in fitness. *Genetics* 54: 1337–1351.

Kimura, M. and Ohta, T. (1971) *Theoretical Aspects of population genetics.* Princeton University Press, Princeton.

Kimura, M., Maruyama, T. and Crow, J. F. (1963) The mutation load in small populations. *Genetics* 48: 1303–1312.

King, J. L. (1965) The effect of litter culling – or family planning – on the rate of natural selection. *Genetics* 51: 425–429.

Kondrashov, A. S. (1988) Deleterious mutations and the evolution of sexual reproduction. *Nature* 336: 435–440.

Lande, R. and Schemske, D. W. (1985) The evolution of self-fertilization and inbreeding depression in plants. *Evolution* 39: 24–40.

Malécot, G. (1948) *Les Mathématiques de l'hérédité.* Masson, Paris.

Nagylaki, T. (1992) *Introduction to theoretical population genetics.* Springer-Verlag, Berlin–New York.

Nei, M. and Murata, M. (1966) Effective population size when fertility is inherited. *Genet. Res.* 8: 257–260.

Ralls, K. and Meadows, R. (1993) Breeding like flies. *Nature* 361: 689–690.

Simmons, M. J. and Crow, J. F. (1977) Mutations affecting fitness in *Drosophila* populations. *Annu. Rev. Genet.* 11: 49–78.

Spielman, D. and Frankham, R. (1992) Modelling problems in conservation genetics using captive *Drosophila* populations: Improvement of reproductive fitness due to immigration of one individual into small partially inbred populations. *Zoo Biol.* 11: 343–351.

Wiens, D., Calvin, C. L., Wilson, C. A., Davern, C. I., Frank, D. and Seavey, S. R. (1987) Reproductive success, spontaneous embryo abortion and genetic load in flowering plants. *Oecologia* 71: 501–509.

Wiens, D., Nickrent, D. L., Davern, C. I., Calvin, C. L. and Vivrette, N. J. (1989) Developmental failure and loss of reproductive capacity in the rare palaeoendemic shrub *Dedeckera eurekensis. Nature* 338: 65–67.

Wright, S. (1969) *Evolution and the genetics of populations. Vol. II.* The University of Chicago Press, Chicago.

Conservation Genetics
ed. by V. Loeschcke, J. Tomiuk & S. K. Jain
© 1994 Birkhäuser Verlag Basel/Switzerland

Extinction risk by mutational meltdown: Synergistic effects between population regulation and genetic drift

W. Gabriel[1] and R. Bürger[2]

[1]*Department of Physiological Ecology, Max Planck Institute for Limnology, Postfach 165 D-24302 Plön, Germany*
[2]*Institute of Mathematics, University of Vienna, Strudlhofgasse 4, A-1090 Vienna, Austria*

Summary. The accumulation of deleterious mutations reduces individual and mean population fitness. Therefore, in the long run, population size is affected. This facilitates further accumulation of mutations by enhanced genetic drift. Such synergistic interaction then drives the population to extinction.

This mutational meltdown process is studied primarily for asexual populations. Recombination cannot stop the meltdown in small sexual populations. Independent of the mode of reproduction, the asexual case is relevant for any paternally or maternally inherited trait and for mitochondria and chloroplasts that can be viewed as asexual populations inside cells.

The extinction risk is maximal for an intermediate value of the selection coefficient. Recombination does not destroy this effect, at least for small populations. In the asexual case, group selection is able to overpower individual selection to establish lineages with low repair capabilities, i.e., highly deleterious mutations.

If the expression of deleterious mutations is modified by the environment, changes in the environment can cause an unexpected increase or decrease in the extinction risk because of the pronounced maximum extinction risk at intermediate values of s. It may be that an environmental management treatment that improves individual fitness, counterintuitively enhances the extinction risk of a population.

Introduction

In addition to many ecological factors, e.g., random fluctuations of demographic parameters or externally forced perturbations of the environment, there are also well known genetic effects, e.g., inbreeding depression or loss of genetic variance that can contribute considerably to the extinction risk of populations. Besides these more classical problems of conservation genetics there is another source of genetic deterioration: the continuous input of slightly deleterious mutations. Lynch and Gabriel (1990) studied the consequences under asexual reproduction, and Gabriel et al. (1991) demonstrated that this mutation load considerably enhances the extinction risk for small sexual populations if it acts together with demographic stochasticity. Recent experiments (Houle et al., 1992) confirm the order of magnitude of the mutation load estimated from other data (see Lynch and Gabriel, 1990).

On average, each individual genome seems to incur one slightly deleterious mutation *per* generation. Mutation rate and mutational effect are hard to estimate and the mutation rate might even have been underestimated (Kondrashov, 1988). The consequences of the accumulation of deleterious mutations for asexual (or parthenogenetic) organisms are unquestionable, but for sexual species it is still debatable. For which population sizes is the mutational meltdown (Lynch and Gabriel, 1990) an important force if compared to other risks such as fluctuations in the environment? The smaller the population size, the more likely there is a synergistic interaction between many risk factors, i.e., the overall extinction risk might be much higher than expected from considering single risk factors.

To assess the impact of deleterious mutations for populations, classic population genetics is very helpful but can be misleading because most of this theory has been developed for constant (effective) population sizes. Historically, the main interest has been the change in relative gene frequencies. Complications arising from population dynamics have often been neglected. On the other hand, in ecological theory genetic influences have been ignored for a long time. Besides historical reasons, population genetics and theoretical ecology are already mathematically quite complex so that a combined treatment might often be hopeless with respect to mathematical tractability.

To investigate genetic effects in combination with population dynamics, a reference model without genetics is needed. For this reason, Gabriel and Bürger (1992) developed a purely demographic model. Studying the extinction risk by demographic stochasticity (random fluctuations of birth rate, death rate, and sex-ratio), they detected that the former theory, which has been developed with stochastic concepts that are valid only for large populations, often predicts extinction risks that are much too low for asexual and sexual populations. In addition, in sexual populations the influence of sex-ratio fluctuations has been ignored or drastically underestimated.

The present paper focuses on asexual populations for which the accumulation of deleterious mutations is an irreversible process and concentrates on aspects not described in Gabriel et al. (1991). It should be stressed, especially in the context of conservation genetics, that all aspects studied for asexual species are also of importance for sexual species. In small sexual populations similar problems arise as in asexuals, but also there are several traits that are maternally or paternally inherited. Consequently, sexual species have to be treated as asexual populations with respect to these characters. Inside each cell there are (maternally or paternally inherited) asexual populations: mitochondria and additionally chloroplasts in plants. Therefore, the studies presented here are of general importance to the survival of all higher organisms – and meant as stimuli for further experimental and theoretical investigations in the context of conservation genetics.

Accumulation of deleterious mutations in asexual populations under the assumption of constant population size

In finite populations, natural selection cannot efficiently remove mutants carrying deleterious mutations if fitness differences caused by mutations are small or the mutation rate is very high. In addition, random genetic drift can play an important role for the accumulation of deleterious mutations, especially under asexual reproduction. We use the following definitions:

s	selection coefficient = fractional reduction in viability caused by a single mutation;
$W_n = (1 - s)^n$	fitness of an individual carrying n mutations (which are assumed to act multiplicatively);
μ	mutation rate *per* genome and *per* generation;
mutational class	individuals carrying the same number of mutations (the loci at which mutations occur might differ between individuals).

An asexual population can be subdivided into mutational classes. Let us first consider the mutation-free class. The smaller this class, the higher the probability that by chance this class does not contribute to the survivors in the next generation. If this class is lost, it is lost forever because the present model neglects back mutations and compensatory mutations. (Backmutations are believed to occur at too low a rate, at least for population sizes below 10^8. However, compensatory, fitness increasing mutations may decrease the extinction risk considerably, cf. Lynch and Gabriel (1990).) Therefore, the mutation-free class cannot be re-established from the higher mutational classes. After the loss of the mutation-free class, the class carrying one mutation becomes the least loaded class and it will meet the same fate of being lost, and so on. With each loss of the actual least-loaded class, the population mean fitness declines. This ratchet-like process was first described by Muller (1964) and has been called Muller's ratchet (Felsenstein, 1974).

The speed of the ratchet critically depends on the size of the least loaded class C_0 in relation to population size N. For the order of events "reproduction-mutation-selection" the ratio C_0/N after selection is expected to be

$$\frac{C_0}{N} = e^{-\mu(1 - s)/s},$$

(1)

72

if N is very large and μ is small (see Gabriel et al., 1993). Eq. (1) differs slightly from $\exp(-\mu/s)$, which is given by Haigh (1978), because of another life cycle. If s is small then the difference between the formulae is negligible. If the population is censured after selection, Haigh's formula cannot be appropriate for large s; this is obvious for s = 1 (lethal mutations) when only non-mutants survive selection. The proportion of the least loaded class must converge to one as the selection coefficient s approaches 1 as shown in Fig. 1.

This figure gives only a hint of the dynamics of the ratchet. If N individuals are drawn randomly after selection to constitute the next generation, then the probability of losing the least loaded class is approximately

$$p = \left(1 - \exp\left[-\frac{\mu(1-s)}{s}\right]\right)^{N}. \tag{2}$$

This probability can be used to calculate the speed of the ratchet in a deterministic fashion if one assumes that after each loss of the least loaded class (after each "turn" of the ratchet) the deterministic equilibrium distribution of mutation classes is the same as before apart from being shifted by one class. But to re-establish the equilibrium distribution takes time. If the ratchet turns too fast, the distribution of mutation classes will never correspond to the deterministic equilibrium distribution that is derived without taking into account an operating ratchet.

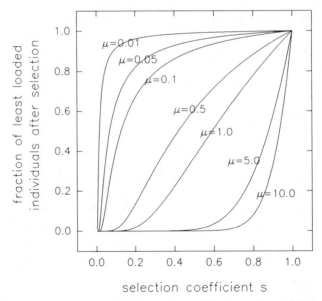

Fig. 1. Deterministic expectation of the proportion of the fittest (least loaded) mutational class as a function of the selection coefficient s, for various genomic mutation rates μ.

Consequently, even for a constant finite population size, the distribution of mutation classes can deviate drastically from the deterministic expectation if the time between successive losses of the least loaded class is of the same order of magnitude or shorter than the time needed to approximate a stable mutation class distribution.

There are only a few detailed studies on the dynamics of the ratchet that go beyond Haigh's (1978) analysis. Bell (1988) explored the dynamics of the ratchet process and the role of recombination in halting the ratchet. He predicted that the optimal class of size n_0 will be lost in approximately $10 \, n_0$ generations, but recent numerical simulations (Charlesworth et al., 1993) yield quantitatively and qualitatively different results. Melzer and Koeslag (1991) studied possible effects of fertility selection. Stephan et al. (1993) derive two diffusion approximations for the speed of the ratchet in asexual populations that work for certain parameter combinations.

The classic models for Muller's ratchet (see Maynard Smith, 1978; and above) keep population size constant. This implicitly implies infinite fecundity, otherwise population size decreases if the ratchet operates and, therefore, mean fitness declines steadily. In reality, one can expect that mean population fitness will affect population size in the long run. Reduced population size, however, facilitates the operation of the ratchet. Therefore, especially in small populations, accumulation of deleterious mutations and extinction might occur much faster than estimated under the unrealistic assumption of constant population size and, eventually, the population will become extinct.

The first study of the ratchet without the assumption of constant population size was performed by Lynch and Gabriel (1990). Recent extensive simulation studies of the ratchet as well as mathematical treatments were performed by Gabriel et al. (1993) and Lynch et al. (1993). Some of their results are discussed later in this paper.

Muller's ratchet with density-dependent population regulation

The incorporation of finite fecundity of organisms into models of Muller's ratchet can be achieved by implementing a finite number of offspring *per* individual or by a density-dependent population regulation. Deleterious effects of mutations may show up in a complicated mixture of reduction in fertility and viability. We will restrict our analysis to viability selection.

Most of the following results have been obtained by means of a demographic model with Poisson-distributed number of offspring and non-overlapping generations. Various properties of this and related demographic models were studied in Gabriel and Bürger (1992). The extinction risk for small populations caused by the combined action of demographic stochasticity and deleterious mutations is studied in Gabriel et al. (1991, 1993) and Lynch et al. (1993).

As a reference model, we will calculate the extinction risk with a "deterministic" null-model by removing all stochasticity. The time-course of the expected population size is then calculated without any noise. Family size is determined exactly by the assumed density dependence; fitness loss corresponds to the average mutation load when genetic drift is neglected. Therefore, in such a deterministic model, the ratchet cannot operate because the least loaded class will never be lost even if its frequency becomes very small. Starting from an unloaded population, the dynamics of the mutational classes can be calculated.

The formulae for the intrinsic growth rate $r = 0$, which corresponds to one surviving offspring *per* individual and implies a constant population size in the absence of other forces that could reduce offspring fitness, are given in Gabriel et al. (1993, Appendix D). This approach can be extended by including population regulation assuming that population regulation does not influence the distribution of mutational classes. Starting with a mutation-free population, the mutation load converges to $e^{-\mu}$ for $t \to \infty$. With the density dependent offspring production F the population size at each time step is then given by

$$N_{t+1} = F(N_t)\, e^{-\mu}\, e^{\mu(1-s)^{t+1}}, \qquad (3a)$$

where

$$F(N_t) = \frac{e^r N_t}{(1 + aN_t)^\beta}, \qquad a = \frac{e^{r/\beta} - 1}{K}. \qquad (3b)$$

Here, K is the carrying capacity and r the intrinsic growth rate (on the time scale of generations^{-1}) so that (in the limit $K \to \infty$) e^r is the maximal reproductive rate (offspring *per* individual). The strength of population regulation is parameterized by β; for example, with $\beta = 1$ the density dependence is equivalent to the classic Verhulst model. (This is a discrete version of the usual logistic growth equation that does not lead to complicated dynamics like chaos; for details see Gabriel and Bürger (1992).) The population becomes extinct if N falls below 1. Survival is possible only if $N(t + 1) \geq 1$ for $N(t) = 1$. In the limit $t \to \infty$, this leads to the condition:

$$K \frac{e^{(r - \mu)/\beta} - 1}{e^{r/\beta} - 1} > 1. \qquad (4)$$

Therefore, necessary conditions for population survival are

$$r > \mu$$
$$K > e^{\mu/\beta}. \qquad (5)$$

The lower limit of K is obtained by allowing infinite growth rates $r \to \infty$. Only if the conditions of Eqs (4) and (5) are fulfilled can the population compensate for the loss due to mutational load, otherwise it becomes extinct even without demographic stochasticity and without genetic drift. The mean time to extinction decreases with random variation in demography and genetic drift. Therefore, this deterministic expectation imposes only a lower limit on the extinction risk. Simulations that include drift and demographic stochasticity are expected to agree with the deterministic model only if the growth rate is small $r \ll \mu$ (if $r > \mu$ the time to extinction becomes infinite for the deterministic model) and for at least moderately large carrying capacity K (increasing K slows down the speed of the ratchet and reduces noise from family size variation).

Fig. 2 shows how the mean time to extinction increases and how extinction risk decreases as growth rate increases. The extinction risk for $s = 0.01$ is much smaller than for $s = 0.1$ although the ratchet turns much faster at the lower s value. The approximate probability that none of the K offspring are drawn from the least loaded class is given by Eq.

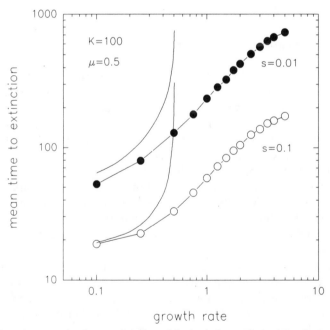

Fig. 2. Mean time to extinction as a function of the intrinsic population growth rate r for the selection coefficients $s = 0.01$ and $s = 0.1$ with a genomic mutation rate $\mu = 0.5$ *per* generation (strength of density dependence of $\beta = 1$). The connected data points show simulation results allowing for genetic drift (Muller's ratchet can operate). For comparison, the "deterministic" solutions without genetic drift and without noise in family size are shown; they become infinite as r approaches μ.

(2). Even if the actual population size is smaller than K, K offspring are always drawn if fecundity is high enough to compensate for the reduced mean fitness caused by the ratchet. For $\mu = 0.5$ and $K = 100$, this probability of losing the least loaded class is $p \approx 1$ for $s = 0.01$ and $p = 0.327$ for $s = 0.1$. Therefore, although the ratchet turns more than three times faster, the time to extinction is larger at the lower s value; this contrasts with the usual view of Muller's ratchet. This effect has been demonstrated already by Lynch and Gabriel (1990) in a model with a very simple population regulation. There it was shown that a lowered speed of the ratchet as s increases can be more than compensated for by the population fitness reduction *per* each turn of the ratchet.

The deterministic model converges to the stochastic simulation model as r becomes very small but gives zero extinction risk when $r > \mu$ as predicted by Eq. (5). The stochastic simulations show a considerable extinction risk even for $r > \mu$. This nicely demonstrates the effect of genetic drift that causes continuous accumulation of deleterious mutations. The extinction risk due to demographic stochasticity alone is negligible for $K \geq 100$ unless r is in the order of zero or below (see Gabriel and Bürger, 1992).

Mutational meltdown in asexual populations

As a consequence of finite fecundity, in the long run the population size N will be affected by the accumulation of deleterious mutations for almost any kind of demographic model. If accumulation of deleterious mutations reduces the actual population size, then subsequent mutations accumulate faster because the chance of losing the least-loaded mutation class is enhanced at reduced population size. Each loss of the actual fittest mutation class further reduces population size. This again speeds up the ratchet. Therefore, mutation accumulation and random genetic drift synergistically intensify each other and drive populations to extinction. This positive feedback mechanism was first described by Lynch and Gabriel (1990) and denoted as "mutational meltdown".

The meltdown has been studied using two different demographic models. In the first model, it is assumed that the maximal number of adults is bounded by the carrying capacity K, and the number of offspring produced *per* individual (fecundity rate) is independent of population size. At high fecundity, the total number of offspring might be much higher than K. If the new generation of adults ($N \leq K$) is drawn from the surviving offspring after selection, then population size N is not reduced immediately by the mutational load. But if the load continues to increase, after a sufficiently long time-span the total number of surviving offspring will fall below K and the number of adults becomes smaller than K. Such a population regulation is studied in detail in Lynch et al. (1993). These models put upper limits on the mean extinction time (especially with high fecundity) because all other additional effects will further increase the extinction risk.

For a second demographic model, let us assume that carrying capacity acts for the number of offspring produced and selection occurs after density-dependent population regulation. Then the number of reproducing adults can be reduced immediately by deleterious mutations (without mutation load the number of offspring and adults would both stay near K). The number of reproducing adults decreases as the accumulation of mutations proceeds. Thereby, the number of offspring *per* individual increases towards the maximum reproductive rate (studied in Gabriel et al., 1993 and Lynch and Gabriel, 1990). The disadvantage is that this model – like all specific demographic models – assumes a more or less arbitrary population regulation. In reality, deleterious mutations might affect fecundity, viability, and population regulation in a complicated manner. Nevertheless, it is helpful to have a model formulated with the parameters r and K which are used widely in ecology. Of course, one has to be careful when applying these parameters. But as an analogy to "effective population size" in population genetics, this model can be used as a null-model with "effective r and K" values.

The meltdown effect is independent of the kind of population regulation and the order of events (Lynch et al., 1993). The time at which the meltdown becomes the dominant force might depend on the demographic model. Also, if the fitness reduction s varies between mutations and some mutations are beneficial, the meltdown effect still occurs (Lynch and Gabriel, 1990). Epistatic fitness effects between single mutations are unlikely to prevent the meltdown (Lynch et al., 1993; Butcher, personal communication).

Fig. 3 gives an example of mean times to extinction for a density-dependent population regulation according to Eq. (1). The striking feature that an intermediate s value minimizes the mean time to extinction is discussed in detail in Gabriel et al. (1993). The position of the minimum can be predicted roughly for moderate and large K values. The position of the minimum shifts to higher s values if the genomic mutation rate increases; but the minimum appears at lower s values if K (see Fig. 3) or r increase (see Gabriel et al., 1993, where approximate formulae to estimate the position of the minimum can be found). With this demographic model, the extinction times for s = 0 and s = 1 can be calculated (see Gabriel and Bürger, 1992; Gabriel et al., 1993). A useful substitution formula (see Gabriel et al., 1993) to estimate the effect of lethal (s = 1) mutations is

$$r_{s=1} = r - \mu$$

$$K_{s=1} = K \frac{e^{r-\mu} - 1}{e^r - 1}.$$

(6)

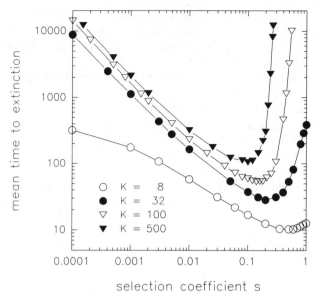

selection coefficient s

Fig. 3. Mean time to extinction shows a minimum for intermediate selection coefficients. Genomic mutation rate $\mu = 0.5$, $r = 1$, $K = 8, 32, 100, 500$.

This means that a population with $s = 1$ behaves like a population with $s = 0$ if r and K are reduced according to Eq. (6). Note that the minimum time to extinction, at $s = s^*$, is many orders of magnitude lower than the extinction time at $s = 1$, unless K is very small. This minimum is not a consequence of the particular demographic model (compare with Fig. 5 where another demography was used).

For conservation genetics, this pronounced minimum may have severe consequences. The selection coefficient of slightly deleterious mutations might differ between species and between habitats. In a real population with genetic diversity the s-values will differ between individuals so that one has to consider a distribution of selection coefficients rather than a single s value. External forces might shift this distribution. This can produce unpredictable changes in the extinction risk. To illustrate the possible effect, let us consider the simplified case of equal s values for all individuals of a population. Imagine that in such a locally adapted population, deleterious mutations cause a fitness reduction that corresponds to s values above s^*, the value at which the extinction time shows the minimum. After a change in the environment – or a management treatment to improve individual fitness – the selection coefficient might decrease. The unexpected consequence would be a dramatic increase in extinction risk.

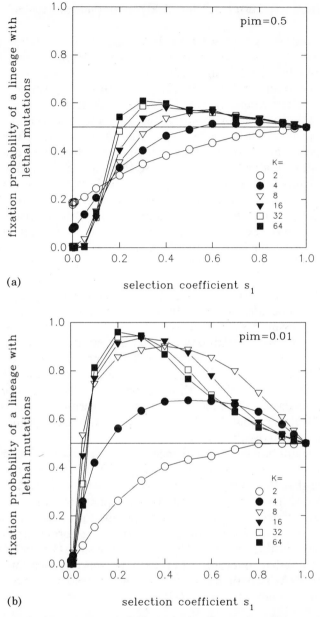

(a)

(b)

Fig. 4. Group selection overpowers individual selection. At start, two lineages with different selection coefficients are separated into two demes in a simple metapopulation that allows reciprocal immigration with probability pim. The fixation probability of the lineage with $s = 1$ is shown as a function of the selection coefficient of the second lineage for different K values. Immigration probabilities are a) $pim = 0.5$ b) $pim = 0.01$.

A population-level advantage for highly deleterious mutations

On an evolutionary time scale, the minimum of the extinction time as a function of s evokes other questions. In relation to the problems mentioned above, one might ask which value of slightly deleterious mutations would be favored during evolution – assuming that the deleterious effect cannot be removed, e.g., by perfect repair mechanisms. On the level of individual selection, any decrease in s would be favored, but on the population level there is a critical amount of repair. For example, compared to accepting a mutation as lethal, one would have to repair until s values are small enough so that the extinction time is larger than for s = 1. In one step, the repair would have to cross a very deep valley of selection coefficients that would otherwise lead to increased extinction risk. This argument, of course, is only valid if selection on the population level (i.e., group selection) is fast or efficient enough to overpower individual selection which favors any reduction in s.

This problem can be evaluated by a simple metapopulation approach. We start with two demes. Initially, each deme consists of a single lineage, but the lineages differ between demes with respect to s; for example, in one lineage any mutation is lethal (s = 1), for the other lineage we assume s < 1. Further, we allow reciprocal immigration between demes with probability p_{im} after successful reproduction. Then, we let both lineages in the metapopulation reproduce until one lineage becomes fixed, i.e., the other lineage is removed from the metapopulation. (The time to fixation is short compared to the extinction time of the metapopulation.) Simulating many such experiments, we then calculate the fixation probability. Fig. 4 shows that, indeed, group selection can overpower individual selection.

As an evolutionary consequence, in many cases it seems to be advantageous not to repair damaged DNA if the repair does not guarantee that the valley of critical s-values will be crossed.

Because of its maternal inheritance, mitochondria (and also chloroplasts) can be viewed as small asexual populations within a cell. Mitochondria do not have genes for gene repair. Therefore, the minimum in s could explain this absence of repair mechanisms. One must carefully determine whether mutations in mtDNA influence replication efficiency of the mtDNA itself or only influence the fitness of the host cell (for discussion see Gabriel et al., 1993).

Does meltdown occur in spite of segregation and recombination?

Even if the mutation-free class is empty, recombination can produce mutation-free individuals as long as the loci, at which the mutations occurred, differ between the individuals, i.e., before specific mutations

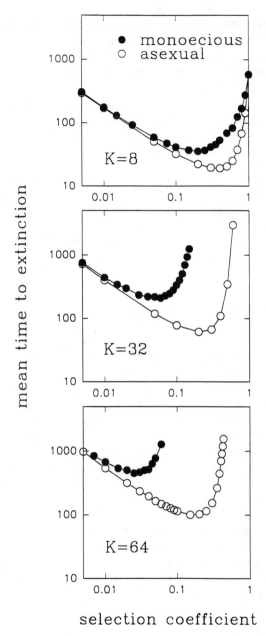

Fig. 5. Comparison of extinction times for asexual and monoecious populations with simple density dependence and reduced demographic stochasticity for carrying capacities K = 8, 32 and 64. A genomic mutation rate of $\mu = 0.6$ is assumed. In the diploid case fitness is multiplicative within and between loci.

are fixed in the population. Recombination should in principle be able to stop Muller's ratchet, but it is debatable how effective and under which conditions this occurs. Bell (1988) predicts that the ratchet cannot be stopped by recombination, but Charlesworth et al. (1993) arrive at different conclusions.

Evidence for the impact of slightly deleterious mutations for small sexual populations is given by Gabriel et al. (1991). In that model, demographic and sex-ratio fluctuations are the dominant sources for extinction if population sizes are very small; at population sizes above 20, however, the genetic effects become quite influential. It is not yet known how large a sexual population has to be in order to effectively purge the mutation load in comparison to an asexual population.

In the study of Gabriel et al. (1991), the extinction process is heavily determined by sex-ratio fluctuations. There is a high risk that no male or no female is left, especially after temporary bottlenecks. In addition, sex-ratio fluctuations reduce effective population size drastically and this implies large effects of genetic drift.

To study the influence of segregation and recombination with a minimum of non-genetic stochasticity, we have performed simulations for small (monoecious) sexual and asexual populations with reduced demographic stochasticity by using the simple population-regulation model of Lynch and Gabriel (1990). (The order of events was: (1) zygotes; (2) mutation; (3) selective mortality; (4) check for extinction; (5) next generation by drawing randomly K zygotes, in the monoecious population, by random mating and free recombination.) There is again an intermediate value of s* that minimizes the expected time to extinction (see Fig. 5). The critical value s* is substantially smaller for sexual than for asexual populations of the same size. The expected time to extinction due to mutation accumulation in sexual populations is higher than in asexuals, but not greatly so for small K or small s. In the neutral (s = 0) and lethal (s = 1) case the mean time to extinction is the same for sexual and asexual populations.

These preliminary results show that for small populations, the mutational meltdown cannot be stopped by recombination. Further investigations with larger population sizes have to be performed to study the general impact of the meltdown on sexual species. Such investigations are currently in progress (Lande; Lynch et al., personal communications).

Conclusions

In this paper, mutations are considered as unconditionally deleterious. There are other kinds of mutational effects that cannot be treated in this way, e.g., a different approach is necessary if fitness is determined by

Table 1. Main effects of the model parameters

s	selection coefficient; fitness reduction *per* mutation: the extinction risk is maximal at intermediate levels of s. The influence on extinction risk is strong. The ratchet turns more slowly when s increases. The damage to the population per turn of the ratchet increases with s and this can overcompensate for the reduction in the speed of the ratchet.
s*	s value that maximizes the extinction risk, resp. minimizes the mean time to extinction.
μ	genomic mutation rate *per* generation: it strongly enhances the speed of the ratchet and drastically increases the extinction risk. Increased μ moves s* to higher values.
K	carrying capacity, related to population size N: extinction occurs if $K < e^{\mu}$ for any s > 0. Larger K reduces the speed of the ratchet and shifts s* to lower values.
r	growth rate, fecundity: If r is too small ($r < \mu$) then extinction is unavoidable for any s > 0. It has strong impact on the onset of the meltdown process. An increase if fecundity above a medium sized number of offspring ($e^r > 10$) has little further effect. Increased r shifts s* to lower values.

several quantitative traits under stabilizing selection. Gabriel and Wagner (1988) and Wagner and Gabriel (1990) studied conditionally deleterious mutations by allowing mutations to compensate phenotypic effects of deleterious mutations. Such compensatory mutations for quantitative characters are as effective as recombination in halting the decline of mean fitness otherwise caused by Muller's ratchet. Extinction was not studied in these papers but it would be interesting to look at how these mutations interact with extinction.

Even if many details still have to be worked out, the importance of the meltdown process for extinction is unquestionable. Tab. 1 summarizes some of the effects of the different model parameters. There remains a large field of necessary theoretical and experimental investigations, for example, we assumed that mutations act multiplicatively and we did not allow for epistatic interactions. Further, there is not enough information on mutation rates and selection coefficients for different species and little is known about how a selection coefficient measured in the laboratory transforms into fitness reduction in the field. The expression of deleterious mutations – even if independent of the genetic background – could be strongly dependent on the environmental conditions. Studies like this paper intend to estimate minimum levels of extinction risk. Additional effects like environmental stochasticity (cf. Lande, 1993) that might themselves synergistically interact with the processes described in this chapter would increase the extinction risk considerably. Hopefully, this study stimulates further experimental and theoretical investigations and increases the awareness of interaction between environmental and genetic problems that we have to deal with in conservation biology.

84

Acknowledgements
Our work has been supported by a grant from Deutsche Forschungsgemeinschaft to WG. We thank Nancy Zehrbach for improving the English and Michael Lynch for helpful comments.

References

Bell, G. (1988) Recombination and the immortality of the germ line. *J. Evol. Biol.* 1: 67–82.

Charlesworth, D., Morgan, M. T. and Charlesworth, B. (1993) Mutation accumulation in finite outbreeding and inbreeding populations. *Genet. Res.* 61: 39–56.

Felsenstein, J. (1974) The evolutionary advantage of recombination. *Genetics* 78: 737–756.

Gabriel, W. and Wagner, G. P. (1988) Parthenogenetic populations can remain stable in spite of high mutation rate and random drift. *Naturwissenschaften* 75: 204–205.

Gabriel, W., Bürger, R. and Lynch, M. (1991) Population extinction by mutational load and demographic stochasticity. *In:* Seitz, A. and Loeschcke, V. (eds), *Species conservation: a population biological approach.* Birkhäuser, Basel, pp. 49–59.

Gabriel, W. and Bürger, R. (1992) Survival of small populations under demographic stochasticity. *Theor. Pop. Biol.* 41: 44–71.

Gabriel, W., Lynch, M. and Bürger, R. (1993) Muller's ratchet and mutational meltdowns. *Evolution*, (in press).

Haigh, J. (1978) The accumulation of deleterious genes in a population. *Theor. Pop. Biol.* 14: 251–267.

Houle, D., Hoffmaster, D. K., Assimacopoulos, S. and Charlesworth, B. (1992) The genomic mutation rate for fitness in Drosophila. *Nature* 359: 58–60.

Kondrashov, A. S. (1988) Deleterious mutations and the evolution of sexual reproduction. *Nature* 334: 435–440.

Lande, R. (1993) Risks of population extinction from demographic and environmental stochasticity, and random catastrophes. *Am. Nat.* 142: 911–927.

Lynch, M. and Gabriel, W. (1990) Mutation load and survival of small populations. *Evolution* 44: 1725–1737.

Lynch, M., Bürger, R., Butcher, D. and Gabriel, W. (1993) The mutational meltdown in asexual populations. *J. Heredity* 84: 339–344.

Maynard Smith, J. (1978) *The evolution of sex.* Cambridge Univ. Press, Cambridge.

Melzer, A. L. and Koeslag, J. H. (1991) Mutations do not accumulate in asexual isolates capable of growth and extinction – Muller's ratchet reexamined. *Evolution* 45: 649–655.

Muller, H. J. (1964) The relation of recombination to mutational advance. *Mutation Res.* 1: 2–9.

Stephan, W., Chao, L. and Smale, J. G. (1993) The advance of Muller's ratchet in a haploid asexual population: approximate solutions based on diffusion theory. *Genet. Res.* 61: 225–231.

Wagner, G. P. and Gabriel, W. (1990) Quantitative variation in finite parthenogenetic populations: What stops Muller's ratchet in the absence of recombination? *Evolution* 44: 715–731.

Part III

Inbreeding, population and social structure

Conservation Genetics
ed. by V. Loeschcke, J. Tomiuk & S. K. Jain
© 1994 Birkhäuser Verlag Basel/Switzerland

Introductory remarks

Two classic treatments of the genetic aspects of conservation, namely, Frankel and Soulé (1981) and Schonewald-Cox et al. (1983), clearly established the potential detrimental roles of genetic drift and inbreeding in the present and future viability of small populations. Elementary population genetics theory has dealt with this topic only under limited conditions of weak selection, more or less stable environments, highly idealized population size or growth parameters, and with only a few models of subdivision-migration. The chapters in this part remind us of the need for further and more careful developments of this theory. At least three different lines of thought should be mentioned here.

The fact that inbreeding can occur in several ways and needs different measures depending on its different genetic consequences on heterozygosity, allelic diversity or pedigrees, has not always been recognized. Estimates of inbreeding coefficients, fixation index (a measure of departure from panmictic expectations), or F-statistics (for hierarchical geographical differentiation of groups, subpopulations, races, etc.) require caution about the appropriate experimental design and analysis. Relative loss of individual and population fitness under inbreeding, leading to more homozygosity, and under joint effects of drift and inbreeding due to loss of genetic variation, should be identified separately; selection and management might succeed in overcoming inbreeding depression even in the evolution of inbreeders. Likewise, the so-called outbreeding depression resulting from the mixing of two different gene pools (or hybridization of locally adapted genotypes), which is often recommended for replenishing genetic variation in small isolates, could be better managed with a clearer understanding of the genetic processes. Several other developments in the theory of mating systems include models of multilocus evolution, biparental inbreeding under spatial substructuring, and evolution of mixed mating systems. Social behavior involving mate choice, polygamy, male *versus* female migration differential, and sex ratio variance affects mating systems so that remaining inbreeding levels would require further extensions of the common textbook concepts. Furthermore, Charlesworth and Charlesworth (1987), Lynch (1991), and Ouborg (1993), among others, emphasized that measures of inbreeding and outbreeding depression require careful experimental work in terms of control of the breeding scheme, environmental conditions and traits to be measured. Altogether, one can readily see a closer interaction

88

between theory and experiment; so much in conservation biology is to be validated from such studies.

The second area of exciting new developments in population genetics theory involves the study of so-called metapopulations (partially isolated populations with drift, migration, and local extinction-recolonization dynamics). Ecological genetics of the colonizing process is incorporated through several models of dispersing colonists and by varying population growth features among and within patches. Holsinger (1993) provides an excellent review of the dynamics of fragmented plant populations. McCauley (1993) concluded that both dispersal and successful colony establishment features of these metapopulation models should be empirically tested. Metapopulation dynamics is naturally very appealing to both ecologists and geneticists, but since there already exists numerous different genetic, demographic or behavioral subsets of models, it is important to develop clear and consistent definitions and ways to estimate effective population size, migration rate, and turnover rate.

A third area of important developments in conservation genetics theory relates to predicting extinction risks under changing (deteriorating) environments. This is important at least for the idea of extinction vortices and interactive (synergistic) genetic demographic stochasticity; for example, inbreeding depression might seriously endanger a population with deteriorating ecological conditions. Lynch and Lande (1993) have initiated a theoretical treatment of genetic, demographic, and environmental stochasticity. Their predictions about avoidance of extinction seem optimistic, at least for polygenic traits, moderate evolutionary change, and a favorable reproductive system. Mangel and Tier (1993), on the other hand, developed a metapopulation model using the MacArthur-Wilson approach to include stochastic births and deaths, immigration, and catastrophes. For even modest catastrophes, extinction risks are high; however, genetics and evolutionary dynamics need to be introduced into this model. (See also Gilpin and Hanski, 1991, for a thorough review of metapopulation dynamics.) Lynch and Lande (1993) also noted that empirical work in this area might be impractical for many species so that a mathematical understanding is well warranted as a start. It does appear very likely that many species recovery programs and restoration projects would have the longterm capability to carry out this work. These and other new directions in population genetic advances will undoubtedly have more realistic applications in conservation.

References

Charlesworth, D. and Charlesworth, B. (1987) Inbreeding depression and its evolutionary consequences. *Annu. Rev. Ecol. Syst.* 18: 237–268.
Frankel, O. H. and Soulé, M. E. (1981) *Conservation and evolution.* Cambridge Univ. Press, Cambridge.

Gilpin, M. and Hanski, I. (1991) *Metapopulation dynamics: empirical and theoretical investigations*. Academic Press, London.

Holsinger, K. E. (1993) The evolutionary dynamics of fragmented plant populations. *In:* Kareiva, P. M., Kingsolver, J. G. and Huey, R. B. (eds), *Biotic interactions and global change*. Sinauer, Sunderland, pp. 198–215.

Lynch, M. (1991) The genetic interpretation of inbreeding depression and outbreeding depression. *Evolution* 45: 622–629.

Lynch, M. and Lande, R. (1993) Evolution and extinction in response to environmental change. *In:* Kareiva, P. M., Kingsolver, J. G. and Huey, R. B. (eds), *Biotic interactions and global change*. Sinauer, Sunderland, MA, pp. 234–250.

Mangel, M. and Tier, C. (1993) Dynamics of metapopulations with demographic stochasticity and environmental catastrophes. *Theor. Pop. Biol.* 44: 1–31.

McCauley, D. E. (1993) Genetic consequences of extinction and recolonization in fragmented habitats. *In:* Kareiva, P. M., Kingsolver, J. G. and Huey, R. B. (eds), *Biotic interactions and global change*. Sinauer, Sunderland, MA, pp. 217–233.

Ouborg, N. J. (1993) *On the relative contribution of genetic erosion to the chance of population extinction*. Ph.D. Thesis, State University Utrecht, Utrecht.

Schonewald-Cox, C. S., Chambers, S. M., MacBryde, B. and Thomas, L. (1983) *Genetics and conservation: A reference for managing wild animal and plant populations*. Benjamin-Cummings, London.

Conservation Genetics
ed. by V. Loeschcke, J. Tomiuk & S. K. Jain
© 1994 Birkhäuser Verlag Basel/Switzerland

Inbreeding: One word, several meanings, much confusion

A. R. Templeton[1] and B. Read[2]

[1]*Department of Biology, Washington University, St. Louis, MO 63130, USA*
[2]*St. Louis Zoological Park, Forest Park, St. Louis, MO 63110, USA*

Summary. Because conservation biologists must frequently deal with small populations, inbreeding (a frequent consequence of small population size) has played a central role in many genetic management programs. However, the word "inbreeding" has several, often contradictory meanings, and a failure to distinguish among these meanings has caused much misunderstanding on the role of inbreeding in genetic management. Three different biological meanings of inbreeding are discussed in this paper: (1) inbreeding as a measure of shared ancestry in the paternal and maternal lineages of an individual; (2) inbreeding as a measure of genetic drift in a finite population, and (3) inbreeding as a measure of system of mating in a reproducing population. The distinction and use of these different measures of inbreeding are discussed and illustrated with a worked example, the North American captive population of Speke's gazelle (*Gazella spekei*). It is shown that these different meanings of the word inbreeding must be kept separated, otherwise erroneous management recommendations and evaluations can occur. On the positive side, the different measures of inbreeding when used jointly can be a powerful management tool precisely because they measure different biological phenomena.

Introduction

Jacquard (1975) wrote an article entitled: "Inbreeding: one word, several meanings." In that article, he described five different genetic and evolutionary phenomena that are measured by "inbreeding coefficients" coupled with three different and mutually incompatible mathematical definitions (as probabilities, as correlations, and as coefficients of deviation from a reference population). As Jacquard correctly pointed out, this diversity of biological and mathematical meanings of the single word "inbreeding" has caused much confusion in the population genetic literature. Attempts to introduce more terms in order to discriminate among these many meanings have been unsuccessful and, as a consequence, confusion over inbreeding still persists, even among professional population geneticists. Unfortunately, this confusion is having a negative impact on the design, implementation, and evaluation of many species management programs in the area of conservation biology.

The aim of this chapter is not to reiterate all the meanings discussed by Jacquard (1975), but rather to focus upon three different meanings of the word "inbreeding coefficient" that are particularly relevant for conservation biology. A clarification of these diverse meanings of inbreeding also demands a clarification of two other terms that have a *plethora* of biological and mathematical meanings: heterozygosity and effective population size (also see Gliddon and Goudet, this volume). The clarification of these terms and their proper use in designing management programs will be illustrated by the example of the Speke's gazelle management program (Templeton and Read, 1983, 1984; Templeton, 1991a), a program in which "inbreeding" in its various manifestations has played a central role.

"Inbreeding" as a measure of coancestry for individuals

Perhaps the simplest and most common use of the word "inbreeding" is to describe the biological phenomenon of two mating individuals sharing ancestors. Because DNA, the genetic material, can make copies of itself and be passed on to future generations through reproduction, there is a finite probability that a specific gene passed on by the father to his offspring is an identical copy of the homologous gene passed on by the mother if the father and mother share common ancestors. Many algorithms exist for calculating these probabilities from pedigrees, and such algorithms are commonly included in the computer packages used for managing captive populations. The resulting probability of identity-by-descent is known as an "inbreeding coefficient" and is often symbolized by "F". There are several properties of this F (which hereafter will be called "pedigree inbreeding" and symbolized by F_p) that should be noted before proceeding to other definitions of inbreeding.

First, F_p is applied to an *individual* of known pedigree. If an individual's pedigree is not known, it is impossible to calculate F_p. In this regard, the standard procedure is to regard all founders for which pedigree information is unknown as unrelated. For example, the captive herd of Speke's gazelle established at the St. Louis Zoological Gardens in the early 1970s had one male and three females as wild-caught founders (Templeton and Read, 1983). As nothing was known about the relationships among these animals, they were regarded as unrelated individuals in the pedigrees of all the individuals subsequently born into this captive population.

Because F_p is applied to specific individuals, different individuals within the same breeding population can have different pedigree inbreeding coefficients. For example, at the onset of the Speke's gazelle captive management program in 1979, 13 adult females and 6 adult males lived in the captive population at two zoos. Under the assumption that the four founders were unrelated, the inbreeding coefficients ranged from 0 (i.e., some animals had no common ancestors in their maternal and paternal lineages) to 0.375 (a level of inbreeding higher than brother-sister matings and indicative of much shared ancestry). Thus, by calculating F_p, a manager can infer which individuals share common paternal and maternal ancestors and which do not. Moreover, among those animals with $F_p > 0$, the value of F_p quantifies the amount of shared ancestry on an individual basis.

Second, because F_p is a probability, its mathematical range is limited from 0 to 1, with 0 corresponding to no shared ancestry in the known pedigree, and non-zero values corresponding to some shared ancestry. Hence, it is impossible to measure avoidance of inbreeding with F_p (i.e., negative F_p's are not defined). With F_p, an individual is either not inbred or is inbred, it can *never* be measured as being produced by a mating that avoided or minimized inbreeding relative to the mating options available in the population. This is perhaps one of the greatest misuses of F_p in the conservation biology literature. For example, the system of mating in the Speke's gazelle program has been described as "intentional inbreeding" designed to "rapidly increase" the inbreeding coefficient (Hedrick and Miller, 1992) and as mating "near relatives deliberately" as opposed to "maximal avoidance of inbreeding" (Simberloff, 1988). Yet, as will soon be discussed, the actual system of mating in this herd minimizes the rate of increase of the inbreeding coefficient through strong avoidance of breeding near-relatives (i.e., "maximal avoidance of inbreeding"), as stated explicitly in Templeton and Read (1983). The confusion obviously stems from the fact that because all captive-born individuals were related (an unavoidable constraint in a population founded by one male and three females), *all* possible matings between captive-born animals produce offspring with $F_p > 0$. However, this fact says nothing about whether or not animals were paired in such a way as to minimize, maximize, or ignore F_p, and this is the information needed to describe a system of mating (as will be discussed later).

Third, because F_p is an indivdual-level measure, it tells one nothing directly about population-level attributes, such as the population's system of mating (as mentioned above) or levels of genetic diversity. However, by using the F_p's of several different individuals in a joint analysis, some population-level phenomena can be studied with pedigree inbreeding coefficients. One of the more important of these is quantification of inbreeding depression. By performing a regression of measures of an individual's viability or health upon the same individual's F_p, it is possible to quantify the impact of shared ancestry upon the health of individuals in the population as a whole. For example, when regressions were performed upon 30-day viability (including stillborns), 1-year viability, and birthweight in Speke's gazelle (using the animals that had been bred up to and including 1982), statistically significant negative regressions were obtained in all cases (Templeton and Read, 1983, 1984), thereby indicating that as pedigree inbreeding increased, the average health of the individuals declined.

As mentioned above, all captive-born animals are related, so further pedigree inbreeding could not be avoided in this herd. Hence, the easy solution to inbreeding depression – namely, avoid pedigree inbreeding – was not an option in this case. Accordingly, Templeton and Read (1983) instituted a breeding program that was designed to maximize genetic variation in the herd while selectively eliminating the inbreeding depression. Inbreeding depression has a genetic basis, and therefore, in principle, it can be eliminated or reduced by evolutionary changes at the contributing loci. Unfortunately, the genetic basis of inbreeding depression is treated in an overly simplistic fashion in much of the conservation literature, being attributed in part to heterotic loci and in part to recessive deleterious alleles (Soulé, 1980). The recessive deleterious alleles are generally believed to be the more important. For example, Lande (1988) claimed that the experimental literature on *Drosophila* showed that all the inbreeding depression was due to recessive, deleterious alleles (half being recessive lethal and semi-lethal mutations, and the remainder being slightly deleterious recessives). Accordingly, the elimination of inbreeding depression is equated to the purging of the population of recessive deleterious and lethal genes (Lande, 1988). However, the *Drosophila* literature actually indicates that a major cause of inbreeding depression lies in the synergistic or epistatic interactions among loci (Hildreth, 1956; Kosuda, 1972; Lucchesi, 1968; Thoday, 1963; Thompson, 1986). Moreover, the Speke's gazelle program was explicitly designed upon the basis of earlier experimental work on *Drosophila mercatorum* that indicated that epistasis was the major factor in allowing populations to adapt to high levels of pedigree inbreeding (Templeton et al., 1976; Templeton, 1979).

The Speke's gazelle program was designed with the idea that the adaptation to pedigree inbreeding is due both to selection *for* alleles or gene combinations as well as to selection *against* recessive, deleterious alleles. As a consequence of using this more complex genetic basis of inbreeding depression, the maintenance of high levels of genetic variation played the primary role in this program because, without variation, it would be impossible to select for the appropriate alleles and gene combinations. Moreover, to increase the amount of genetic variation at the combinatorial level, Templeton and Read (1983) deliberately bred the animals in such a way as to maximize the effectiveness of genetic recombination, and this was monitored through the "hybridity coefficient" (Templeton and Read, 1984).

After this breeding program was initiated, the inbreeding depression was indeed significantly reduced as this population of Speke's gazelles evolved in captivity (Templeton and Read, 1983, 1984; Templeton, 1991a). Moreover, the hybridity coefficient had a significant, positive effect on this response to selection that could be separated from other, potentially confounding factors (Templeton and Read, 1984; Templeton et al., 1987; Templeton, 1991a). Recombination in the previous generation should have no effect if the genetic basis of inbreeding depression was due simply to independently acting recessive deleterious mutations. The opposite result, particularly when coupled with the rich *Drosophila* literature, indicates that models of inbreeding depression and its elimination should take epistasis into account, and that the simple "purging" models of recessive deleterious alleles are unrealistic and could lead to erroneous recommendations.

The detection of inbreeding depression in the Speke's gazelle, the monitoring of its reduction, and the separation of the factors responsible for that reduction were all accomplished by analyses involving the pedigree inbreeding coefficient. This illustrates that pedigree inbreeding coefficients can be used to make inferences about the population, but only with additional information and statistical analyses. Without this extra effort, no population-level inference is justified from pedigree inbreeding coefficients.

"Inbreeding" as a measure of genetic drift in a population

The simplest way of converting the individual F_p's into a population measure is to take the arithmetic average over all individuals. This average probability of identity-by-descent of the population will be symbolized by F_d in this paper, with the "d" signifying that its primary utility is as a population or deme level measure of the evolutionary force of genetic drift.

Genetic drift is the evolutionary force that is associated with the random sampling of genes in going from one generation to the next in a finite population. There are many ways of measuring the evolutionary impact of genetic drift upon a population, and one way is through the use of the average probability of identity-by-descent. Wright (1969) considered an idealized finite population of size N in which all individuals are monoecious or hermaphroditic, self-compatible, have a Poisson-distributed number of offspring, discrete generations, and random mating. There is much confusion about this last assumption as well. Random mating does not imply an absence of inbreeding in the pedigree sense. Instead, it simply means that mates are chosen irrespective of their degree of relationship so that inbreeding in the pedigree sense is neither sought out nor avoided. When the population is finite (as all real populations are), random mating implies that some individuals mate with biological relatives and hence inbreeding in the pedigree sense is *always* a consequence of random mating. The intensity of pedigree inbreeding that is caused by random mating depends upon the size of the population. For example, in the idealized monoecious, self-compatible population, the probability of an individual mating with itself is

1/N, the same probability that it has of mating with any other individual. As the size of the population decreases, the probability of an individual mating with itself (the most intense inbreeding) increases.

The increase of the average probability of identity-by-descent can now be calculated for this idealized population. Producing N offspring requires drawing a total of 2N gametes. Under the idealized assumptions, each gamete is paired at random and is replaced with another gamete. The probability of picking the second gamete as an identical copy of the first gamete drawn at a particular locus is simply 1/2N. In this case, the probability of identity-by-descent is 1 at the locus. With probability $[1 - 1/2N]$, a gamete is paired with a gamete that is not an identical copy from the previous generation. However, let $F_{d(t-1)}$ be the average probability of identity-by-descent at generation $t-1$ due to sharing of ancestors in previous generations. Thus, even when genes are not identical copies from the previous generation, they still could be identical copies due to earlier shared ancestry with probability $F_{d(t-1)}$. Hence, the average probability of identity-by-descent $t(F_{d(t)})$ at generation is given by:

$$F_{d(t)} = 1/2N + F_{d(t-1)}[1 - 1/2N]. \tag{1}$$

So,

$$1 - F_{d(t)} = [1 - 1/2N](1 - F_{d(t-1)}). \tag{2}$$

Now, suppose that no pedigree information is available past generation 0. As with F_p, we assume no past shared ancestry, so $F_{d(0)} = 0$. Using this initial condition, and the recursion formula (2), it follows that

$$1 - F_{d(t)} = [1 - 1/2N]^t, \tag{3}$$

or

$$F_{d(t)} = 1 - [1 - 1/2N]^t. \tag{4}$$

Note that $F_{d(t)}$ is an increasing function of t (time in generations) for any finite N. Ultimately, $F_{d(t)}$ goes to 1, and how fast it goes to 1 depends upon the inverse of population size. The quantity $1 - F_d$ is called the "heterozygosity" in the population genetic literature. Hence, another way of stating Eq. (3) is that heterozygosity ultimately goes to 0 in this idealized finite population.

"Heterozygosity," like "inbreeding," is one word with several meanings. Note that in the context of Eq. (3), heterozygosity simply (and only) means the average probability of non-identity by descent in the population. This meaning of heterozygosity may or may not be related to the actual amount of genetic variation in the population, but it is commonplace to equate this "heterozygosity" with "genetic variation" in the conservation biology literature. This was lead to another great misunderstanding, namely, the false idea that inbreeding causes a loss of genetic variation (Hedrick and Miller, 1992; Senner, 1980; Shaffer, 1981; Soulé, 1980). This idea has lead to some wrong choices in conservation programs. For example, because the Speke's gazelle breeding program inevitably caused an increase in F_d, Simberloff (1988) argued that the goal of eliminating inbreeding depression of Templeton and Read (1983) would lead to "decreasing genetic diversity." However, the primary stated goal of the Speke's gazelle program was indeed to maintain genetic diversity (Templeton and Read, 1983), and as we shall soon see, this goal was met. Thus, there is no validity to the idea that inbreeding causes a loss of genetic variation.

What is true is that genetic drift can cause the average probability of identity by descent of a population to increase, and genetic drift can cause the loss of genetic variation in a population; hence, both an increase in F_d and a decrease in genetic variation can be *correlated* because of a common causal factor – genetic drift. As with most correlations, this relationship is not universal. For example, subdividing a population into small breeding units can cause F_d to go to 1 very rapidly in all subpopulations and hence in the total population, but genetic variation will be preserved at *higher* levels and for longer periods of time than if the total population were randomly mating or even avoiding inbreeding (as a system of mating) (Chesser, 1991; Gliddon and Goudet, this volume; Templeton, 1991b).

As shown above, inbreeding, in the senses considered so far, is not a cause of loss of genetic variation, and is only a correlate of the loss of genetic variation due to genetic drift. As will soon be shown, inbreeding when used as a measure of system of mating has no impact at all upon the amount of genetic variation (numbers of alleles and their frequencies) in a population. Yet, many management programs seek to avoid inbreeding in order to preserve genetic variation. However, the real cause for the loss of genetic variation is genetic drift and selection, not inbreeding in any of the senses of the word. An effective management program for genetic diversity must be based on an accurate identification of the factors that actually erode genetic diversity in managed populations. Otherwise, the program faces the possibility of utilizing management procedures that have the opposite of the desired effect.

Eqs (3) or (4) are also the sources for another widely misunderstood concept in the conservation biological literature; the concept of effective population size. Recall that these equations were derived under the assumption of a monoecious, self-compatible, random mating (including selfing), discrete generation population with a Poisson distributed number of offspring from each individual. It is unlikely that any real population will satisfy all these assumptions. The concept of effective population size was invented in order to measure the strength of genetic drift in populations other than this idealized one. The strength of drift can be monitored by its impact on various genetic parameters that characterize the population, and the average level of identity-by-descent relative to some base population is one such parameter. If a real population after t generations from its base generation had an average probability of identity-by-descent of, say, F_t, then the idealized hypothetical population would have the same average probability of identity-by-descent after t generations if the idealized population had (from Eq. 4) the following population size:

$$N = \frac{1}{2[1 - (1 - F_t)^{1/t}]} \tag{5}$$

Thus, relative to its reference generation, the real population has accumulated through genetic drift the same amount of average identity-by-descent as an idealized population of the size given by Eq. (5). Hence, Eq. (5) provides a way of quantifying the evolutionary impact of genetic drift upon the real population relative to the inbreeding parameter F_t. Eq. (5) is known as the inbreeding effective population size of the real population. Note that N is determined exclusively from F_t and t; it has no direct dependency upon the actual population size. Population geneticists (e.g., Crow and Denniston, 1988) have derived several equations that relate inbreeding effective size to actual population size under a variety of assumptions. However, the primary definition of inbreeding effective size is that given by Eq. (5), and all other equations for inbreeding effective size are based on certain restrictive assumptions that may or may not be appropriate for a given real population (Simberloff, 1988).

If one has pedigree data on a managed population so that F_t can be calculated, then it is possible to determine the inbreeding effective size without the use of secondary equations that may be quite inaccurate in situations commonly arising in conservation biology. For example, the average probability of identity by descent in the Speke's gazelle breeding herd in 1979, regarding the four founders as unrelated (and hence the reference generation), was 0.1283. The average number of generations of these animals from the founders was 1.7237 generations. Then, from Eq. (5), the inbreeding effective size of the 1979 herd is 6.5305, a number which is considerably less than the census size of 19 breeding animals. This low size is attributable to the founder effect: although there were 19 breeding animals available in 1979, their genes were derived from only four founders and therefore they accumulated inbreeding at a very fast rate, as indicated by the low inbreeding effective size. In 1979, Templeton and Read (1983) instituted a new management program that included avoidance of breeding near biological relatives. The first generation bred from the 19 animals available in 1979 consisted of 15 offspring with an average probability of identity-by-descent of 0.1490. Using this value for F_t and augmenting t by 1, Eq. (5) yields an inbreeding effective size of 8.6932. Note that the inbreeding effective size is still smaller than the census size, once again showing the persistent effects of the initial founding event. However, note that the inbreeding effective size increased from 6.5305 to 8.6932 even though the census size *decreased* from 19 to 15 when our inference is confined to the animals bred under the new program. This increase in inbreeding effective size reflects the impact of the avoidance of inbreeding (in a system of mating sense). In order to focus more directly upon the genetic impact of the Templeton and Read (1983) breeding program, one can define the 1979 herd as the reference population rather than the four founders. Note that the difference in average probability of identity-by-descent of the first generation born into the Templeton and Read (1983) program from that of their parents is 0.0207. Since the parental herd is now the reference, $t = 1$, therefore Eq. (5) yields the inbreeding effective size of the first generation of Speke's gazelle born under this breeding program to be 21.0556. Note that the inbreeding effective size is *greater* than the census size of 15, reflecting the stated goal of decreasing (not increasing) the rate of increase of F_d (Templeton and Read, 1983). This reveals yet another widespread fallacy: that effective population sizes are smaller than census sizes (Soulé, 1980). Quite the contrary, inbreeding effective sizes are commonly larger than census sizes in many well managed populations.

Note also that we now have two very different inbreeding effective sizes (8.6932 and 21.0556) for the same animals. The first size tells us how rapidly inbreeding has occurred both before and during the Templeton and Read (1983) program, the later size tells us about the effects on average probability of identity-by-descent after the inception of the breeding program. Hence, both numbers give valuable information. However, as Simberloff (1988) pointed out, it is commonplace in the conservation genetic literature to refer to "the effective population size" as if there were only one effective size for a population. As the Speke's gazelle example clearly shows, there can be many different effective sizes for the same population as a function of different (and meaningful) reference populations.

However, the situation is actually much worse than merely forgetting to specify a reference population in most invocations of effective size. Genetic drift is a powerful evolutionary force that influences many genetic parameters of a population in addition to average probability of identity-of-descent. For example, genetic drift will cause the frequency of an allele to fluctuate at random around its value from the previous generation. This random fluctuation can be measured through the variance in allele frequency. In the same idealized population used to define Eqs (3) and (4), this theoretical variance is $pq/2N$, where p is the frequency of the allele in the parental generation ($q = 1 - p$). Suppose that in a real, non-idealized population the variance of an allele frequency was σ^2. Then, the variance effective size is defined as the number that simply expresses this real variance in the form of the idealized variance; that is,

$$\sigma^2 = pq/2N_{ev}, \tag{6}$$

or

$$N_{ev} = pq/2\sigma^2, \tag{7}$$

where N_{ev} is the variance effective size.

Another important genetic parameter is the rate at which alleles become fixed or lost; that is, the rate at which allelic variability is lost at a locus. In an idealized population, this occurs at a rate of $1/2N$ *per* generation. In real populations, this rate of loss is expressed as $1/2N_{e\lambda}$ where $N_{e\lambda}$ is the eigenvalue effective size (Ewens, 1982; Gliddon and Goudet, this volume).

Unfortunately, much of the conservation genetic literature treats effective size as a unitary concept, failing to acknowledge that effective size varies as a function of reference population and of the genetic parameter of interest (Simberloff, 1988; Gliddon and Goudet, this volume). However, by keeping the different concepts distinct, a manager has a potentially powerful monitoring device for tracking the genetic impact of a breeding program. Indeed, in the situations commonplace in conservation biology, effective sizes can differ by large amounts (Simberloff, 1988). When population sizes are declining (as for many endangered natural populations), N_{ev} and $N_{e\lambda}$ are usually much smaller than N_{ef}. When populations are expanding (as is generally recommended for the initial stages of a captive breeding program starting from a small founder base), N_{ev} and $N_{e\lambda}$ are usually much larger than N_{ef} (Templeton, 1980). Consequently, equating these concepts is misleading and can lead to erroneous management recommendations (Gliddon and Goudet, this volume). Genetic drift is a powerful evolutionary force, and the only way to effectively manage for its consequences is to realize that those consequences are diverse and not unitary.

Effective management also requires a clear distinction between causal relationship (e.g., genetic drift causes a reduction in genetic variability, as measured by $N_{e\lambda}$) *versus* correlational relationships (e.g., genetic drift reduces genetic variability, as measured by $N_{e\lambda}$, and increases inbreeding in the sense of average probability of identity by descent, as measured by N_{ef}, leading to a negative correlation between genetic variation and inbreeding). Because correlational relationships can be violated in particular instances, management decisions should be based on causal relationships whenever possible. For example, it would be incorrect to avoid inbreeding in order to preserve genetic variation; instead, one should institute policies that directly increase $N_{e\lambda}$, as was done for the Speke's gazelle population (Templeton and Read, 1983, 1984).

"Inbreeding" as a measure of system of mating in a population

The final sense of the word inbreeding to be considered in this chapter is as a measure of a system of mating for a local breeding population. There have been several inbreeding coefficients designed for this purpose (Jacquard, 1975), but only one will be considered here, and it will be called the panmictic index and symbolized by f. The panmictic index measures inbreeding as a deviation from a reference population. In this case, the reference population has a system of mating known as random mating in which alleles at a locus are paired together in proportion to their frequencies in the overall population. For an autosomal locus with two alleles (say A and a) at frequencies p and q ($=1-p$), the expected genotype frequencies under random mating are given by the well-known Hardy-Weinberg formulae: the frequency of $AA = p^2$, the frequency of $Aa = 2pq$, and the frequency of $aa = q^2$. The panmictic index measures a population's system of mating as a deviation from the heterozygosity frequencies expected under random mating; that is,

$$f = 1 - H_o/H_e, \tag{8}$$

where H_e is the expected heterozygosity under random mating (i.e., $2pq$ for an autosomal locus with two alleles) and H_o is the observed heterozygosity. If the observed heterozygosity is the same as the expected, $f = 0$, the population is randomly mating. If the observed heterozygosity is less than expected, $f > 0$, the population is regarded as having an inbreeding system of mating. If the observed heterozygosity exceeds the expected, $f < 0$, the population has a system of mating in which inbreeding is avoided.

Note also that three different meanings of the word "heterozygosity" have now been given in this paper (recall Eq. (3)). Heterozygosity, like inbreeding, is one word with several meanings, and the failure to discriminate among the various usages of the word has caused much confusion. For example, heterozygosity in the sense of Eq. (3) is simply the inverse of F_d. Hence, by mathematical necessity, that type of heterozygosity declines as F_d increases. However, one can easily have a population with a high F_d that nevertheless has high levels of either expected and/or observed heterozygosity. For example, a genetic survey on the Speke's gazelle herd (Templeton et al., 1987) revealed that this herd had higher levels of expected heterozygosity than most large ungulate natural populations or mammalian populations in general, despite having a very high value of F_d. Moreover, the observed heterozygosity was even higher, as will now be shown.

Tab. 1 gives the observed and expected heterozygosities for the animals from 1982 (the first generation of the Templeton and Read breeding program), from the genetic survey data given by Templeton et al. (1987). From these observed and expected heterozygosities, the

Table 1. Observed and expected heterozygosities in the Speke's gazelle herd of 1982, for four isozyme loci, phosphoglucose isomerase (PGI), general protein (GP), malic dehydrogenase (MDH), and colin esterase (CE). The uncorrected and corrected (see text) panmictic indices, f and f_c, respectively, are also given for each of these loci

Locus:	PGI	GP	MDH	CE
Observed het.:	0.765	0.500	0.235	0.300
Expected het.:	0.472	0.375	0.208	0.255
f:	−0.619	−0.333	−0.133	−0.176
f_c	−0.600	−0.308	−0.103	−0.151

panmictic index can be calculated for each locus, as also given in Table 1. In a finite population with separate sexes, random mating is actually expected to yield a negative f because of random differences in allele frequency between the sexes. One can correct for this sex bias by solving the equation (Gliddon, personal communication):

$$f = \frac{1 - H_o/H_e + 1/(2N/(1+f) - 1)}{1 + (2N/(1+f) - 1)}. \tag{9}$$

Taking the average of these corrected indices across loci, the overall f is −0.291. Hence, the system of mating of the Speke's gazelle herd is one of strong avoidance of inbreeding. Since avoidance of inbreeding was an explicit management policy (Templeton and Read, 1983), these results are not at all surprising. Although these animals are avoiding inbreeding in a system of mating sense, recall that $F_d = 0.1490$ for these same animals. Thus, the same animals are inbred (in the pedigree and drift senses) and avoiding inbreeding (in the system of mating sense) at the same time. There is no contradiction here, merely the fact that there are different inbreeding coefficients that are designed to measure different features of the population.

Because many management programs have as a major goal the preservation of genetic variability, it is important to investigate the impact of system of mating upon genetic variation. To do this, one first of all has to have a measure of genetic variation. Heterozygosity in the sense of probability of non-identity by descent is inappropriate because it is not related even in theory to the actual levels of genetic variation present in a population and is completely redundant in its information content to F_d. Observed heterozygosity is better, but it is mathematically a function of f, so using observed heterozygosity yields no meaningful biological insights concerning the role of system of mating. Expected heterozygosity is a function only of allele frequencies in the gene pool of the population, and hence there is no mathematical constraint upon expected heterozygosity and system of mating. Therefore, expected heterozygosity provides the most biologically meaningful measure of genetic variability when exploring the consequences of system of mating.

Expected heterozygosity is strictly a function of allele frequencies. Allele frequencies are calculated from genotype frequencies by taking the frequency of the genotype homozygous for the allele of interest and adding to it one half the sum of the frequency of heterozygotes that also bear the allele of interest (the half enters because only half the alleles born by heterozygotes are of the allele type of interest). For example, in the two allele, random mating population, the frequency of the A allele is $p^2 + pq = p(p + q) = p$. Under inbreeding (in the sense of f), the genotype frequencies are given by $p^2 + pqf$ for AA, $2pq(1 - f)$ for Aa, and $q^2 + pqf$ for aa. Hence, the frequency of the A allele in non-randomly mating populations is given by:

$$p^2 + pqf + \tfrac{1}{2}[2pq(1 - f)] = p^2 + pqf + pq(1 - f) = p^2 + pq = p. \quad (10)$$

As Eq. (10) shows, the allele frequencies (and hence the expected heterozygosities) are not influenced by f. Hence, regardless of whether the population has a system of mating of inbreeding, avoidance of inbreeding or random mating, the level of expected heterozygosity is the same. Obviously, it is genetic drift, not the system of mating, that causes an erosion of genetic variation in populations.

Conclusions

Because "inbreeding" receives so much attention in the genetic management of small populations, it is critical to distinguish among the various definitions of inbreeding and to use them in a proper fashion. This, unfortunately, is rarely done in the literature (e.g., all three definitions of inbreeding given here appear in three consecutive equations on page 37 in Hedrick and Miller (1992), but the only formal definition of inbreeding coefficient in that paper is F_p on p. 34). By explicitly acknowledging these different definitions and by calculating individual inbreeding coefficients, demic averages of probabilities of identity by descent, and panmictic indices, a genetic manager has a powerful set of statistics with which to monitor a population and from which to make management recommendations. By confusing these concepts, erroneous management decisions are likely, cause and effect are confused with correlation, and management for other genetic factors, such as genetic diversity, will suffer. Population geneticists in particular have a responsibility to treat inbreeding in an accurate fashion because their writings are often the source of information for zoo and wildlife managers. They must also take their responsibilities much more seriously in the future than they have in the past because, in the words of Hedrick and Miller (1992), "To err in the application of genetics to an endangered species may result in its extinction." Only in this way can the confusion, erroneous statements, and misguided management plans and evaluations of the past be avoided.

Acknowledgements
We would like to thank Subodh Jain and Volker Loeschcke for their excellent suggestions on an earlier draft of this manuscript. This work was supported by NIH grant R01 GM31571.

References

Chesser, R. K. (1991) Influence of gene flow and breeding tactics on gene diversity within populations. *Genetics* 129: 573–583.

Crow, J. F. and Denniston, C. (1988) Inbreeding and variance effective population numbers. *Evolution* 42: 482–495.

Ewens, W. J. (1982) On the concept of the effective population size. *Theor. Pop. Biol.* 21: 373–378.

Hedrick, P. W. and Miller, P. S. (1992) Conservation genetics: techniques and fundamentals. *Ecol. Appl.* 2: 30–46.

Hildreth, P. E. (1956) The problem of synthetic lethals in *Drosophila melangaster*. *Genetics* 41: 729–742.

Jacquard, A. (1975) Inbreeding: One word, several meanings. *Theor. Pop. Biol.* 7: 338–363.

Kosuda, K. (1972) Synergistic effect of inbreeding on viability in *Drosophila virilis*. *Genetics* 72: 461–468.

Lande, R. (1988) Genetics and demography in biological conservation. *Science* 241: 1455–1460.

Lucchesi, J. C. (1968) Synthetic lethality and semi-lethality among functionally related mutants of *Drosophila melanogaster*. *Genetics* 59: 37–44.

Senner, J. W. (1980) Inbreeding depression and the survival of zoo populations. *In:* Soulé, M. E. and Wilcox, B. A. (eds), *Conservation biology: an evolutionary-ecological perspective*. Sinauer, Sunderland, MA, pp. 209–224.

Shaffer, M. L. (1981) Minimum population sizes for species conservation. *BioScience* 31: 131–134.

Simberloff, D. (1988) The contribution of population and community biology to conservation science. *Annu. Rev. Ecol. Syst.* 19: 473–511.

Soulé, M. E. (1980) Thresholds for survival: maintaining fitness and evolutionary potential. *In:* Soulé, M. E. and Wilcox, B. A. (eds), *Conservation biology: An evolutionary-ecological perspective*. Sinauer, Sunderland, MA, pp. 151–169.

Templeton, A. R. (1979) The unit of selection in *Drosophila mercatorum*. II. Genetic revolution and the origin of coadapted genomes in parthenogenetic strains. *Genetics* 92: 1265–1282.

Templeton, A. R. (1980) The theory of speciation via the founder principle. *Genetics* 94: 1011–1038.

Templeton, A. R. (1991a) Genetics and conservation biology. *In:* Seitz, A. and Loeschcke, V. (eds), *Species conservation: A population biological approach*. Birkhäuser Verlag, Basel, pp. 15–29.

Templeton, A. R. (1991b) Off-site breeding of animals and implications for plant conservation strategies. *In:* Falk, D. A. and Holsinger, K. E. (eds), *Genetics and conservation of rare plants*. Oxford University Press, Oxford, pp. 182–194.

Templeton, A. R., Davis, S. K. and Read, B. (1987) Genetic variability in a captive herd of Speke's gazelle (*Gazella spekei*). *Zoo Biol.* 6: 305–313.

Templeton, A. R. and Read, B. (1983) The elimination of inbreeding depression in a captive herd of Speke's gazelle. *In:* Schonewald-Cox, C. M., Chambers, S. M., MacBryde, B. and Thomas, L. (eds), *Genetics and conservation: A reference for managing wild animal and plant populations*. Benjamin/Cummings, Menlo Park, CA, pp. 241–261.

Templeton, A. R. and Read, B. (1984) Factors eliminating inbreeding depression in a captive herd of Speke's gazelle (*Gazella spekei*). *Zoo Biol.* 3: 177–199.

Templeton, A. R., Sing, C. F. and Brokaw, B. (1976) The unit of selection in *Drosophila mercatorum*. I. The interaction of selection and meiosis in parthenogenetic strains. *Genetics* 82: 349–376.

Thoday, J. M. (1963) Locating synthetic lethals. *Am. Nat.* 97: 353.

Thompson, V. (1986) Half-chromosome viability and synthetic lethality in *Drosophila melanogaster*. *J. Heredity* 77: 385–388.

Wright, S. (1969) *Evolution and the Genetics of Populations*. University of Chicago Press, Chicago.

Conservation Genetics
ed. by V. Loeschcke, J. Tomiuk & S. K. Jain
© 1994 Birkhäuser Verlag Basel/Switzerland

The genetic structure of metapopulations and conservation biology

C. Gliddon and J. Goudet

Functional and Evolutionary Biology Group, School of Biological Sciences, University of Wales, Bangor, Gwynedd LL57 2UW, U.K.

Summary. A range of models describing metapopulations is surveyed and their implications for conservation biology are described. An overview of the use of both population genetic elements and demographic theory in metapopulation models is given. It would appear that most of the current models suffer from either the use of over-simplified demography or the avoidance of selectively important genetic factors. The scale for which predictions are made by the various models is often obscure. A conceptual framework for describing metapopulations by utilising the concept of fitness of local populations is provided and some examples are given. The expectation that any general theory, such as that of metapopulations, can make useful predictions for particular problems of conservation is examined and compared with the prevailing 'state of the art' recommendations.

Introduction

Many species of animals and plants are distributed patchily at some spatial scale. That is, they occur in a series of more or less isolated groups which are themselves distributed over the landscape. This patchy distribution may be due to a number of factors including habit disconti-nuities and social groupings. In the 1940s, population geneticists started to attempt to determine the consequences of population subdivision (Wright, 1940, 1943), usually focusing on the rate of loss of genetic variability. In the 1950s, the emphasis changed to asking how popula-tion subdivision affected the ease with which genetic polymorphism could be maintained (Levene, 1953; Dempster, 1955). However, in most of this early work, although populations were distributed patchily in space or time, the phenomena of local extinction and recolonization were not taken into account. The concept of the metapopulation was introduced by Levins (1970) and can be defined as a population of local populations which are established by colonists, survive for a while, send out migrants and eventually disappear. That is, a metapopulation is a set of local populations connected through migration and recoloniza-tion. If we contrast this last definition with that of a Mendelian population (a set of individuals connected through parenthood and mating), we can see the great similarity between the two concepts, the difference mainly being one of the scale of biological complexity that is being addressed.

Why study metapopulations?

A commonly stated reason for studying metapopulations is that they represent more of an 'ecological reality' than many other classes of model. In the sense that they can describe more exactly the spatial and temporal distribution of many species, this is a valid reason. In particular, if metapopulation models include both elements of population dynamics and population genetics, then they represent a more complete description of the ecological reality than most other classes of model. However, the criterion of 'greater reality' is only of use as a reason to prefer one class of model over another if there are either major quantitative or qualitative differences in their predictions of the behavior of biological systems. That is, increasing the complexity of models is only useful when there are emergent properties (the more complex model predicts behavior that is not predicted by any simpler model or sets of models applied independently).

Metapopulation theory has become very popular over the past decade, primarily for the reason that it appears to offer a theoretical framework that may be of help to conservation biology (see, for example, Hanksi and Gilpin, 1991). Since many of the forces that may have acted to bring about a decline in numbers of certain species often result in habitat fragmentation, the structure of a metapopulation has an immediate apparent relevance both to assessing the impact of this habitat fragmentation and to devising ways of changing the patterns of connection between local populations so as to increase the chances of survival of a species.

If the classical conservation dilemma of whether an arrangement of a *Single Large Or Several Small (SLOSS)* patches is the better means of preserving a species is examined from the perspective of pure demography (in the sense that the probability of extinction of a patch is a convex increasing function of decline in patch size) then it is apparent that the solution *Single Large* is the better. On the other hand, if conservation of genetic diversity is deemed to be of prime importance, a population genetic approach would suggest that extreme subdivision of the population (*Several Small*) is the optimal solution. Since there are two opposing solutions, which depend upon whether demography or genetics is deemed to be the major consideration, it would appear desirable to examine the part of metapopulation theory that combines both demography and genetics, that is, a population biological theory of metapopulations.

Problems with current models

Most current models of metapopulations can be divided into two, more or less exclusive, categories: those which emphasise demographic aspects of population structure, and those which emphasise genetical aspects.

In the first category, the local populations frequently have no real dynamics. For example, after foundation by a specified number of individuals, a local population increases instantly to its carrying capacity (e.g., Wade and McCauley, 1988). Alternatively, where local populations have explicit dynamics, they are usually assumed to be logistic with r and k defined as constants (i.e., independent of genotype). That is, the possibility of an effect of change of genotype frequencies within local populations on growth rate or extinction probability is excluded from the model. In many senses, these models could be considered to focus on relatively short-term ecological effects at the expense of ignoring possibly longer term evolutionary consequences.

In the second category where fitness of genotypes within local populations are explicitly defined, that fitness is constant (i.e., independent of density and/or frequency). That is, possible ecological effects tend to be ignored to allow evolutionary consequences to be studied.

In addition, the majority of models (for some notable specific exceptions see the review by Olivieri et al., 1990) tend to divorce both colonising ability of individuals and probabilities of extinction of local populations from any genotypic effects.

Given the relative lack of incorporation of genotypic effects on parameters of clear importance for conservation such as colonising ability and extinction probability, it should come as no surprise that the majority of models predict that demographic (genotype independence) effects are of major concern in designing conservation strategies. It is incumbent upon the developers of metapopulation theory and models to examine the conditions (if any) under which greater concern for conservation of genetic variability over time is required. That is, there is a need for measures which combine both demographic and population genetic effects.

A 'Neutralist' approach

A measure that combines both the census size of a population and the genetic variability contained therein is the genetically effective size, N_e. It might be reasonable to assume that it is the genetically effective size that is inversely related to the probability of extinction rather than the census size. However, there are a number of problems which are associated with this approach.

Which genetically effective size should be used?
There are a number of genetically effective sizes which have been defined in population genetic theory (see e.g., Crow and Denniston, 1988). Three relatively common varieties are:

(i) The inbreeding effective size (Wright, 1931, 1939). Measures the rate of loss of heterozygous individuals from a defined set of individuals;
(ii) The variance effective size (Wright, 1931, 1939; Crow, 1954). Measures the rate of loss of genetic variance from a defined set of individuals;
(iii) The extinction effective size (Haldane, 1939). Measures the asymptotic rate of loss of segregating loci (also known as the eigenvalue effective size (Ewens, 1979, 1982)).

While many population genetics texts suggest that all these measures of genetically effective size yield essentially the same results under most circumstances, it is important to emphasise that with population subdivision and/or when population census number is changing over time, there may be orders of magnitude differences in their values. In the context of metapopulations, it is extremely important, therefore, to decide which is the appropriate type of genetically effective size.

Inbreeding effective size
Gilpin (1991) calculated the inbreeding effective size for a metapopulation and concluded that it may be several orders of magnitude lower than the census size of the metapopulation. That is, in Gilpin's model, metapopulations with extinction and recolonization lose heterozygous individuals far faster than a control with no extinction or recolonization. One would conclude from this result that, if the inbreeding effective size is an appropriate measure, population subdivision with extinction and recolonization is extremely undesirable from a conservation perspective. However, in Gilpin's model, the probability of local extinction is related solely to the local carrying capacity and is, therefore, independent of the genetic constitution of the local population. It is difficult to see, in the context of this model, what the relevance of any genetically effective size is, since it is bound to have no effect on the behavior of the model and, presumably, should be assumed to have no effect on the real situations that are being modelled! Additionally, one still needs to ask whether the inbreeding effective size is the most appropriate combination of census size and genetic variation for conservation biology.

Variance effective size
As stated above, the variance effective size measures the rate of loss of genetic variability from a population, independent of the way in which

that genetic variability is 'packaged' (i.e., within or among individuals). Crow and Denniston (1988) stated:

"If one is interested in conserving genetic variance,... it [variance effective size] is the most appropriate effective number."

Wright (1940) stated in a verbal model that extinction/recolonization will enhance the genetic differentiation of local populations because the number of founders of a local population is likely to be substantially smaller than the local carrying capacity. That is, the variance effective size of the set of local populations would be increased by the extinction/recolonization process.

Slatkin (1977, 1985, 1987) incorporated extinction/recolonization into the classical Island models, thereby formalising Wright's (1940) verbal model. He found that, under some circumstances, extinction and recolonization effectively increase gene flow, thereby reducing among-population differentiation and, hence, the global variance effective size.

Wade and McCauley (1988) re-cast Slatkin's models in terms of F_{st}. They considered two different origins for founders: the 'Propagule Pool' in which there is one (large) source population from which migrants originate; and the 'Migrant Pool' in which migrants are drawn at random from the extant local populations. In the 'Propagule Pool' model, their results were in accord with those of Wright (1940) in that F_{st} is increased, relative to a no-extinction control, and, therefore, the variance effective size is increased, providing the number of founders is lower than the local carrying capacity. In the 'Migrant Pool' model, F_{st} is increased and, hence, variance effective size, providing $(4Nm + 1)$ is greater than twice the number of founders, where N is the local carrying capacity and m the rate at which local populations exchange migrants (see McCauley, 1991, for a fuller review).

Extinction effective size
The extinction (or eigenvalue) effective size is not used as frequently as either variance or inbreeding effective size. However, in conservation genetics, a strategy for rescuing a species will often involve a breeding programme designed to maximize the number of founder equivalents and this approach will also maximise the extinction effective size (see Templeton and Read, this volume).

Problems with all genetically effective sizes
The conclusions that can be drawn from the various models that describe the variance effective size of metapopulations is that there are circumstances in which the global variance effective size is increased relative to the census size which is in direct contrast to the findings for inbreeding effective size by Gilpin (1991). The reason for this is that the inbreeding effective size focuses on the number of heterozygous individ-

uals and, therefore, is determined primarily by the rate of loss of genetic variability *within local populations*. In contrast, the global variance effective size is determined by the rate of loss of genetic variability *over all populations*. That is, the two measures of genetically effective size refer to different levels of population structure and this should be borne in mind when deciding which is the more relevant for particular conservation purposes. However, there is another major problem which arises with the use of genetically effective size as a measure of the likelihood of extinction. All measures of effective size require that the genetic variability, which is used in their calculation, is selectively neutral. For the purposes of metapopulation theory, it is necessary, therefore, that selective neutrality at the individual level is transformed into differences, at the level of among populations, that affect the rates of extinction and/or colonisation. While this may seem a somewhat far-fetched scenario, there are some particular circumstances under which it may not be an unreasonable assumption. For example, if a major cause of extinction were a pathogen to which there were a variety of resistance genes in a population, providing the carrying of a resistance gene in the absence of the pathogen did not cause a reduction in individual fitness, those populations or species which carried a variety of resistance genes would have a lower likelihood of extinction than those populations or species which were monomorphic.

However, given the problems pointed out above, coupled with the lack of local dynamics included in those models for which genetically effective sizes have been calculated, it is very difficult to propose any one genetically effective size as an important factor to consider when designing a conservation programme.

A 'Selectionist' approach

Given the similarity between the definition of a metapopulation and that of a Mendelian population described in the Introduction, it is tempting to consider the possibility of defining the equivalent of individual survivorship and fecundity as components of the 'fitness' of a local population. That is, it might be possible to measure colonising ability of a local population as the equivalent of individual fecundity and resistance to extinction as the equivalent of survivorship. These components of fitness would be functions of both size and genotypic constitution of local populations and the modelling of the fate of a metapopulation could then become relatively straightforward. For example, if a population which contains genetically variable individuals is 'fitter' than a monomorphic population, then this is a sufficient condition for the maintenance of genetic polymorphism at the level of the metapopulation and is likely to render the metapopulation less prone to global extinction. This increase in fitness of the local population may be achieved in at least two ways:

(i) If heterozygous individuals are fitter than homozygotes then polymorphic local populations will be fitter than similarly sized monomorphic populations. This is simply a restatement of the classical 'balancing selection' argument.

(ii) Populations containing individuals of differing genotypes may be 'fitter' than monomorphic populations if intra-genotypic competition is stronger than inter-genotypic competition. This does not require individual 'heterozygote advantage'.

The latter scenario is similar to the concept of niche selection and there are certainly many empirical observations which support the view that genotype mixtures may be 'fitter' than monocultures. However, the application of this type of approach to conservation genetics is still in its infancy and it is difficult to assess its importance at this time.

Conclusions

The usefulness of metapopulation theory to conservation biology is rather difficult to assess at this point in time. There have been some rather unrealistic expectations on the part of conservation biologists as to of what such a general theory might be capable. From another perspective, the development of metapopulation theory, particularly where both genetics and demography are included in the models, has been driven by analytic tractability and has obscured some of the underlying assumptions that are made. In particular, very few authors have pointed out what type of genetically effective size they are calculating for particular metapopulation models and what the implications would be for conservation.

We are clearly at a point in time when pressing questions are being posed to population biologists by conservationists with the expectation of receiving clear advice. It is essential that the answers given clearly refer to the appropriate level and time-scale. For example, in situations where inbreeding depression may be expected to be important, strategies which maximise inbreeding effective size are more appropriate than those which use global variance effective size. Short-term solutions may well be best based on purely demographic criteria, although genetic effects can occur on time scales which are similar to demographic changes.

While it is unrealistic to expect that any one general theory will provide adequate answers to specific conservation problems, this lack of generality is equally a problem for the sets of guidelines currently advocated by, for example, the International Union for the Conservation of Nature (IUCN). Strict adherence to IUCN recommendations for re-introduction of locally extinct species would have prevented the apparently successful re-introduction of the Arabian Oryx. In particular, many existing conservation guidelines suffer from what could be called the 'Cute and Cuddly' syndrome. That is, they are based on the presumption that what is thought to be good for a number of mammals, such as inbreeding avoidance, is universally good for all species in danger of extinction. It is necessary, therefore, the continue to examine classes of models dealing with fragmented habitats with the particular emphasis being to describe any novel properties that emerge as a result of the combination of both demography and population genetics, in order to improve the quality of the advice that can be given to conservation biologists.

References

Crow, J. F. (1954) Breeding structure of populations. II. Effective population number. *In:* Kempthorne, O., Bancroft, T. A., Gowen, J. W. and Lush, J. L. (eds), *Statistics and mathematics in biology.* Iowa State College Press, Ames, pp. 543–556.

Crow, J. F. and Denniston, C. (1988) Inbreeding and variance effective population numbers. *Evolution* 42: 482–495.

Dempster, E. R. (1955) Maintenance of genetic heterogeneity. *Cold Spring Harbor Symp. Quant. Biol.* 20: 25–32.

Ewens, W. J. (1979) *Mathematical population genetics.* Springer Verlag, Heidelberg & New York.

Ewens, W. J. (1982) On the concept of effective population size. *Theor. Pop. Biol.* 21: 373–378.

Haldane, J. B. S. (1939) The equilibrium between mutation and random extinction. *Ann. Eugen.* 9: 400–405.

Hanski, I. and Gilpin, M. (1991) Metapopulation dynamics – brief history and conceptual domain. *Biol. J. Linn. Soc.* 42: 3–16.

Gilpin, M. (1991) The genetic effective size of a metapopulation. *Biol. J. Linn. Soc.* 42: 165–175.

Levene, H. (1953) Genetic equilibrium when more than one ecological niche is available. *Am. Nat.* 87: 331–333.

Levins, R. (1970) Extinction. *Lect. Math. Life Sci.* 2: 75–77.

McCauley, D. E. (1991) Genetic consequences of local population extinction and recolonization. *Trends Ecol. Evol.* 6: 5–8.

Olivieri, I., Couvet, D. and Gouyon, P-H. (1990) The genetics of transient populations – research at the metapopulation level. *Trends Ecol. Evol.* 5: 207–210.

Slatkin, M. (1977) Gene flow and genetic drift in a species subject to frequent local extinctions. *Theor. Pop. Biol.* 12: 253–262.

Slatkin, M. (1985) Gene flow in natural populations. *Annu. Rev. Ecol. Syst.* 16: 393–430.

Slatkin, M. (1987) Gene flow and the genetic structure of natural populations. *Science* 236: 787–792.

Wade, M. J. and McCauley, D. E. (1988) Extinction and recolonization: their effects on the genetic differentiation of local populations. *Evolution* 42: 995–1005.

Wright, S. (1931) Evolution in Mendelian populations. *Genetics* 16: 97–159.

Wright, S (1939) *Statistical genetics in relation to evolution. Exposé de Biométrie et de Statistique Biologique.* Herman & Cie, Paris.

Wright, S. (1940) Breeding structure of populations in relation to speciation. *Am. Nat.* 74: 232–248.

Wright, S. (1943) Isolation by distance. *Genetics* 28: 114–138.

Conservation Genetics
ed. by V. Loeschcke, J. Tomiuk & S. K. Jain

Effects of inbreeding in small plant populations: Expectations and implications for conservation

T. P. Hauser, C. Damgaard and V. Loeschcke

Department of Ecology and Genetics, University of Aarhus, Ny Munkegade, Building 540, DK-8000 Aarhus C, Denmark

Introduction

Man-made changes of the landscape have reduced the habitats of many species, causing a diminished size and increased isolation of populations, which may in many ways threaten their survival. The increase in inbreeding expected in small populations relative to large ones may be one such threat. As matings in small populations occur among a lower number of mates, this increases the chance of matings among relatives, and as inbreeding is known to cause decreased fitness in populations of many species (e.g., reviewed by Charlesworth and Charlesworth, 1987), the fitness of individuals could thus be expected to drop after a reduction in population size.

Inbreeding, on the other hand, seems to be a permanent and stable trait of many species. Small populations seem to survive for long periods of time, and inbreeding mating systems with intermediate selfing rates and apomixis are widespread in plants. Ecological and population genetic studies have tried to explain the stability and evolution of these mating systems (Lande and Schemske, 1985; Charlesworth and Charlesworth, 1987; Holsinger, 1991a,b). In the genetic models two main hypotheses on the cause of inbreeding depression have been proposed: the over-dominance model, in which the decrease in fitness with inbreeding is caused by the reduced heterozygosity of overdominant loci under inbreeding, and the dominance model, where the increased homozygosity of recessive or partly recessive deleterious alleles causes the inbreeding depression. In this chapter, we will consider the consequences of inbreeding under the dominance model, which most data seem to support (Charlesworth and Charlesworth, 1987; Savolainen, this volume). The increased expression of deleterious recessive alleles under inbreeding that is observed as inbreeding depression, exposes at the same time these alleles to selection. The frequency of deleterious alleles may thus be reduced over some generations, which has been termed "purging of the genetic load". The success of agricultural procedures utilizing hybrids among inbred lines, can be partly ascribed to selection among the homozygous lines reducing the frequency of deleterious alleles in the final hybrid population (Allard, 1960).

A reduction in population size might thus lead to a reduction of fitness due to inbreeding depression, but the inbreeding may, on the other hand, reduce the genetic load, and thereby maintain the mean fitness in the population at the level of the large population. The expectations for the fitness consequences of increased inbreeding in smaller populations are thus not clear, and may, among other factors, depend on the time elapsed since the drop in population size. For conservation purposes, it is important that the processes of inbreeding depression and population size are well understood, in order to evaluate whether negative fitness effects of small population size in nature reserves and habitat fragments may threaten the survival of the populations.

The objective for the present study was therefore to examine the fitness effects of a population size reduction by using genetic computer simulations of a plant population. The fitness of a population was estimated before and after a drop in size, and this was compared to the inbreeding depression in the large population. Our results imply that a stronger focus on the variance in response to inbreeding is essential for predictions of the consequences of a diminished population size. Analogous considerations apply for outbreeding depression, where selection may result in a higher hybrid fitness than suggested by the mean response estimated in an experiment, if there is genetic variation for fitness among the hybrids. Finally, we suggest a management procedure for controlled gene flow from one plant population into another.

Simulations

In order to evaluate the effect of a reduction of population size on the mean population fitness and the inbreeding depression, we performed a computer simulation in which we modeled genomes in a plant population subjected to mutation and selection processes. The fitness of individuals can be followed in a population that drops in size from large to small, and the inbreeding depression in the same population can be estimated at given generation numbers before and after the size reduction.

The model

A random-mating Fisher-Wright population of diploid genomes with an infinite number of mutable sites is simulated on the computer using Monte-Carlo simulations. A genome consists of a pair of autosomal chromosomes of a fixed length (set to 1000 map units). Every new mutation "creates" a locus at a randomly determined site, and the rate of recombination between any two loci is determined by the map distance between them. When the mutation at a given locus is either lost or fixed by chance or selection, the site is removed from the genomes, but the information saved as a population variable. The number of new mutations in the population *per* generation is given by a normal distribution with mean and variance of $N \cdot U$ (an approximation to the Poisson distribution), where N is the number of individuals in the population, and U is the deleterious mutation rate for the diploid genome. The mutations are distributed randomly among the 2N chromosomes. The selection coefficients for homozygotes of the new mutations are distributed between 0 and 1 by a left-skewed distribution (i.e., the uniform distribution to the fourth power), and given the relative fitness of $1 - hs$ for heterozygotes and $1 - s$ for homozygotes of a mutation, the degree of dominance, h, is negatively linearly dependent on the selection coefficient (s), i.e., $h = h_0 \cdot (1 - s)$, where h_0 is the dominance for $s = 0$ (in these simulations set to 0.3). The fitness is multiplicative among loci.

A generation cycle is started with N seedlings (among which mutations occur) that survive to reproductive age. A zygote for the next generation is found by combining two gametes randomly drawn from the plants in the previous generation, after recombination and segregation has taken place. The fitness of the zygote is calculated and it survives according to the selection mode used, which is either proportional (for each zygote a random number is drawn, and its fitness has to be above that value to survive), or truncational (survival of the zygote if its fitness is above a certain fraction of the maximum individual fitness in the previous generation). This procedure takes place until N zygotes have survived.

In each run of the model, the population is allowed to reach a quasi-equilibrium of mutation and selection at a size of 200 individuals (after approx. 1200 generations), and thereafter the number of individuals is reduced to 20 and the population followed for 300 further generations. For each generation the mean fitness after mutation (relative to the fitness of a population without any deleterious mutations) and the intensity of selection (the proportion of rejected zygotes to the total number tried) are determined. At specified generation numbers before and after the drop in size, the inbreeding depression is estimated by letting a sample of all the individuals go through 10 consecutive generations of selfing and monitoring the fitness parameters for each line.

For this study, the following sets of parameters were analyzed: Proportional selection at genomic mutation rates of 0.5, 1.0, and 2.0, and truncation selection at 50% of the maximal fitness with a genomic mutation rate of 1.0.

The distribution of pseudo-random numbers used for the simulations was checked for several intervals, all of which behaved similarly. The program was checked by using neutral alleles, and the results were in agreement with neutral expectations.

Results and Discussion

For all mutation rates and selection modes studied, the mean population fitness decreased in the generations after the size reduction (Fig. 1). In most cases, the fitness decreased gradually over the 300 generations, but did not do so soon after the size reduction as could be expected if the effect was due to inbreeding depression caused by the increased homozygosity of the recessive deleterious alleles. Instead, the fitness decrease can be ascribed to fixation of deleterious alleles in the small population, as shown in Fig. 2, which gives the mean population fitness calculated from polymorphic loci only, thereby eliminating the effect of fixation. In that case, the fitness did not decrease after the drop in population size. In another series of similar runs, but using a uniform distribution of selection coefficients assigned to new mutations, we found no decrease in population fitness after the size reduction, even when including fixed loci in the fitness calculation. The skewed distribution of selection coefficients used for the present runs thus resulted in a higher number of fixed deleterious alleles, causing the fitness decrease shown in Fig. 1. A few highly fit populations showed a serious decrease in fitness right after the size reduction (see, e.g., Fig. 1c), which may be an effect of the increase in inbreeding. The level of mean population fitness was dependent on the mutation rate and the selection mode, but as the values are relative to a hypothetical population without deleterious alleles, only the relative response of the fitness during the size transition is of interest for this study.

The results from the "inbreeding experiments" performed in the population before the drop in size and at several times after this, are shown in Fig. 3. In case of proportional selection, there was a pronounced inbreeding depression in the base population for all three mutation rates (bold line), and the inbreeding depression was less severe for the later generations. A different response was found for the truncation selection, where fitness is only slightly affected by the inbreeding.

The strong inbreeding depression found in the large populations (Fig. 3a,b,c) suggests that the populations would experience a pronounced drop in fitness when subjected to a sudden reduction in population size. As stated before, the mean population fitness does not generally decrease in response to the increased inbreeding level due to size, but only

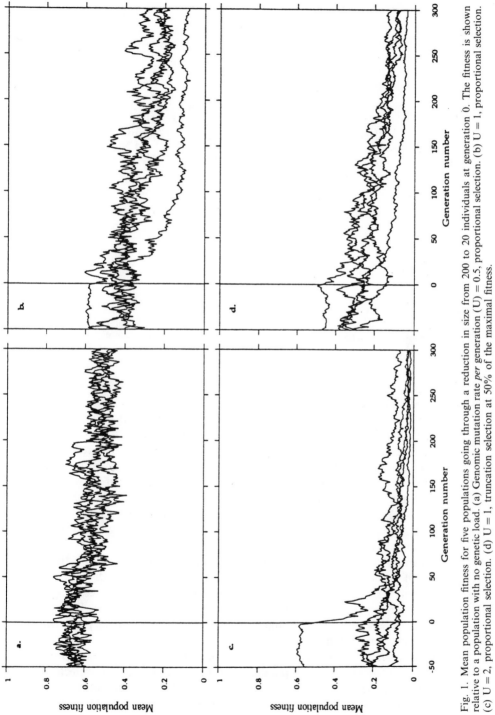

Fig. 1. Mean population fitness for five populations going through a reduction in size from 200 to 20 individuals at generation 0. The fitness is shown relative to a population with no genetic load. (a) Genomic mutation rate *per* generation (U) = 0.5, proportional selection. (b) U = 1, proportional selection. (c) U = 2, proportional selection. (d) U = 1, truncation selection at 50% of the maximal fitness.

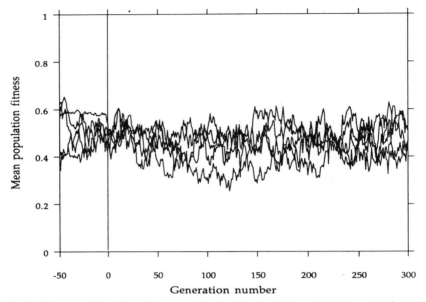

Fig. 2. Mean population fitness for segregating loci (i.e., loci that are fixed for a mutation are not included in the calculation) for five populations going through a reduction in size from 200 to 20 individuals at generation 0. The fitness is shown relative to a population with no genetic load. Genomic mutation rate *per* generation = 1, proportional selection.

as a consequence of the fixation of deleterious alleles. Selection against less fit inbred individuals is thus efficient in purging the harmful alleles at the polymorphic loci from the populations, whereas slightly deleterious alleles are constantly being fixed. In order for the selection to work, there must be genetic variation among the individuals for the response to inbreeding. This is indicated in Fig. 4, which shows the inbreeding response of one of the populations from Fig. 3b (generation 0), but for each selfing line separately. A large variance in response to the inbreeding is found among the 20 lines, with some lines dropping to near zero in the first generations of selfing, and other lines maintaining a relatively high fitness even under severe inbreeding.

It is important to point out that our conclusions do not imply that there will be no harmful genetic effects of a population size drop. They only imply that the inbreeding depression caused by changes in the genotype proportions due to the increased inbreeding may have little or no effect, whereas fixation of harmful alleles may seriously depress the fitness over generations.

The magnitude of the fitness decrease due to fixation of deleterious alleles was found to be dependent on the distribution of selection coefficients assigned to the mutations, with a more severe fitness decrease if the distribution was skewed towards the mildly deleterious

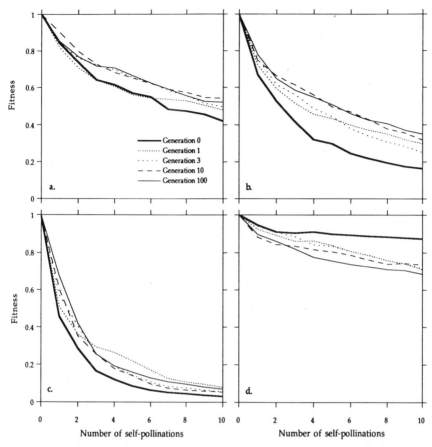

Fig. 3. Inbreeding response before (generation 0) and after (generation 1, 3, 10, and 100) a drop in population size from 200 to 20 individuals, averaged over three populations. At the given generation number, individuals in the populations are forced to self-pollinate for 10 sequential generations, and the mean fitness is shown relative to the fitness of non-inbreds. (a) Genomic mutation rate *per* generation (U) = 0.5, proportional selection. (b) U = 1, proportional selection. (c) U = 2, proportional selection. (d) U = 1, truncation selection at 50% of the maximal fitness.

alleles. This is because these slightly deleterious alleles are almost neutral in a small population, and are thus more likely to become fixed than the more harmful alleles.

The inbreeding depression decreased in magnitude in the generations after the size reduction, as expected from the purging hypothesis. Ouborg (1993), van Treuren et al. (1993b), and Hauser and Loeschcke (submitted) compared the inbreeding depression among large and small natural populations, but no clear trend was found in any of the studies. This may be because the populations have been small for too

122

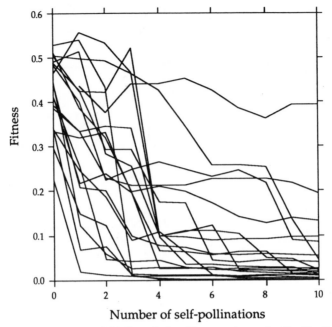

Fig. 4. Inbreeding response of 20 lines during 10 generations of self-pollination ($U = 1$, proportional selection). Same data as in Fig. 3b (generation 0), but shown for each line separately.

short a time to change the genetic system, but could also be because the differences in inbreeding depression are too small and variable to detect in an experiment. Analysis of the data from our simulations show that only for a high genomic mutation rate ($U = 2$) could a difference in inbreeding depression (one-time selfing versus non-inbred) be detected between the large and small population (generation 0 and 100, tested as a significant interaction in a two-way Anova between inbreeding level and population size). Data were used from all 20 individuals in the small population, a number that is above the number of plants commonly analyzed in such experiments. Under most experimental conditions, the environment will additionally increase the variances, making it even less likely to detect an eventual difference.

When truncation selection was used in the simulations, the inbreeding depression was not as severe as with proportional selection. Truncation selection is more effective in removing deleterious alleles from the population, as only zygotes with a relatively high fitness survive. Under proportional selection, more zygotes carrying deleterious alleles will be recruited to the new generation, and these alleles will then be expressed during inbreeding and cause a depression of fitness. These results suggest that different responses to inbreeding among natural populations may result from differences in the selection mode imposed by the environment.

Our simulation study obviously reduces the genetical and ecological parameters studied to a manageable suite. Other factors are known to influence the fitness changes in response to small population size. Demographic stochasticity may threaten the population if reduced to the size used in this study, and if the population size decreases in response to the fixation of deleterious alleles, interactions among these two factors may make things worse (Gabriel and Bürger, this volume). In the model used here, population size was kept constant in order to concentrate on the study of inbreeding. An indication that our results may not differ much if the population size was allowed to fluctuate comes from the selection intensities calculated during the simulations. The selection intensity never exceeded an average of eight zygotes to the one surviving (Fig. 5), which is a low number compared to the large number of seeds produced by most flowering plants during their lifetime. Enough offspring would thus be produced each generation to replenish the sites in the population. However, this might be different for species with a different demography. If only few offspring are produced during a lifetime, the replenishment may be more vulnerable to a lack of genetically fit offspring.

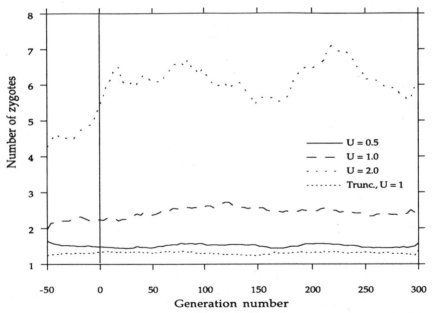

Fig. 5. Mean selection intensity at different mutation rates and selection modes for five populations dropping in size from 200 to 20 individuals at generation 0. Shown as the average number of zygotes tried to obtain one that survives at a given position in the population. The curves are smoothed.

The environment may obviously be very different for a small population, which could in itself threaten the survival of a given population. For plants, the pollinator or seed disperser fauna may be very sensitive to changes in the environment, and a diminished population size or a larger degree of isolation may thus alter the mating system of a plant population through effects from mutualists (Jennersten, 1988; see also scenario D, Olesen and Jain, this volume).

Implications

Inbreeding depression and experiments

The main conclusion that we draw from our simulation study is that selection against deleterious alleles in a population of strongly reduced size may be efficient in preventing a decrease in fitness due to the increased inbreeding, even in cases when a strong inbreeding depression can be found in the large original population.

This has important implications for conservation biology with respect to the focus of studies and experiments, and for management procedures. Whereas studies aimed at inferring past evolution, e.g., inbreeding in small populations, may compare the average inbreeding response in populations of different sizes, a prediction on the future effects of inbreeding must additionally consider the genetic variance for inbreeding response in the population, as recognized by the genetic literature. For many conservation projects, such a prediction of possible effects of inbreeding under managed conditions is essential.

In the following, we discuss some commonly used designs for studying inbreeding depression, and suggest ways in which to improve these in order to estimate the genetic variance for inbreeding depression.

Most studies of plant inbreeding depression have compared fitness components between selfed and outcrossed progeny, either within the same plant or with self and outcross pollinations performed on different plants.

Unless controlled crossings are made in advance, the exact inbreeding level of the parental plants will be unknown. If the maternal plants have different inbreeding levels, the selfed offspring group will also span a range of inbreeding coefficients, thus introducing a variance in the estimate of inbreeding depression that cannot be resolved. (Selfed offspring from an outbred maternal plant have an inbreeding coefficient of 0.5, whereas, for example, selfed offspring from a selfed plant will have an inbreeding coefficient of 0.75).

Different inbreeding levels among maternal plants may introduce another source of unwanted variance in the estimate of inbreeding depression due to maternal effects. These are most pronounced in the early life stages of plants (e.g., Biere, 1991a,b), but selfed plants have been shown to give rise to overall less fit offspring than outcrossed maternal plants (Wolfe, 1993; Hauser and Loeschcke, submitted). Parts of the variance in fitness among offspring may thus be caused by variation in the inbreeding level among their parents.

We therefore recommend that studies on inbreeding response and variance be made with parental families of known inbreeding status, and that offspring of several inbreeding levels are compared, as this will add valuable information on the complicated relationship between inbreeding level and fitness. As was shown by Damgaard et al. (1992), an inbreeding depression estimated from only self *versus* outcross may cover a spectrum of different evolutionary potentials with respect to selfing rate, dependent on the function of fitness to progressive inbreeding.

A commonly used experiment in plant breeding practices is the monitoring of fitness components for several generations of self-pollina-tion, or any milder form of inbreeding (see, e.g., Lynch, 1988). The speed at which the inbreeding accumulates may influence the selection against harmful alleles, and if the interest is in the consequences of small population size, a slow rate of inbreeding may be preferred, as it will better simulate the processes during a size transition. The crossing design could be an extension of a diallel cross with several inbreeding groups, or include reciprocal crossings to a defined population. These extensions make it possible to estimate the maternal effects of inbreeding.

The inbred families may finally be crossed among each other after the wanted number of generations, and the fitness and the inbreeding depression be estimated in this hybrid population. If compared to the fitness and inbreeding depression in the original population, a sugges-tion of the amount of selection that has occurred during the experiment can be gained (El-Nahrawy and Bingham, 1989; Barret and Charlesworth, 1991).

In the analysis of the experiments, the variance among families for response to inbreeding should be included. The heritability of inbreed-ing response can be estimated from the correlation in fitness among parents and descendant, and if several generations of selfing are per-formed, the proportion of the variance among inbred lines to the total phenotypic variance (the repeatability) can be used as a maximum estimate for the heritability of the fitness component in that population (Falconer, 1989). We are well aware that there are serious problems connected with the variance estimation of quantitative genetic experi-ments, and its application to the non-experimental conditions. Even the most "natural" experiments will inherently interfere with the environ-ment, and results from one year may not apply under different condi-tions another year. Still, we believe that much is to be learned on the dynamics of inbreeding and inbreeding depression from more detailed experiments, and that this knowledge is central for insights on the fate of threatened populations.

Outbreeding depression

In our study, we found that the fitness consequences of the increased inbreeding after a drop in population size were not as harmful as might be expected from the mean inbreeding depression estimated in the large population.

Analogous considerations may apply for the fitness effects of crossings among populations. The mean fitness of the hybrids may be higher than the mean of the two parental populations if the hybrids are heterozygous for loci that are fixed for deleterious alleles in the parental populations. Alternatively, the hybrids may have a reduced relative fitness, if population-specific adaptations to the environment or among genes in the genomes are lost in the hybrids ("outbreeding depression"). A large variance in fitness can be expected among individuals of the crosses, and even in case of a relatively low average fitness of the hybrids, selection may result in a higher fitness of the descendants (Templeton, 1986).

The discussion of outbreeding depression has been somewhat unclear with respect to the kind of hybridizations discussed. The fitness consequence of a hybridization event is different for a population composed of descendants of the crossing only, and for the descendants of the hybrids if these are back-crossed to one of the parental populations. The first hybrid generation (F_1) among inbred lines or populations is expected to have a relatively high fitness, as all individuals will be heterozygous for parentally fixed alleles, and as parental adaptations are still partly intact in the two parental genomic contributions. In subsequent hybrid generations (F_2 and F_3), the fitness is expected to decrease, as the individuals are no longer all heterozygous, and as the adaptations are partly lost by recombination and segregation (Falconer, 1989; Lynch, 1991). If the crosses subsequent to the hybridization are back-crosses to the parental population instead, parts of the local adaptations may be restored, with the result that a smaller fitness decrease can be expected than for the pure hybrid population. If the descendants from the back-cross survive, they may introduce genetic variation to loci fixed for deleterious alleles.

Conservation management

In a situation where a population is supposed to suffer from inbreeding depression or negative effects from fixation of deleterious alleles, or where a drop in population size is supposed to create such problems, two management strategies may be envisioned.

One strategy would be to increase the effective number of individuals in the population in order to slow down the increase in general inbreeding level and the rate of fixation of deleterious alleles. An increase in effective size of plant populations can be achieved by, for example, managing populations to ensure a high outcrossing rate (Van Treuren et al., 1993a). Seeds can be harvested evenly from a large sample of individuals and dispersed in the population in order to minimize possible effects of isolation by distance and uneven fecundity of individuals. If some life stages are more sensitive to inbreeding than others, it may be possible to relieve the inbreeding effects by reducing the impact of these, e.g., by mowing the surrounding vegetation if inter-specific competition is severe.

Another strategy would be to create controlled gene flow between populations that could introduce variation to loci fixed for deleterious alleles. As already mentioned, such import might simultaneously cause outbreeding depression in the hybrids, and in the following, we therefore suggest a procedure for controlled gene flow among plant populations.

Instead of transplanting adult plants from one population into another, we suggest that pollen be transferred, if possible. The style may act as a sieve against either less fit pollen tubes, or against pollen tubes that interact less well with the maternal plant (inbreeding and outbreeding depression) (Stephenson et al., 1992). Further selection on gene combinations may occur during the seed stage, and the harvested seeds may thus have eliminated some of the less fit genetic combinations of the hybridization. Pollen transfer will additionally minimize the risk of introducing diseases and pests, as pollen carries a much lower number of diseases than whole plants.

Pollinations should be performed from several plants in the donating population to several in the recipient population, and the flowers must be kept bagged to prevent unintentional secondary transfer of the migrant pollen. After harvesting, the F_1 seeds are sown out in controlled seed lots in the recipient population. The resulting F_1 plants are monitored for survival, vigor, and flowering phenology. At flowering they are bagged to prevent natural pollination, and pollen is transferred from the hybrids to flowers on local plants, to create back-crosses. The direction of the cross is important, as the pollen and its interactions should be tested under the conditions of the local plants. The resulting seeds may thereafter be dispersed in the population for natural growth, or an additional round of controlled pollination be performed.

The proposed design can be viewed as a "safe strategy" with respect to the differing theories on the cause of inbreeding and outbreeding depression. Possible negative effects of outbreeding depression are minimized due to the back-cross design, and the hybrid gene combinations are exposed to several stages of selection under local conditions before letting them into the local gene pool. As the hybrid descendants are known and marked, the introduction may work as a natural experiment in which several fitness related parameters can be followed. It is important to stress, though, that this procedure does not replace proper assessments of ecological and genetical differences between the involved populations that might cause maladaptedness in the hybrids. This information would instead be useful as a guide to which traits and life stages to monitor in the hybrid generations.

Our reason for proposing this procedure is also to show that our concepts of inbreeding and outbreeding depression may influence the suggestions we make for management procedures. If a strong outbreeding depression was found on average for hybrids between two populations, this might lead to the conclusion that gene flow management would not be a beneficial strategy. For the same set of data, but considering a large genetic variation in fitness found among the hybrids, gene flow may still be considered a method of choice, especially if a back-cross design is utilized.

Conclusion

A simulation study was performed in order to examine the fitness effects of inbreeding due to a drop in population size. For three different mutation rates and two modes of selection, we found that the mean population fitness only rarely decreased in response to the increase in inbreeding after a drop in population size, but that the fitness decreased gradually over time as genetic drift fixed loci for deleterious alleles. Purging of deleterious alleles may thus be efficient in eliminating negative fitness effects of inbreeding, and a mean inbreeding depression estimated in a large population has therefore little predictive power with respect to the possible effects of a drop in size. As selection operates on the variance among individuals, estimates of the genetic variance in response to inbreeding are essential for predictions on the fitness effects of a population size reduction. We therefore discuss experimental procedures commonly used for inbreeding studies and propose ways in which to improve these.

Outbreeding depression can under certain circumstances be expected for hybrids among populations, but, as argued for the inbreeding depression, the mean fitness response may tell little about the consequences of a particular hybridization event. The genetic variance in response to outbreeding is therefore of interest for conservation studies.

A procedure for managing controlled gene flow among plant populations is proposed which utilizes pollen transfer instead of transfer of

adult plants. It relies on a back-cross design of hybridization to minimize possible outbreeding depression, and on a large variation
of genetic combinations to be screened under local environmental
conditions.

Acknowledgements
We want to thank Denis Couvet and Chris Cliddon for valuable suggestions in the development of the computer program, and Bernt Guldbrandtsen for discussions and technical
assistance. Jürgen Tomiuk and Subodh Jain commented on earlier versions of the manuscript.
We are also grateful to the Danish Natural Science Research Council for supporting this
project (grant no. 11-9025-1 and 11-9634-1).

References

Allard, R. W. (1960) *Principles of plant breeding*. Wiley & Sons, New York.

Barrett, S. C. H. and Charlesworth, D. (1991) Effects of a change in the level of inbreeding
on the genetic load. *Nature* 352: 522–524.

Biere, A. (1991a) Parental effects in *Lychnis flos-cuculi*. I. Seed size, germination and seedling
performance in a controlled environment. *J. Evol. Biol.* 4: 447–465.

Biere, A. (1991b) Parental effects in *Lychnis flos-cuculi*. II: Selection on time of emergence and
seedling performance in the field. *J. Evol. Biol.* 4: 467–486.

Charlesworth, D. and Charlesworth, B. (1987) Inbreeding depression and its evolutionary
consequences. *Annu. Rev. Ecol. Syst.* 18: 237–268.

Damgaard, D., Couvet, D. and Loeschcke, V. (1992) Partial selfing as an optimal mating
strategy. *Heredity* 69: 289–295.

El-Nahrawy, M. A. and Bingham, E. T. (1989) Performance of S_1 Alfalfa lines from original
and improved populations. *Crop. Sci.* 29: 920–923.

Falconer, D. S. (1989) *Introduction to quantitative genetics*. 3rd ed. Longman, New York.

Holsinger, K. E. (1991a) Inbreeding depression and the evolution of plant mating systems.
Trends Ecol. Evol. 6: 307–308.

Holsinger, K. E. (1991b) Mass-action models of plant mating systems: the evolutionary
stability of mixed mating systems. *Am. Nat.* 138: 606–622.

Jennersten, O. (1988) Pollination in *Dianthus deltoides* (*Caryophyllaceae*): Effects of habitat
fragmentation on visitation and seed set. *Cons. Biol.* 2: 359–366.

Lande, R. and Schemske, D. W. (1985) The evolution of self-fertilization and inbreeding
depression in plants. I. Genetic models. *Evolution* 39: 24–40.

Lynch, M. (1988) Design and analysis of experiments on random genetic drift and inbreeding
depression. *Genetics* 120: 791–807.

Lynch, M. (1991) The genetic interpretation of inbreeding depression and outbreeding
depression. *Evolution* 45: 622–629.

Ouborg, N. J. (1993) *On the relative contribution of genetic erosion to the chance of population
extinction*. Ph.D. Thesis, Univ. of Utrecht.

Stephenson, A. G., Lau, T.-C., Quesada, M. and Winsor, J. A. (1992) Factors that affect
pollen performance. *In:* Wyatt, R. (ed.), *Ecology and evolution of plant reproduction.*
Chapman & Hall, New York, pp. 119–136.

Templeton, A. R. (1986) Coadaptation and outbreeding depression. *In:* Soulé, M. E. (ed.),
Conservation biology, the science of scarcity and diversity. Sinauer, Sunderland, MA, pp.
105–117.

Van Treuren, R., Bijlsma, R., Ouborg, N. J. and Van Delden, W. (1993a) The effects of
population size and plant density on outcrossing rates in locally endangered *Salvia
pratensis*. *Evolution* 47 (4).

Van Treuren, R., Bijlsma, R., Ouborg, N. J. and Van Delden, W. (1993b) Inbreeding
depression and heterosis effects due to selfing and outcrossing in *Scabiosa columbaria*.
Evolution 47 (5).

Wolfe, L. M. (1993) Inbreeding depression in *Hydrophyllum appendiculatum*: role of maternal
effects, crowding, and parental mating history. *Evolution* 47: 374–386.

Conservation Genetics
ed. by V. Loeschcke, J. Tomiuk & S. K. Jain

The interaction of inbreeding depression and environmental stochasticity in the risk of extinction of small populations

A. J. van Noordwijk

Netherlands Institute of Ecology, P.O. Box 40, NL-6666 ZG Heteren, The Netherlands

Summary. Current population genetic and population dynamic models are inappropriate to judge the risk of extinction of small populations due to the combined effects of inbreeding, genetic drift, demographic stochasticity, and environmental stochasticity. Instead, a model based on the aggregated fates of individuals is advocated. The unequal distribution of resources over individuals is an essential part of this model. The model allows the incorporation of the mutation-selection dynamics of alleles leading to inbreeding effects and to fixation of slightly deleterious mutations as a result of genetic drift. The slightly deleterious mutations lower the conversion of resources into offspring. Whereas lethal alleles are rapidly eliminated by selection in small populations, the selection against mild deleterious effects depends strongly on effective population size and on the social system, that is, on the division of resources among individuals. The model allows for the study of rates at which processes occur while far away from equilibrium, which is crucial in understanding the extinction risks of threatened populations. One example of the latter is illustrated in simulations in which small populations become extinct between approximately 100 and 200 generations after they became small populations, due to a gradual accumulation of mildly deleterious mutations.

Introduction

Over 1000 of the world's 9000 species of birds are considered to be threatened with extinction (Collar and Andrew, 1988). Many of these species have world populations well below 1000 individuals (Green and Hirons, 1991). Extrapolating from the species for which census data are available, about 200 species have world populations of less than 100 individuals (Green and Hirons, 1991), while actual breeding populations may even be considerably smaller. In this range of population sizes, one expects the effects of inbreeding and loss of genetic variation through random drift to cause real problems. Further, stochastic processes may greatly enhance the likelihood of extinction. First, because of demographic stochasticity (Gilpin and Soulé, 1986) which is the stochastic effect of reproduction with low numbers that may result in a zero population size or individuals of a single sex only (Lynch and Gabriel, 1990). Second, because fluctuating environments causing fluctuating population sizes will reduce the effective population size and thereby increase the effects of inbreeding and drift. Yet, even though inbreeding depression can be demonstrated in virtually all species (e.g., Ralls et al., 1986), I am not aware of data suggesting that species on the border of extinction show very low reproductive success and low viability in juvenile offspring that are typical for inbreeding depression (e.g., Ballou and Ralls, 1982; van Noordwijk and Scharloo, 1981). Unfortunately, the interaction between stochastic elements in population dynamics and genetic effects of small population size have so far not been investigated in a single framework. Moreover, previous work on the population dynamic aspects of risk of extinction have been based on classical population dynamic theory (Boyce, 1992; Mangel and Tier, 1993; Tomiuk and Loeschcke, 1994; Witting, 1993), which gives a reasonable approximation for large populations, but which is probably inappropriate for small populations.

The model presented here includes the dynamic interaction between new alleles arising through mutation and their elimination or fixation through their consequences for fitness or through chance events. The alleles are lethal recessives, or alleles with a small additive negative effect on reproduction or wild-type alleles. I will use the term inbreeding for the cases in which homozygosity is important, and the term genetic drift for the gradual replacement of wild-type alleles by additive mutants. The interaction between the deleterious effects of mutant accumulation and the increased elimination of harmful recessives from the population caused by the stronger selection through the increased homozygosity will appear as emergent properties in the model.

Population models with both complex population dynamics and realistic genetics have many parameters. It is therefore quite understandable that the customary simplifying assumptions made in either population dynamics or genetics bar the inclusion of the other parameters (Loeschcke and Tomiuk, 1991). Whereas inbreeding affects the variance of the distribution of alleles over individuals, but not their mean frequency, most population dynamic models ignore the variance among individuals. When individual variation in, for example, reproductive performance is taken into account (e.g., Gabriel et al., 1991; Gabriel and Bürger, this volume) it is in the form of the net effect of a probability distribution. However, analogous to developments in economics, where macro-economic properties are derived from the aggregation of simulated households, computing facilities are now sufficiently powerful to derive population dynamics from the sum of the fates of individuals over many generations. The advantage of such an approach is that complex interactions between macro-scale parameters can emerge from simple interacting processes at the individual level.

Estimating the extinction risk in small populations due to a combination of genetic and stochastic processes is an area in which the relation between an individual's genotype and its fitness plays an important role. This relation is very complex, but the energy allocation within individuals and the division of resources among individuals give a satisfactory description at an intermediate level of complexity and can easily be included in the model. Thus, the model in which individuals are simulated provides an opportunity to study genetic effects and non-genetic stochastic effects in the context of a life-history that resembles those of the organisms whose extinction concern us most, rather than in the context of bacteria life histories, implicit in population dynamic formulas. In this model, changes in allele frequencies and their distribution over individuals lead to changes in the reproductive output of individual with its population dynamic consequences.

One of the main ideas to help explain the considerable variation observed in reproductive traits is the trade-off (e.g., Lessells, 1991; Stearns, 1992). A larger clutch should automatically lead to smaller offspring and/or a larger effort by the parents, thus resulting in their lower survival. Through these compensatory effects on its components, the variation in the net reproduction over a lifetime is expected to be less than the variation in each of its components. One important aspect in the consideration of trade-offs is the interaction between the total amount of resources that individuals have at their disposal and the allocation to the various components. If the variation among individuals in the acquisition of resources is large, one may find that the amounts allocated to competing traits is positively correlated and no trade-off is visible (van Noordwijk and de Jong, 1986). The interaction between the acquisition and allocation of resources provides an alternative model to study the micro-evolution of genetically correlated traits (de Jong and van Noordwijk, 1992; de Jong, 1993). Although the explicit consideration of variation in the acquisition of resources provides insight in the evolution of allocation strategies, we do not yet have a theoretical framework in which the division of resources over individuals is linked to these allocation properties. Apart from its use in the study of extinction, the model presented here also provides a framework for linking the evolution of life history strategies to their physiological basis and their population dynamic consequences. Here, I will concentrate on a simple version in which the mutation selection balance of lethal recessives and of slightly deleterious mutations is studied in connection with their population dynamic consequences.

The distribution of resources over individuals

Lomnicki (1988) has provided an alternative formulation of population dynamics, based on the distribution of resources over individuals. One of his key motives for developing this formulation is that population extinctions are much rarer events than the classical population dynamic equations suggest. He distinguishes three basic social systems: an egalitarian distribution (NT = Non Territorial) of resources over individuals, a strictly territorial system ST, with a fixed territory size and an intermediate system of compressible territories CT. In fact, the compressible territorial system can be subdivided into a fully compressible territorial system FCT and a partly compressible territorial system PCT; the difference between these two types lies in the distribution at high densities. In the PCT system, there is a triangular resource distribution with one pair still getting the maximum amount and all pairs beyond a certain number getting no resources at all. In the FCT system, there is a triangular resource distribution with all pairs getting some resources and the maximum equal to twice the average amount. It is thus possible in the FCT system that even the pair with the maximum amount of resources has insufficient resources for reproduction. The key element in the ST and PCT systems is that at high population densities a number of individuals get no resources at all, which leaves sufficient resources for a number of individuals to reproduce normally. This leads to a stable population dynamics with a potentially high reproductive rate that would lead to chaos in the traditional population dynamic models, based on an equal distribution of resources over individuals (May, 1974). Lomnicki's formulation uses a simple allocation of resources within individuals; he defines a minimum requirement for maintenance (= survival) and a surplus that is allocated to reproduction, bounded by a maximum capacity to process resources (see Fig. 1). The actual reproduction then depends on the amount of resources available for reproduction and the efficiency with which this is converted into offspring.

Individuals (or breeding pairs) can be ranked according to the total amount of resources available to them. The amount needed for maintenance may vary from pair to pair, depending on their individual properties. In the example, in Fig. 1a, the first 11 pairs each get the maximum amount of resources, while pairs 12–18 get nothing; in Fig. 1b, the first 11 pairs have sufficient resources for survival and have some left over for reproduction. Pairs 12–18 have some resources for survival, but too little to survive until reproduction. In the simulations reported here, the division of resources among individuals is independent of their genotype, but the model structure would allow the division of resources among individuals to be affected by properties of the individuals. It is implicit in the model that there is no fixed proportion allocated to both survival and reproduction, but instead a fixed amount of resources necessary for survival is first subtracted from the available total. This makes it easy to incorporate effects due to alleles with a small effect realistically, namely, by a reduction in the reproductive efficiency, more resources being needed *per* offspring produced. This automatically leads to a small selection pressure against such alleles, depending on the composition of the population. The net effect of the reduction in

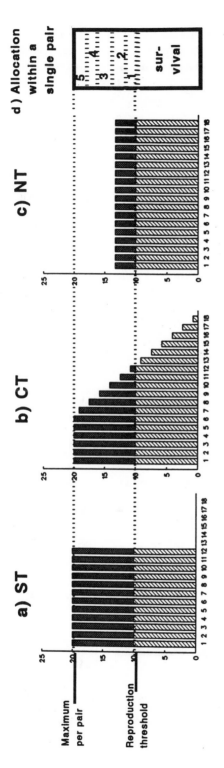

Pairs according to social rank

Fig. 1. Three different ways of dividing resources over individuals (after Lomnicki (1988)) at intermediate population densities. (a) In the strictly territorial system ST, a number of pairs gets the maximum amount while the other pairs get nothing. (b) In the compressible territorial system CT, some pairs still get the maximum amount, but many pairs get different amounts of resources less than the maximum. (c) In the non-territorial system NT, all pairs get the same amount of resources, which may allow none of the pairs to reproduce. (d) For each pair, the first x units of resources available are necessary for survival. The remaining resources are converted into offspring. The number of offspring produced depends on resources availability and conversion efficiency.

offspring number of a pair depends on the reproduction of other pairs. In this manner, a soft selection component that acts against mutations with a small negative effect on fitness comes about through the relative reproductive performance of the pairs in the populations.

The model used

The model is based on the flow chart in Fig. 2. The diploid organism forms breeding pairs and has non-overlapping generations. With each generation a given amount of resources is available and is divided over the pairs according to one of the four social systems: NT, FCT, PCT, and ST. The use of the term social system and the reference to territoriality is no more than a label. The essence lies in the shape of the unequal resource distribution. Each pair reproduces according to its available resources and the efficiency of transforming resources into offspring. This efficiency is reduced by the number of mutant alleles in both members of the pair.

Offspring receive one copy from each parent at 20 loci. In the simulations presented, the loci were unlinked. At each locus, 100 alleles are possible: 10 of these are "wild type," 50 of these are mutants that

Flow chart of simulations

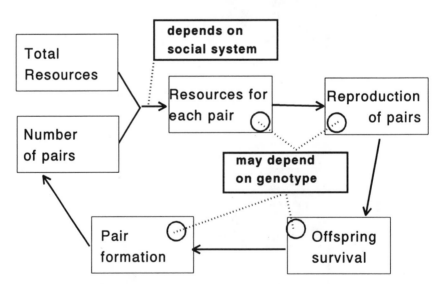

Fig. 2. The flow chart of the model used. In the simulations reported here, reproductive efficiency of pairs was dependent on the genotypes of both male and female, and offspring survival depended on the absence of homozygous lethal recessives. No effects of genotype on the resource distribution among pairs or on the pair formation were incorporated.

only have a small negative additive effect on reproductive efficiency, and 40 alleles have the same negative effect if heterozygous, but these are lethal in homozygous condition. Mutation is independent of the previous alleles. Thus, the back-mutation rate is about 10% of the mutation rate from wild type to mutant. At the start of the simulations, populations were generated with random mixtures of all 10 wild-type alleles. In choosing the number of loci to be modeled, it is important to note that the Mendelian transfer of alleles consumes by far the most computing time. This leads to a choice of a low number of loci. The risk of obtaining the same allele through independent mutations can be lowered by including many potential alleles per locus. The mutation rate used in the results presented here is 0.01 *per* locus *per* gamete *per* generation, giving a total mutation rate of 0.2 mutations *per* haploid genome *per* generation, or 0.4 *per* zygote *per* generation. The effect of the additive mutations was modeled by multiplying the amount of resources available for the production of offspring by $(1.3 - F_{\female}) \times (1.3 - F_{\male})$, where F is the frequency of mutant alleles in females and in males respectively. One additional mutation thus leads to a reduction in reproduction of about 2%. Thus, the resulting load, μs, is about 0.008, which is well within the range of $0.0002 < \mu s < 0.02$ estimated by Lynch and Gabriel (1990) for non-lethal additive mutations in eukaryotes.

A simulation program was written in Pascal. Usually, 50 runs were made with each parameter combination for each of the four social systems. One line of output was produced for every generation in each simulation. The resulting 500 (max generation number used) × 50 (runs) × 4 (social systems) output files were then used to calculate means and standard deviations *per* generation across runs for every social system. The pseudo-random generator was tested for evenness and for the distribution of differences between subsequent draws. In all the runs presented here, the initial population was given sufficient resources to have several hundred pairs for the first 100 generations. Then the amount of resources was linearly reduced from generations 101 to 120. The first 120 generations are included in the figures to illustrate the fact that the frequency of lethal mutants had reached equilibrium values and that the non-lethal mutations were usually approaching an equilibrium.

Results

The exploration of the parameter space with the help of this model is still at an early stage. Nevertheless, a number of interesting points can be illustrated. First of all, the social systems with more unequal distribution of resources lead to more stable populations. In fact, the more

138

strictly territorial social systems allow high per capita reproductive rates without causing instability. Therefore, the reproduction can still be sufficient to maintain population size in a very poor year. The effect of the four social systems on the extinction rate is illustrated in Fig. 3. For one of the parameter combinations, the NT and FCT system lead to

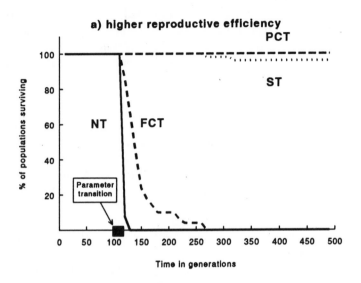

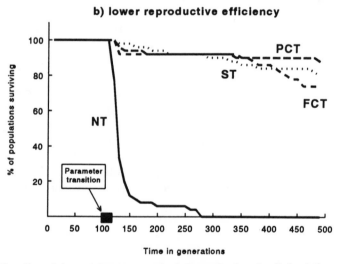

Fig. 3. The effect of the social system on population extinction. In all simulations, the total resources *per* generation allowed for several hundred individuals during the first generations and were reduced linearly from generations 101 to 120 to an amount allowing for populations of 3 to 10 breeding pairs. In (a) reproductive efficiency allowed pairs with maximum resources to produce up to 20 offspring, while in (b) the maximum number of offspring *per* pair was about 10 individuals.

mass extinctions shortly after the reduction in available resources. The lower panel of Fig. 3 illustrates that a reduction in reproductive rate (by 50%) leads to a situation in which extinction in the fully compressible territory system is much rarer and occurs much later. This indicates that the overshooting of the population size is a major cause for extinction. This result is a consequence of non-overlapping generations and the allocation rule of first allocating to survival and only then to reproduction which, compared to a traditional proportional allocation, leads to a zero reproduction instead of a low reproduction at high population densities. The result is that extinction may result at net reproductive rates lower than the chaotic domain in the classical difference equations (May, 1974).

Inbreeding depression and genetic drift

In all runs in which the population size is decreased gradually between generations 100 and 120, the frequency of lethal alleles is quickly reduced to a very low level. If populations are reduced in size instantaneously, extinction due to inbreeding may occur after two to five generations. In these cases a substantial proportion of all offspring produced in the last few generations are classified as lethal homozygotes. It therefore appears that inbreeding depression due to lethal recessives is rarely a cause of extinction.

The frequency of the less harmful mutations often rises very gradually after the reduction in population size. At the time of population size reduction, the frequency of the additive mutations has reached an equilibrium value that is maintained at the larger population size. The rise in frequency of the additive mutations results from wild-type alleles being lost from the population through random drift rather than from inbreeding. This involves mutations arising after the reduction in population size, since the increase in frequency depends strongly on the mutation rate applied. The accumulation of these mutations seems to reach a pseudo-equilibrium when populations become extinct (see Figs 4 and 5). This is caused by the fact that the populations with the higher frequencies become extinct first. Thus, the average frequency given in the figure is the mean of a censored distribution. These extinctions occur after a long delay and, even in populations of less than 10 reproducing pairs, may only occur after more than 100 generations! The importance of this aspect can be demonstrated by varying the mutation rate. The value of this result depends critically on the realism of the implementation of the reduction in reproductive output due to accumulated mutations as well as on the realism of the selection operating against these alleles (see discussion).

The effects of environmental stochasticy

In the less stochastic regime used, the total available resources *per* generation was multiplied by a random factor ranging from 0.8 to 1.2, following a uniform probability distribution. In the more stochastic environment, the random factor for year quality varied from 0.4 to 1.6. In most cases this was irrelevant to the pattern of extinctions, although the extinctions occurred slightly earlier in the more stochastic environment. However, in the egalitarian type and in the entirely compressible territory type, extinction due to overshooting the carrying capacity was more likely in the more stochastic environment. Fig. 4 shows a comparison between two runs that only differ in having the less (Fig. 4a) and the more stochastic environment (Fig. 4b). The most obvious difference is that the variance of the population sizes across runs is greater in the more variable environment leading to longer error bars. Extinctions tend to occur slightly earlier in the more stochastic environment. This effect is greatest around generation 250, when half the populations in the more stochastic environment have become extinct *versus* 20% in the less stochastic environment, but in both cases the last out of 50 becomes extinct just before generation 500.

Interaction among parameters

The most interesting feature of the model that emerged so far is the possibility that populations become extinct many generations after the severe reduction in population size. In the example given in Fig. 5, one of 50 populations becomes extinct in the first 100 generations after the reduction of population size at generation 120, but more than half the replicates become extinct in the next 100 generations and no population survives until generation 500. There are two clues to what goes on. On the one hand, the population size, which includes non-breeders in the strictly territorial system, slowly declines. At the same time, the frequency of the non-lethal mutations increases, even though it had reached an equilibrium level in the larger population before the reduction in population size. Apparently, nonlethal mutations get fixed through random drift, which leads to a lower efficiency of converting resources into offspring. After one to a few hundred generations the productivity then drops below the level necessary to maintain the population under adverse environmental conditions. Of course, the time span after which these extinctions occur depends on the potential productivity of offspring, on the amount of environmental stochasticity, and especially on the mutation rate and the way in which mutations decrease the rate of offspring production. However, a reduced vigor is a frequently observed phenomenon in inbred populations. Whereas the

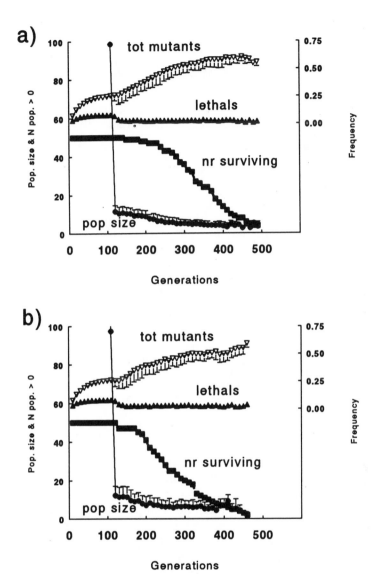

Fig. 4. The effect of environmental stochasticity. (a) In the less stochastic environment the total amount of resources available varies uniformly from 0.8–1.2 × average. (b) In the more stochastic environment resources vary from 0.4–1.6 × average. All other parameters are equal: a PCT social system, a very low amount of resources, a high potential reproductive rate, and an intermediate mutation frequency. The amount of resources was reduced linearly between generations 100 and 120. Population size and the number of surviving populations out of 50 replicates refer to the left axis, while the frequency of lethal mutants and of total mutants refer to the right axis.

142

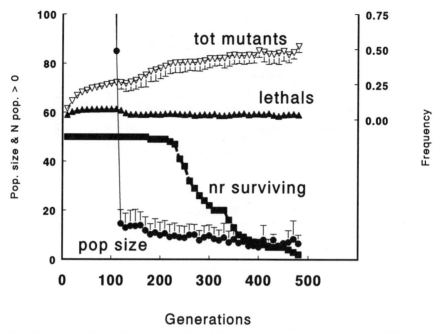

Fig. 5. An example of delayed extinctions. The amount of resources was reduced linearly between generations 100 and 120. Population size and the number of surviving populations out of 50 replicates refer to the left axis, while the frequency of lethal mutants and of total mutants refer to the right axis. This example has an ST social system, a low amount of resources, an intermediate reproductive conversion efficiency, an intermediate mutation frequency, and a less stochastic environment.

selection against mutations with a major effect on vigor is sufficiently strong to eliminate such mutations quickly after they arise, this is not the case for mutations with a small effect. The message is clearly that it is quite possible that small populations apparently survive well for a hundred generations or so, but while they remain small, they may nevertheless accumulate many slightly deleterious mutations that will inevitably lead to extinction. These phenomena occur in types PCT and ST population dynamics. Under the type NT (egalitarian and traditional) population dynamics, the very small populations could never exist. This phenomenon is similar to the mutational meltdown (Gabriel and Bürger, this volume; Gabriel et al. 1991; Lynch and Gabriel, 1990).

Discussion

The fact that the Lomnicki population dynamics allows very small populations to exist with little risk of extinction in a territorial system is not new (Lomnicki, 1988). One of the interesting features of this population dynamics is that it allows for high reproductive rates and, thus, for potentially strong selective mortality before any stage at which density regulation occurs, which allows a high amount of lethal recessives to be continuously eliminated. This raises the question at which stage in the life-cycle inbreeding depression occurs. Most studies on inbreeding depression have been performed in the laboratory or under breeding conditions in a favorable environment with little or no competition. In these studies, most inbreeding depression will occur at a very early stage in the development, with the important exception of traits affecting the fertility (Hauser et al., this volume). This implies that most mortality due to inbreeding depression is likely to occur before stages with strong competition and density-dependent mortality. Thus, lethal recessives are unlikely to cause population extinction, unless either the primary reproductive rate is low or the frequency of these lethals is much higher than expected for the population size, e.g., after a sudden, very sharp reduction in population size. This is not true for the more subtle effects of mutations with an effect on reproductive efficiency or competitive ability. In this context, it will make a substantial difference how the unequal distribution of resources comes about: Is it affected by individual quality, for instance, as a result of competition among individuals for a territory, or is it independent of individual properties, e.g., seeds landing on rich or poor micro-habitat patches? In the latter case, the selection against the mildly deleterious mutants will be far weaker.

Although these aspects have been insufficiently explored, the examples given show that mutations slightly reducing reproductive efficiency may lead to a slowly ticking time bomb. This phenomenon depends only on population size and on the inclusion of mutations that slightly lower reproductive efficiency. Is this likely to be realistic? I believe that it is, since a more or less gradual reduction in reproductive efficiency is very common in selection lines in all sorts of organisms (e.g., Falconer, 1981). The crucial point here is the fact that such selection lines tend to be maintained with few pairs and their effective population size may even be smaller.

The possibility to draw conclusions relevant to the real world depends on the realism of the model structure and parameter ranges used. I have chosen non-overlapping generations and extremely small populations to illustrate the fact that under particular social systems populations of down to three breeding pairs could exist for hundreds of generations, even with non-overlapping generations and with a fourfold fluctuation of resources available to them. In this, the odds have been stacked in favor of rapid extinction and the interesting aspect is the persistence of these small populations. I believe that the range of parameter values in terms of resources needed per pair for maintenance and reproduction are more or less realistic for a passerine bird over a lifetime, although I am not aware of any bird species with non-overlapping generations. The one choice made that might be unrealistic is to model a small number (20) of loci with high *per* locus mutation rates and many alleles *per* locus. This choice was made for practical reasons, since the genetic component uses most of the computing time and is linearly dependent on the number of loci. However, by using a large number of possible alleles *per* locus, the chance of creating the same allele through two independent mutation events has been made low. The major difference with a system with more loci and a lower *per* locus mutation rate is that a saturation with mutations occurs more quickly in this model with few loci. However, the interaction among various loci with a slightly delete- rious effect is probably far more important for the overall realism in this parameter range. The assumption of additivity is, although customary, probably unrealistic. The fact that the delayed extinction emerging from the model is analogous to frequently occurring problems in maintaining small selection lines is therefore reassuring.

The deleterious effects of mutations have been modeled as a reduction in the efficiency of converting resources into offspring. There are many different ways in which this could come about, but at the same time the net effect of many different mutations can be described as such. This leads to an interesting interaction with competition. On the one hand, intra-specific competition between individuals with fewer and with more mutations will automatically lead to selection against the mutations. On the other hand, a population consisting only of individuals that could never compete successfully may nevertheless survive in an empty niche. Koelewijn (1993) gives a good example in which inbred *Plantago coronopus* have the same fitness when competing against each other, but perform very poorly when competing against crossbred individuals from the same lines. This is directly relevant for the importance of intra- specific *versus* inter-specific competition. Inter-specific competition is absent from the model, but it could well lead to a much faster extinction if the distribution of resources among the competing species was affected by the mutations in the small population. Including competing species in a model would make it hopelessly complicated. However, by giving the model an ecophysiologically realistic structure, i.e., modeling mutations as reducing reproductive conversion efficiency, it becomes possible to estimate the effects that inter-specific competition could have.

A further area that should be investigated with the help of this model is the reason why a population has become small and threatened in the first place. There are two main categories of causes, habitat shrinkage and habitat deterioration. In the first, the quality of the habitat has not changed, but there is much less of it. The simulations shown here, in which the total amount of resources is drastically reduced, but the maximum amount that can be processed by a pair and the conversion efficiency are left unchanged, is a representation of this cause for creating a small population. In contrast, habitat deterioration could lead to an unchanged territory in terms of area, but a reduced territory in terms of the maximum amount of a critical resource that a pair could convert into offspring. It is likely that in such a situation more energy would be needed by the parents for foraging and the net conversion efficiency of transforming resources into offspring would be lowered. The various causes for small populations have received almost no attention in the modeling literature (but see Witting, 1993), but is a major concern in the practice of conserving particular species. Both trajectories for threatening a population, or any combination of them, can easily be included in the model here, because the biological interpretation of the parameters in the model is straightforward. Together with the inclusion of a physiologically realistic effect of deleterious mutants, this is a step forward in the direction of considering risks of extinction in their ecological context.

Conclusions

The social system is crucially important for the stability of small populations and their resistance to demographic and environmental stochasticity (Lomnicki, 1988). Small breeding populations with a pool of non-reproducing floaters are at low risk of extinction. The interaction of density-dependent processes and the physiological effects of mutant alleles are poorly known and have considerable consequences for the risk of extinction of small populations. In this model, the increase in frequency of the mutations by which populations eventually become extinct occurs after the reduction in population size and is directly related to population size. This implies that the risk of extinction can always be reduced by increasing the size of the breeding population. In assessing the extinction risk of small populations, the social system and the occurrence of a decline in reproductive output in absolute terms, however slow and gradual, are the most important parameters for predicting the time until extinction. This calls for basic ecological data on individual variance in reproductive rates that are not too difficult to collect.

References

Ballou, J. and Ralls, K. (1982) Inbreeding and juvenile mortality in small populations of ungulates: a detailed analysis. *Biol. Cons.* 24: 239–272.
Boyce, M. S. (1992) Population viability analysis. *Annu. Rev. Ecol. Syst.* 23: 481–506.

Collar, N. J. and Andrew, P. (1988) *Birds to watch: a checklist of the world's threatened birds.* ICBP, Cambridge.

Falconer, D. S. (1981) *Introduction to quantitative genetics,* 2nd edition. Longman, London.

Gabriel, W., Bürger, R. and M. Lynch (1991) Population extinction by mutational load and demographic stochasticity. *In:* Seitz, A. and Loeschcke, V. (eds), *Species conservation: a population-biological approach.* Birkhäuser, Basel, pp. 49–59.

Gilpin, M. E. and Soulé, M. E. (1986) Minimum viable populations: processes of species extinction. *In:* Soulé, M. E. (ed.), *Conservation biology: the science of scarcity and diversity.* Sinauer, Sunderland, MA, pp. 19–34.

Green, R. E. and Hirons, G. J. M. (1991) The relevance of population studies to the conservation of threatened birds. *In:* Perrins, C. M., Lebreton, J.-D. and Hirons, G. J. M. (eds), *Bird population studies, relevance to conservation and management.* Oxford University Press, Oxford, pp. 594–633.

de Jong, G. (1993) Covariances between traits deriving from successive allocations of a resource. *J. Funct. Ecol.* 7: 75–83.

de Jong, G. and van Noordwijk, A. J. (1992) Acquisition and allocation of resources: genetic (co)variances, selection and life histories. *Am. Nat.* 139: 749–770.

Koelewijn, H. P. (1993) *On the genetics and ecology of reproduction in Plantago coronopus.* Thesis, University of Utrecht.

Lessells, C. M. (1991) The evolution of life histories. *In:* Krebs, J. R. and Davies, N. B. (eds), *Behavioural ecology.* Blackwell, Oxford, pp. 32–68.

Loeschcke, V. and Tomiuk, J. (1991) Epilogue. *In:* Seitz, A. and Loeschcke, V. (eds), *Species conservation: a population-biological approach.* Birkhäuser, Basel, pp. 277–281.

Lomnicki, A. (1988) *Population ecology of individuals.* Princeton Univ. Press, Princeton NJ.

Lynch, M. and Gabriel, W. (1990) Mutational load and the survival of small populations. *Evolution* 44: 1725–1735.

Mangel, M. and Tier, C. (1993) Dynamics of metapopulations with demographic stochasticity and environmental catastrophes. *Theor. Pop. Biol.* 44: 1–31.

May, R. M. (1974) *Stability and complexity in model ecosystems.* Princeton Univ. Press, Princeton NJ.

van Noordwijk, A. J. and de Jong, G. (1986) Acquisition and allocation of resources: their influence on variation in life history tactics. *Am. Nat.* 128: 137–142.

van Noordwijk, A. J. and Scharloo, W. (1981) Inbreeding in an island population of the Great Tit. *Evolution* 35: 674–688.

Ralls, K., Harvey, P. H. and Lyles, A. M. (1986) Inbreeding in natural populations of birds and mammals. *In:* Soulé, M. E. (ed.), *Conservation biology: the science of scarcity and diversity.* Sinauer, Sunderland, MA, pp. 35–56.

Stearns S. C. (1992) *The evolution of life histories.* Oxford University Press.

Tomiuk, J. and Loeschcke, V. (1994) On the application of birth–death models in conservation biology. *Cons. Biol.* (in press).

Witting, L. (1993) Environmental factors in population dynamics. Nord. Seminar og Arbejdsrapporter 572: 130–138.

Conservation Genetics
ed. by V. Loeschcke, J. Tomiuk & S. K. Jain
© 1994 Birkhäuser Verlag Basel/Switzerland

Genetic structure of a population with social structure and migration

G. de Jong[1], J. R. de Ruiter[2] and R. Haring[1,2]

[1]*Population Genetics Group, University of Utrecht, Padualaan 8, NL-3584 CH Utrecht, The Netherlands*
[2]*Ethology and Socio-ecology Group, University of Utrecht, Padualaan 14, NL-3584 CH Utrecht, The Netherlands*

Summary. Long-tailed macaques (*Macaca fascicularis*) live in social groups consisting of resident adult females and their offspring, and immigrant males. Subadult males leave their birth group, and might establish themselves as reproducing males in another group. Females do not leave their birth group. Such a social pattern might have consequences for the genetic differentiation between groups and the genetic relationships within groups.

In a field study of long-tailed macaques (*Macaca fascicularis*) in Ketambe, Sumatra, Indonesia, blood samples were taken from individuals in seven adjacent social groups. Electrophoretic analysis showed 17 blood proteins and enzymes to be polymorphic, allowing the computation of heterozygosities and of the F-statistics. Of the F-statistics, F_{IS} indicates the deviation from Hardy-Weinberg equilibrium averaged over local populations, F_{ST} indicates the differentiation in allele frequency between local populations, and F_{IT} indicates the deviation from Hardy-Weinberg equilibrium over the total population.

In a computer simulation of the population of long-tailed macaques using many loci with many neutral alleles, F_{IS} and F_{ST} values proved to be characteristic for a certain demography and life history of the population, and proved not to depend upon the number of alleles or level of heterozygosity. F_{ST} values found in the simulation were compatible to those found in the field; in the simulation, values for F_{IS} and F_{IT} were consistently negative.

The explanation for the negative F_{IS} appears to be that genetic drift causes differentiation in allele frequencies between groups, and that due to this differentiation, allele frequencies differ between resident females and immigrant males, leading to offspring with an excess of heterozygotes (negative F_{IS}) relative to the expectation based upon the overall allele frequency.

The excess of heterozygotes might imply that slightly deleterious alleles are protected from selection. A population with a social structure and differential migration of the sexes is liable to accumulate deleterious recessives and, as a consequence, to be very sensitive to inbreeding on disruption of the social structure, as for instance in zoos.

Introduction

Many mammalian species live in socially structured populations. Several types of social structure exist, ranging from almost free roaming individuals to the tightly knit colonies of the naked mole rat. It seems possible to think of social structure on a graded scale, from independent individuals that interact more or less equally with all other individuals in the population, to loose associations with different probabilities of interactions between individuals, to groups of individuals forming a metapopulation, to local populations with occasional migration. Together with the grouping comes reproduction: is mating for a female equally likely to be with any male in the population, or more likely to be with a male living in her own group? Mating pattern and group relations will often interact to form a social structure. Family groups of females and young together with unrelated immigrant adult males are a recurrent type of mammalian social organization.

The genetic structure of a population is the pattern of genotypic differentiation between groups and genetic relationships within groups. A genetic structure of a population follows from an uneven distribution of genotypes over the area where the population lives. Social structure will highly influence the genetic structure: the social patterns of the population will be reflected in its genetic patterns. But while it is obvious that the social structure will influence the genetic structure of the population, it is not immediately clear *how* social structure influences genetic structure and how strong such an effect can be. A social structure involving resident females and migrating males promotes inbreeding avoidance – but it might retard the purging of deleterious recessives. Both the immediate and long-term consequences of any social structure on the genetic structure are therefore of interest in conservation genetics.

Social structure and genetic structure in *Macaca fascicularis*

The long-tailed macaque (*Macaca fascicularis*; Raffles, 1821) is widespread throughout Indonesia, Malaysia, and the Philippines (Kawamoto and Ischak, 1984). In the Ketambe Nature Reserve Area, in the north of the island of Sumatra, these monkeys live in the rain forest along the Ketambe river in mixed sex groups of about eight to 50 individuals. Several of these groups have been under intensive study since 1976, so that we now have a detailed knowledge of their behavior, ecology, and social structure (e.g., van Noordwijk and van Schaik, 1988; de Ruiter, 1992). Male and female life histories differ, in that females remain in their birth group and males always emigrate. After the hazards of youth, females start reproducing at 6 years of age, and have infants with an interbirth interval of about 18 months. Males migrate at puberty (4 to 6 years of age), typically to an adjacent group and often in peer groups. Reproducing males in any group are always immigrants. The dominant α-male enjoys paternity of about 80% of the infants born during his tenure of about 3 years; practically all the remaining 20% are fathered by the β-male (de Ruiter et al., 1992). There is a tendency that the α-male has more fertilizations in matrilines where the closely related females all have a high social rank, whereas the β-male fathers relatively more offspring in low-ranking matrilines. Individuals can be expected to live until about the age of 15 years.

This social structure clearly leads to a high genetic relationship within groups: all females within a group are sisters, half-sisters or cousins of some degree, or mother and daughter. α-Male and β-male within a group might well be half-brothers. Infants born at the same time have related mothers, and are likely to have the same father. The number of adult reproducing individuals within a group is small, never more than half the census number. The number of reproducing males is much more limited than the number of reproducing females. Therefore, the chance processes inherent in heridity will come predominantly into play: genetic drift will clearly be of importance in such a socially structured population. But the migration of subadult males *prior* to reproduction connects the social groups: after all, it is one viscous population. Genetic drift has to be considered both on the level of the social group, and on the level of the total population.

Blood samples were taken from all individuals of seven social groups in the Ketambe area. By electrophoresis, 17 blood protein and allozyme loci proved to be polymorphic out of 29 loci (Scheffrahn et al., 1987). On the basis of 10 of these loci that gave reproducible results and had sufficient variability for our purpose, F_{ST} and F_{IS} were computed. The average F_{ST} over all loci was 0.045 ± 0.013, and the average F_{IS} 0.016 ± 0.028 (Scheffrahn and de Ruiter, 1993). The F_{ST} value agrees with a value found for eight neighboring social groups of long-tailed macaques on the island of Bali (Aoki and Nozawa, 1984). F_{ST} values of this magnitude or higher have been found for other monkey species, but F_{IS} has often been found negative in other studies of monkey populations with a similar social structure (Aoki and Nozawa, 1984; Pope, 1992).

The genetic structure of a population is often reported as an F_{ST} and an F_{IS} value. This provides a great deal of very useful data: for instance, the survey collected by Pope (1992) shows enough studies reporting negative F_{IS}'s (that is, a heterozygote excess within groups) to realize that this is a real phenomenon, not an unlucky, occasional outcome. The question should then be asked, what do these F_{ST} and F_{IS} values mean? How characteristic is a given F_{ST}? Is it a moment's value, to be replaced in the next change in heterozygosity due to genetic drift, or is it indeed a characteristic value for a given population? What causes a negative F_{IS}? The question therefore is, whether we can identify those aspects of the social structure of the population that influence F_{IS} and F_{ST}, whether we can quantify those aspects, and whether we can predict what effects changes in the social structure would have on the genetic structure.

150

Computer simulation

A computer simulation of a macaque population was set up with two objectives in mind. First, to investigate whether a population with a macaque social structure reaches a characteristic set of F_{ST} and F_{IS} values. Second, if such characteristic values exist, to investigate what consequences changes in the social structure and demography of the population have on the F_{ST} and F_{IS} values. Both objectives can be realized by using individual oriented simulation, a computer simulation in which each individual in the population is actually represented with its age, sex, male status, genotype, and group adherence.

In the simulation an abstracted demography of the long-tailed macaques was followed (Fig. 1). In a group were a fixed number of individuals: 20 males and 20 females. Of the 20 males, 6 were juveniles of ages 0 to 4.5 years, 2 were subadults of ages 4.5 and 6 years bent on migration, 2 were immigrant adults of 6 to 9 years keen on reproducing, 2 were reproducing males (identified as α-male and β-male) of age 6 to

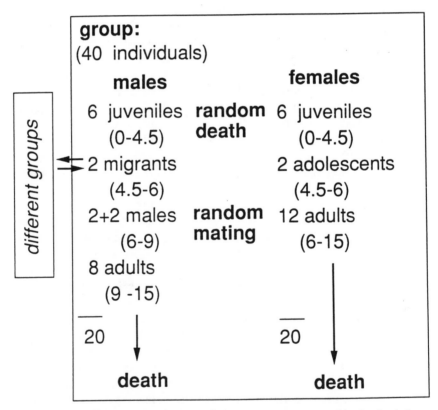

Fig. 1. Abstracted demography of a long-tailed macaque group, as used in the simulation.

12 years, and 8 were post-reproducing males of up to 15 years. After his tenure as α-male, a male becomes the β-male of the group. Of the 20 females, 6 were juveniles of ages 0 to 4.5 years, 2 were adolescents of ages 4.5 to 6 years, and 12 were reproducing females of 6 to 15 years. Females produce an infant every 18 months, the father being the group's α-male with 80% probability and the β-male with 20% probability. The population was surveyed every 1.5 years – subadults emigrating, migrants immigrating, male reproduction status settled between immigrants and tenured reproducing males, infant birth, and juvenile mortality to keep the number within a group at the required total (Fig. 1).

A population was composed of a preset number of groups, most often 35. Male migration by two peers was to a random group. Both the α-male and the β-male mated, respectively, with 80% and 20% probability, with any female from any matriline. The model takes many aspects of long-tailed macaque demography and social structure into account; not included are rank-related female differential reproduction, and rank-related male non-random mating with the females from the different matrilines. This difference between the field and the simulation will tend to increase field F_{IS} over simulation F_{IS}, as a field group is, in fact, itself subdivided and the Wahlund effect comes into play.

The genotype of each individual was given by a single locus. In the starting population, all of the alleles were individually distinguished. A population begins therefore with a locus with a number of alleles that is twice the number of the individuals over all the groups. Apart from their identification number, the alleles are totally identical. The model is of selectively neutral alleles. The population starts fully heterozygotic; identity by descent of every allele can be established over the whole population in any generation. Due to genetic drift, heterozygosity declines with time over the total population, and differentiation in allelic frequencies over the groups develops.

The genetic structure of a population at any one moment can be given by the F_{IT}, F_{ST}, and F_{IS} values; these are computed from the heterozygosities. Definition of five heterozygosities within a group, averaged over groups and between groups gives

H_i = the observed heterozygosity of an individual in group i;

$\bar{H}_I$ = the observed heterozygosity of an individual, averaged over k groups;

$= \Sigma H_i / k;$

H_{Si} = the expected heterozygosity of an individual in group i, based upon the allele frequencies p_j in group i;

$= 1 - \Sigma p_{j,i}^2;$

$\bar{H}_S$ = the expected heterozygosity of an individual based upon the allele frequencies in group i, averaged over k groups,

$$= \Sigma H_{Si}/k;$$

$\bar{H}_T$ = the expected heterozygosity of an individual, based upon the average allele frequencies over all groups;

$$= 1 - \Sigma p_j^2.$$

The F statistics become (Nei, 1977):

$$F_{IS} = 1 - \bar{H}_I/\bar{H}_S;$$

$$F_{ST} = 1 - \bar{H}_S/\bar{H}_T;$$

$$F_{IT} = 1 - \bar{H}_I/\bar{H}_T.$$

F_{IS} indicates the average deviation from Hardy-Weinberg frequencies in the groups ($-1 < F_{IS} < 1$); F_{IS} is positive when over all groups more homozygotes are present than expected *per* group, and negative when more heterozygotes are found. F_{IT} indicates the overall deviation from Hardy-Weinberg frequencies if the group structure is totally ignored ($-1 < F_{IT} < 1$); F_{IT} is positive when more homozygotes are present than expected for the total population, and negative when more heterozygotes are found. F_{ST} indicates the difference in allele frequencies between groups, and between the groups and the average allele frequencies over the total population. F_{ST} can be regarded as a variance ratio, giving the ratio of the observed variance in the allele frequencies over groups, relative to the expected variance. In theory, and for each locus in a single population,

$$(1 - F_{IS})(1 - F_{ST}) = (1 - F_{IT}),$$

as can be seen from the definitions. Average F_{IT} over loci, or average F_{IT} over populations or simulation runs, does not obey this equation. In such cases, F_{IT} has to be computed separately. It is therefore not sufficient to focus on F_{IS} and F_{ST} in such cases.

The F statistics were computed from the simulation data according to the method of Weir and Cockerham (1984). This method has a better behavior under sampling than the method of Nei and Chesser (1983), but the two methods agree in values for large populations (Haring, MSc thesis, Utrecht University). In this computer simulation many replicate runs of the standard situation were used to see whether expected values for F_{IS} and F_{ST} could legitimately be stated. Once it became clear this was possible, subsequently multiple replicate runs, after changing a specific parameter, were used to evaluate the influence of that parameter.

Simulation results

(a) The F values are constants characterizing a social structure
The first result of the computer simulation was that F_{ST} and F_{IS} quickly reached an approximate asymptotic value in any run in a few generations (timesteps of 1.5 years) (Fig. 2), despite the continued fixing of alleles (Fig. 3) and the continued decrease of heterozygosities $\bar{H}_I$, $\bar{H}_S$ and $\bar{H}_T$ (Fig. 4). The heterozygosities decrease, but not their ratios in F_{ST}, F_{IS}, and F_{IT}. Analytically, it has been shown that F_{ST}, as a variance ratio, remains constant, even though the actual observed and expected variances in allele frequency over groups decrease due to genetic drift (Chesser, 1991a,b). In the simulation, runs and generations differ slightly in F_{ST} and F_{IS}; this is used to estimate an average F_{ST} or F_{IS} value and its standard deviation. The fluctuations over runs and over generations are relatively minor. This means that F_{ST}, and F_{IS}, are values characterizing a population, and can be used to evaluate the influence of alternative social structure or life history parameters.

The demography of the long-tailed macaque in the reference model yielded, averaged over five runs, an $F_{ST} = 0.034 \pm 0.004$, an $F_{IS} = -0.045 \pm 0.012$, and an $F_{IT} = -0.0024 \pm 0.0008$. (The averaging over runs causes the relation $(1 - F_{IS})(1 - F_{ST}) = (1 - F_{IT})$ to be invalid). These F-values have to be compared with the field estimates of $F_{ST} = 0.045 \pm 0.013$, $F_{IS} = 0.016 \pm 0.028$, and $F_{IT} = 0.060$. The correspondence for F_{ST} is quite satisfactory, giving confidence in the ap-

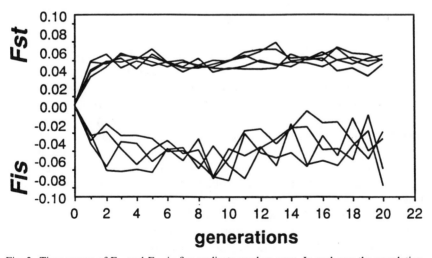

Fig. 2. Time-course of F_{IS} and F_{ST} in five replicate random runs. In each run the population consists of 35 groups of 40 individuals, males migrate, and only one α-male and β-male reproduce. F_{IS} and F_{ST} quickly reach an asymptote, subject to random variation. The average value over 4 runs suffices to characterize a population with a known social structure.

154

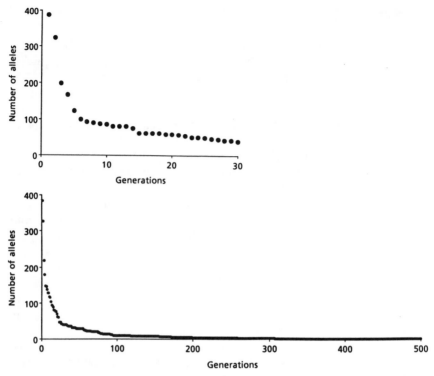

Fig. 3. Time-course of the number of alleles within a population. The initial number of alleles is 400. Comparison of this figure with Figs 2 and 4 shows that the mean values of F_{ST}, F_{IS}, and F_{IT} over time are independent of the number of alleles. The variances of F_{ST}, F_{IS}, and F_{IT} increase with decreasing number of alleles.

proach. The difference between the F_{IS} and F_{IT} from the field data and from the simulation is appreciable. Two considerations are pertinent to this difference. First, it is known from our knowledge of the long-tailed macaque population that a reproductive subdivision between matrilines within groups might possibly exist. A group as used in the simulation might in the real world be a composite of reproductive subgroups. Subgroups within a group would lead to a Wahlund effect for the group, that is, a deficit of heterozygotes on the level of the group. A deficit of heterozygotes at the level of the group due to internal subdivision would lead to a positive F_{IS}. This internal subdivision is not present in the simulation. The real world F_{IS} is therefore expected to have a higher positive value than the simulation F_{IS}. Another consideration follows from basic population genetics: if allele frequencies differ in males and females, their offspring show an excess of heterozygotes (see Appendix A). Since allele frequencies differ between the groups – as $F_{ST} > 0$ testifies – the expectation is that mating between males born in

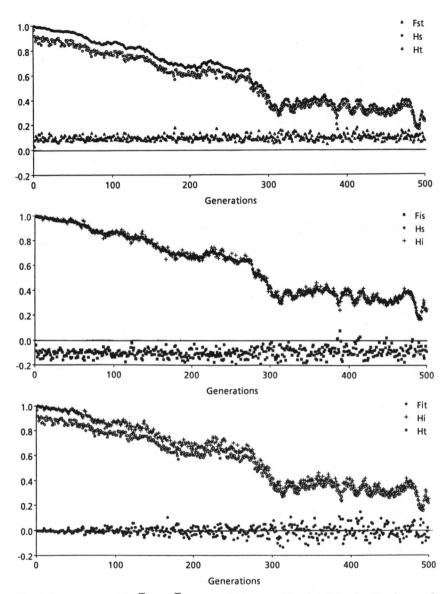

Fig. 4. Time-course of the $\overline{H}_S$ and $\overline{H}_T$, the two heterozygosities that determine F_{ST}, in one of the populations of Fig. 2. Despite the decrease in both heterozygosities in time, their ratio remains constant.

one group and females born in another group leads to an excess of heterozygotes within any group (Prout, 1981). The negative F_{IS} found in the simulation is therefore compatible with *a priori* expectations. While the expectation is not fulfilled in the long-tailed macaques in the parent study, negative F_{IS}-values have been repeatedly found in monkeys with a similar social organization (Pope, 1992). A further subdivision within the groups in the field that is not represented in the simulation would lead to a more positive F_{IT} in the field than in the simulation. For F_{IT} no *a priori* expectation of its sign exists, in contrast to the negative sign expected for F_{IS}. These considerations lead us to accept the simulation model.

(b) F values change with the social structure

Social structure has an influence on genetic structure, as summarized by F-statistics. We can change the parameters of the social structure and study the changes in the F-statistics that follow a change in social structure. The most obvious parameters of social structure are:

– group size: the smaller the group size, the greater the expected effect of chance, genetic drift, within a group.
– group number: the population might consist of a differing number of groups.
– the reproductive pattern: only α-male or β-male reproduce *versus* all adult males reproduce.
– the sex of the migrants: males migrate in the long-tailed macaque, but in some monkey species both sexes migrate, and in a few species only the females migrate. The best-known example of female migration is the chimpanzee; both males and females migrate in some colobus and howler monkeys (Pusey and Packer, 1987). Since the variance in the number of offspring of males is far higher than the variance in the number of offspring of females, it can be expected that this has an influence on the genetic structure of the population.
– migration pattern: do males migrate to a random group, or is there some preference in the direction of migration? A preference in the direction of migration might be dictated by the geography of the terrain, or simply by distance. In a model, a first contrast is between migration of males to random groups, and one-directional migration; in this last case, the groups are arranged in a "circular" design in the model representation of the population.

In each case, only one parameter of social structure was changed at a time from the reference model.

The influence of the number of individuals in a group on F_{ST} values is relatively small (Fig. 5). There are two reproducing males: group size involves number of females per male. F_{ST} decreases with the number of females in the group. The difference in numbers between males and females leads to the pattern of change of F_{ST} with female number: at high number of individuals (that is, of females) per group, F_{ST} reaches a low asymptote, as F_{ST} comes to depend almost solely on the males. Lower number of females implies more scope for genetic drift, and higher random group differentiation. Subdivided populations rapidly lose variability from within each subpopulation, but retain variation across subpopulations better than does a panmictic population (Lacy, 1987). A possible conclusion might be that a social system involving very large harem groups is not expected to have a very high F_{ST}. At low number of individuals *per* group, less than 40, the effect of the number of females in the group on F_{ST} is appreciable.

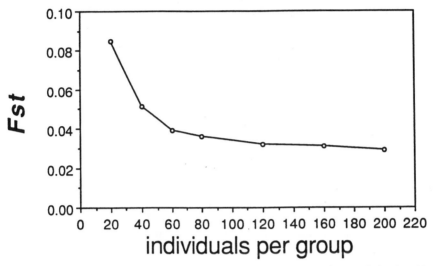

Fig. 5. The influence of the number of females in a group on F_{ST}. The population has 35 groups, male migration, and each group has an α-male and β-male.

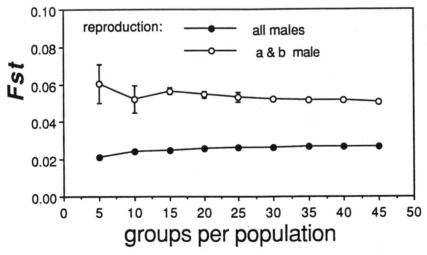

Fig. 6. The influence of the number of groups on F_{ST}. Number of groups has almost no influence on F_{ST}, unless the number of groups is very small. This holds for both mating systems.

158

The influence of the number of groups, each with 40 individuals, is not significant (Fig. 6). The observed and expected variances in allele frequency over groups depend on the drift within a group; and unless the number of groups is less than five, this is fairly independent of the number of groups. For very low numbers of groups, the estimates of F_{ST} become less reliable; the groups had better be regarded as one population.

Interestingly, the number of heterozygotes over the overall population is slightly higher than it would have been had the overall population mated at random: F_{IT} is consistently negative in the simulations (Fig. 7). The mating systems and migration patterns influence F_{IT} as they influence F_{IS}.

The effect of the mating system on the F_{ST} and F_{IS} predicted by the model is relatively large (Fig. 8). When both males and females migrate, the differentiation of allele frequencies between groups is relatively low, and both F_{ST} and F_{IS} are low in absolute value. Male migration has a slightly stronger influence than female migration. The presence of limited reproduction in males (by an α-male and a β-male only, as opposed to reproduction by all males in the group) has a very strong effect. The limited number of reproducing males gives more leeway to chance, leading to strong group differentiation (F_{ST}) and, therefore, to an appreciable difference in allele frequencies between immigrant males and resident females (F_{IS}).

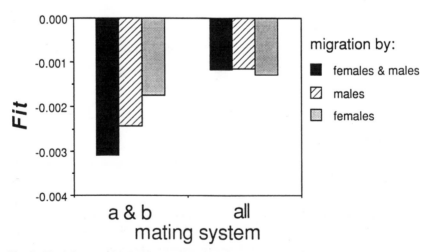

Fig. 7. The influence of migration and mating on F_{IT}, for 35 groups of 40 individuals. The migration comparison is of populations where only two males immigrate into a group, only two females immigrate into a group, or a male and female immigrate. The effect of the mating system comes from comparing a population where all males in a group mate with a population where only an α-male and a β-male mate.

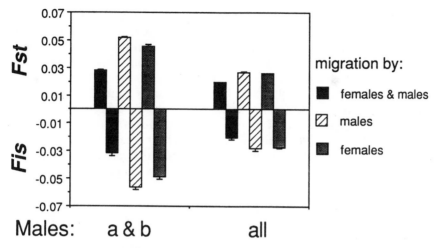

Fig. 8. The influence of migration and mating on F_{ST} and F_{IS}, for 35 groups of 40 individuals. The migration comparison is of populations where only two males immigrate into a group, only two females immigrate into a group, or a male and female immigrate. The effect of the mating system comes from comparing a population where all males in a group mate with a population where only an α-male and a β-male mate.

Random migration of males leads to less differentiation between groups than circular migration: after all, the mean "distance" between any two groups in the random migration model is lower than in the circular migration model. This implies that features of the geography of the habitat that facilitate or impede migration between groups are certain to influence the genetic structure of the population.

The simulations were only pursued for a limited range of changes from the reference model. In all cases, the basic long-tailed macaque life history and demography was maintained. However, the principal findings seem to have a wider applicability.

Discussion

The study of the long-tailed macaques in Ketambe resulted in F_{IS} and F_{ST} values, two statistics often used to describe the genetic structure of a population for social groups (Nozawa et al., 1982; Aoki and Nozawa, 1984; Pope, 1992). The main finding in the simulation study precipitated by the field data (Scheffrahn and de Ruiter, 1994) is that the influence of social structure on the genetic structure of a population is relatively strong. Aspects of the social structure such as group size, reproduction pattern, and migration pattern all influence the genetic structure of the population. In the long-tailed macaques, we are considering groups of resident females reproductively connected by migrating males. The relatively small group size and the fact that at any one time in one group only a few males father all the infants, lead to a relatively large role of genetic drift, and therefore appreciable genetic differentiation between the social groups. On the other hand, the individuals within a group are closely related, due to the fact that the females remain in their group of birth, and the fact that all infants born at the same time might well have the same father. Variation in allele frequencies between groups and genetic relationships between individuals within a group are related concepts (Pamilo, 1989). However, the high degree of relatedness between individuals within a group does not imply increased homozygosity: highly related individuals might well all be heterozygotes.

Small group size, a variance in the number of offspring, and unequal number of males and females all lead to lower effective population size. Effective population size is, however, not one concept, but a conflation of three different concepts that may each yield a different effective population size if applied to the same data (Gliddon and Goudet, this volume). If migration pattern, group size, male reproduction, etc., all should lead to the use of the same concept for effective population size, it would be possible to describe how the social structure influences effective population size, and then determine how effective population size influences the genetic structure. But it is not immediately obvious that the same concept for effective population size is useful to describe the effect of all parameters of the social structure. Moreover, a direct description of the effect of the parameters of the social structure on the genetic structure seems biologically equally informative.

The simulation only distinguishes alleles by an identification number, as in all theoretical studies of the F statistics the alleles are selectively neutral. In conservation genetics (slightly) deleterious alleles that would lead to inbreeding depression upon homozygosity are of more interest. Such alleles will be subject to selection as well as to genetic drift. The relative strength of genetic drift and selection will depend upon group size mating pattern, and migration pattern: exactly on the same features of the social structure of the population that influence the genetic structure even without selection. Strong selection may override genetic drift, and lead to groups that do not vary in their allele frequencies – or only do so due to recurrent mutation. Weak selection implies that "genetic," "selective" deaths are rare; therefore, the pattern of differentiation between groups will be akin to that for neutral alleles. Again, differentiation in allele frequencies between groups implies different allele frequencies between the females resident in a group and the males immigrating into the group and, hence, some excess of heterozygotes in the offspring. An excess of heterozygotes, in turn, will protect a slightly deleterious recessive allele. A population with negative F_{IS} and negative F_{IT} might therefore well have a higher genetic load than a random mating population with genotype frequencies in Hardy-Weinberg proportions (Fig. 9).

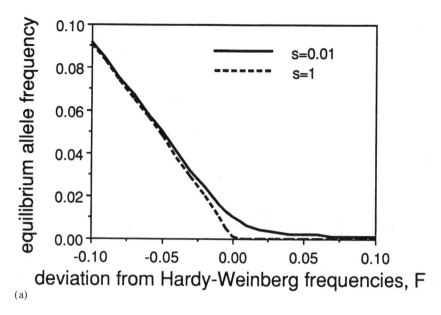

(a)

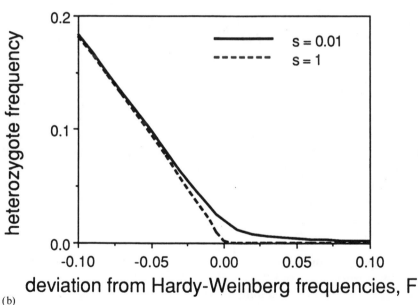

(b)

Fig. 9. A theoretical infinite population under mutation selection balance. (a) The equilibrium allele frequency depends upon the deviation from Hardy-Weinberg equilibrium F. (b) The heterozygote frequency depends upon the deviation from Hardy-Weinberg equilibrium F.

This is only conjecture at this point, but the reasoning is as follows. In any population, the genotype frequency of a homozygote at a locus with two alleles can be written as $q^2 + pqF$, with $-1 < F < +1$, as F is the correlation between alleles within an individual. For any deleterious recessive homozygote the genotype frequency at birth before selection can be written in this way, too. The deleterious recessive allele is assumed to be maintained by a mutation selection balance, the mutation rate to the deleterious allele being u and the selection coefficient against the homozygote being s. If the individuals in the population mate at random, and the population is not structured, $F = 0$ and the textbook solution to the equilibrium frequency of the deleterious allele is: $q = \sqrt{(u/s)}$. In a socially structured population, the deviation from Hardy-Weinberg proportions F might be independent of the allele frequency q, even under selection – in fact, it is so for neutral alleles (Figs 2, 3). Using the same type of approximation as in the textbook case, the equilibrium frequency of the homozygote deleterious recessive is

$$q^2 + pqF = u/s. \tag{1}$$

The equilibrium allele frequency q becomes a function of F (Fig. 9a); the heterozygote frequency at this equilibrium allele frequency is given in Fig. 9b. For a negative F, the equilibrium frequency of the deleterious allele increases as the deviation from Hardy-Weinberg proportions increases. The frequency of heterozygote carriers of the deleterious allele might become fairly high: at $F = -0.05$, the frequency of heterozygote carriers becomes 0.099 to 0.095, for selection coefficients $s = 0.01$ to $s = 1$; and at $F = -0.005$ the carrier frequency becomes 0.010, compared with 0.002 at $s = 1$ and $F = 0$. The equilibrium situation in a socially structured population might well be a higher genetic load than in a totally random mating population.

The evolution of a social structure as found in the long-tailed macaques has been thought to be mediated by the avoidance of incest and the concomitant inbreeding depression (Harvey and Ralls, 1986). Given small groups, and given a certain frequency of a deleterious recessive allele, a male who migrates has a good chance of having all his offspring non-inbred and healthy, while a male who mates within his group of birth risks offspring suffering from inbreeding depression. A female who mates with an immigrant male has a better chance of having a non-inbred, healthy offspring than a female who accepts a mating with a male native to the group. The comparison is at a given allele frequency, and is concerned with the immediate best strategy for a female or male. However, the point made in the previous paragraph is that a migration strategy, once adopted, leads to higher equilibrium frequencies of deleterious alleles. This becomes apparent in the heterozygote carrier frequency: the equilibrium frequency of the deleterious homozygotes is independent of F (expression 1). The number of actual selective deaths therefore depends upon the selection coefficient, and not on F or on the migration strategy adopted by the population.

The potential for accumulation of deleterious alleles by migration and socially prevalent outbreeding has to be further investigated. It is however clear that if genetic load does accumulate at a higher rate in a socially structured population, disruption of the social structure would lead to severe inbreeding depression. The prediction would be that species with a socially structured population in nature are particularly liable to severe inbreeding depressison upon disruption of the social structure, as when brought together into one population in a single zoo.

Acknowledgements
We thank the Indonesian Institute of Sciences (LIPI) and its biological division (Puslitbang Biologi), the Directorate of Nature Conservation (PHPA), and the Universitas Nasional Jakarta (UNAS) for sponsoring the fieldwork and for permission to work in the Ketambe study area. We thank our collaborator W. Scheffrahn, R. D. Martin for stimulating the project, and J. A. R. A. M. van Hooff for his continued interest. The simulation is part of a MSc thesis by Mr. R. Haring for the University of Utrecht. The fieldwork was financed by WOTRO, the Netherlands Foundation for the Advancement of Research in the Tropics.

References

Aoki, K. and Nozawa, K. (1984) Average coefficient of relationship within troops of the Japanese monkey and other primate species with reference to the possibility of group selection. *Primates* 25: 171–184.

Chesser, R. K. (1991a) Gene diversity and female philopatry. *Genetics* 127: 437–447.

Chesser, R. K. (1991b) Influence of gene flow and breeding tactics on gene diversity within population. *Genetics* 129: 573–583.

Harvey, P. H. and Ralls, K. (1986) Do animals avoid incest? *Nature* 320: 575–576.

Kawamoto, Y. and Ischak, T. M. (1984) Genetic variations within and between groups of crab-eating macaques (*Macaca fascicularis*) on Sumatra, Java, Bali, Lombok and Subawa, Indonesia. *Primates* 25: 131–159.

Lacy, R. C. (1987) Loss of genetic diversity from managed populations: interacting effects of drift, mutation, immigration, selection, and population subdivision. *Cons. Biol.* 1: 143–158.

Nei, M. (1977) F-statistics and analysis of gene diversity in subdivided populations. *Ann. Hum. Genet.* 41: 225–233.

Nei, M. and Chesser, R. K. (1983) Estimation of fixation indices and gene diversities. *Ann. Hum. Genet.* 47: 253–259.

Noordwijk, M. A. van and van Schaik, C. P. (1988) Male careers in Sumatran long-tailed macaques. *Behaviour* 107: 24–43.

Nozawa, K., Shotake, T., Kawamoto, Y. and Tanabe, Y. (1982) Population genetics of Japanese monkeys: II. Blood protein polymorphisms and population structure. *Primates* 23: 252–271.

Pamilo, P. (1989) Estimating relatedness in social groups. *Trends Ecol. Evol.* 4: 353–355.

Pope, T. R. (1992) The influence of dispersal patterns and mating system on genetic differentiation within and between populations of the red Howler Monkey (*Alouatta seniculus*). *Evolution* 46: 1112–1128.

Prout, T. (1981) A note on the island model with sex dependent migraiton. *Theor. Appl. Genet.* 59: 327–332.

Pusey, A. E. and Packer, C. (1987) Dispersal and philopatry. *In:* Smuts, B. B., Cheney, D. L., Seyfarth, R. M. Wrangham, R. W. and Struhsaker T. T. (eds), *Primate societies*, University of Chicago Press, Chicago, pp. 250–265.

Ruiter, J. R. de (1992) Capturing wild long-tailed Macaques (*Macaca fascicularis*). *Folia Primatol.* 59: 89–104.

Ruiter, J. R. de, Scheffrahn, W., Trommelen, G. J. J. M., Uitterlinden, A. G., Martin R. D. and van Hooff, J. A. R. A. M. (1992) Male social rank and reproductive success in wild long-tailed macaques: paternity exclusions by blood protein analysis and DNA fingerprinting. *In:* Martin, R. D., Dixson, A. F. and Wickings E. F. (eds), *Paternity in primates: Genetic tests and theories.* Karger, Basel, pp. 175–191.

Scheffrahn, W. and de Ruiter, J. R. (1994) Genetic relatedness between populations of *Macaca fascicularis* on Sumatra. *In:* Fa, J. E. and Lindburg, D. G. (eds), *Evolution and ecology of Macaque societies.* Cambridge University Press, Cambridge (in press).

Scheffrahn, W., Socha, W. W., de Ruiter, J. R. and van Hooff, J. A. R. A. M. (1987) Blood genetic markers in Sumatran *Macaca fascicularis* populations. *Genetica* 73: 179–180.

Weir, B. S. and Cockerham, C. C. (1984) Estimating F-statistics for the analysis of population structure. *Evolution* 38: 1358–1370.

Appendix A

Difference in allele frequencies between males and females leads to excess heterozygotes in offspring.

Let the allele frequency at a locus A with two alleles A_1 and A_2 be p_f in females and p_m in males. After random mating, the allele frequency in the offspring is

$$p = \tfrac{1}{2}(p_f + p_m),$$

while the frequency of the heterozygotes is

$$p_f q_m + q_f p_m.$$

The expected frequency of heterozygotes under Hardy-Weinberg frequencies on the basis of p equals:

$$2 \cdot \tfrac{1}{2}(p_f + p_m) \cdot \tfrac{1}{2}(q_f + q_m) = 2 \cdot \tfrac{1}{4}(p_f q_f + p_f q_m + p_m q_f + p_m q_m)$$

$$= \tfrac{1}{2} \cdot (p_f q_m + p_m q_f) + \tfrac{1}{2} \cdot (p_f q_f + p_m q_m).$$

We are interested whether

$$\tfrac{1}{2} \cdot (p_f q_m + p_m q_f) + \tfrac{1}{2} \cdot (p_f q_f + p_m q_m) < p_f q_m + q_f p_m$$

$$(p_f q_f + p_m q_m) - (p_f q_m + q_f p_m) < 0$$

$$(p_f - p_m)(q_f - q_m) < 0,$$

which is always true for two alleles. The observed heterozygosity is therefore higher than expected under Hardy-Weinberg proportions, and can be written as

$$2pq(1 - F) > 2pq,$$

implying that the F giving the deviation from the expected Hardy-Weinberg frequencies is negative.

Conservation Genetics
ed. by V. Loeschcke, J. Tomiuk & S. K. Jain
© 1994 Birkhäuser Verlag Basel/Switzerland

Guidelines in conservation genetics and the use of the population cage experiments with butterflies to investigate the effects of genetic drift and inbreeding

P. M. Brakefield[1] and I. J. Saccheri[1,2]

[1]Section of Evolutionary Biology, Institute of Evolutionary and Ecological Sciences, University of Leiden, Schelpenkade 14a, NL-2313 ZT Leiden, The Netherlands
[2]Conservation Genetics Group, Institute of Zoology, Regent's Park, London NW1 4RY, U.K.

Summary. Bottleneck experiments were performed on the tropical butterfly *Bicyclus anynana*. The reasons for choosing a species of butterfly are described together with the methodology used in the population cage experiments. Replicated lines founded by single pairs and then measured in generation 2 or 3 showed strong phenotypic differentiation, declines in heritabilities of morphological traits, dramatic reductions in egg fertility (due to inbreeding depression) and an increased sensitivity to insecticide applications (probably due to a loss of alleles conferring tolerance) relative to unbottlenecked control lines. Egg fertility showed some recovery in further generations. An experiment involving repeat bottlenecks provided support for the hypothesis that a purging of deleterious alleles contributed to this "fitness rebound". The preliminary results are examined in the context of developing guidelines in conservation genetics.

Introduction

The central purpose of conservation biology theory is to understand what determines the survival of natural populations or of communities of such populations. It is well established that catastrophes (natural or man-induced), environmental stochasticity, demographic stochasticity, and genetic stochasticity are the major causes of extinction. However, a quantitative description of the importance of each of these processes and how they interact in different species and environments is only at an early stage of development (Nunney and Campbell, 1993), such that the management of rare species is based on largely untested theoretical rules of thumb with little knowledge of the long-term consequences. In particular, the importance of genetic variation remains essentially unquantified for almost all organisms.

The dual environmental pressures of habitat fragmentation and contraction are tending to produce a greater isolation and smaller size of populations. Small population size can lead to genetic change and deterioration either through (1) introducing the stochastic sampling effects of genetic drift to produce a loss of heterozygosity, rare alleles and additive genetic variance; or (2) increasing the incidence of matings between individuals related by descent and thus of inbreeding depression due to higher levels of homozygosity of rare recessive deleterious alleles (Lande and Barrowclough, 1987; Falconer, 1989). Any intermittent reduction in numbers in an already small population will result in a "bottleneck" event and a greatly increased chance of genetic drift and inbreeding. Heterozygosity at polymorphic loci and additive genetic variance for quantitative characters are expected to decline because of drift by one over twice the effective population size *per* generation of a bottleneck. Isolation restricts the possibility of migration and gene flow introducing novel genetic diversity; mutation will only provide slow or very slow recoveries in small populations (Lande and Barrowclough, 1987). Isolation will also lower the chance of recolonization following any local extinctions. Genetic diversity is likely to be directly related to the variability in individual fitness within a population and the ability of a population to produce an adaptive response to an environmental perturbation or change (Brakefield, 1991). Genetic diversity must also play a role in discussions of conservation priorities where choices need to be made in applying limited resources to the preservation and management of scattered populations of a rare species.

Local populations of many organisms are probably commonly established from small numbers of founder individuals (for examples involving butterflies see Oates and Warren, 1990; Pavlicek-van Beek et al., 1992). In these sorts of situations one can ask, and attempt to quantify, how the probability of establishment and persistence is related to the genetic diversity within the founders. Furthermore, an active conservation policy which is becoming increasingly important for butterflies in European countries such as The Netherlands (J. v. d. Made, personal communication) and Britain (Oates and Warren, 1990) is the reintroduction of a species into habitats where it has become extinct; such a policy also involves choices with respect to the founders. More specific questions need to be answered to produce adequate and rigorous guidelines to maximize the chance of success of such introductions; they include: how many, and which individuals (including which source when local adaptation and differentiation have evolved). Another policy which is being pursued particularly actively in The Netherlands is the establishment of landscape corridors and networks. In this case one question is: how much movement should be encouraged?; too much gene flow may swamp any processes of local adaptation. In rare species conservation where only small numbers of fully isolated local populations are extant, individuals could be moved around artificially; in which case, how many and how often? These types of questions all have a genetic component. The theory of population genetics can provide partial theoretical answers or guidelines (Lande and Barrowclough, 1987). Furthermore, computer simulations can examine the consequences of this theory in terms of heterozygosity indices (see de Jong et al., this volume). However, more empirical data are necessary to be able to apply the theory and simulations with any real confidence to populations in the wild. Moreover, the relevance of heterozygosity and molecular-based indices of genetic diversity to variation in ecologically-important quantitative traits and to processes of adaptation in natural populations are essentially unexplored in a conservation context. A reliance solely on estimates of heterozygosity from molecular surveys could lead to incorrect decisions when developing conservation priorities or management policies.

Our laboratory experiments using butterflies in population cages are designed to describe some of the potential processes involved in genetic deterioration and to investigate how they influence variation in fitness within populations. Some of the preliminary results will be outlined here together with the rationale of performing these experiments. The first series of experiments involves examination of the effects of single-generation bottlenecks of differing size on genetic variances and fitness parameters in independent lines. A second series will investigate and describe empirically the effect of gene flow in groups of interacting small subpopulations ("metapopulations"); these will not be discussed further (C. van Oosterhout et al., unpublished).

Aspects of the initial experiments to be covered in this preliminary report are:

(1) why butterflies?;
(2) the methods of working with butterfly populations in cages;
(3) the experimental design and objectives;
(4) a summary of some of the results from bottleneck experiments (Saccheri, unpublished data).

Materials and Methods

Why butterflies?

Butterflies attract a high level of public awareness and conservation activity. The ecology and food plant specialization of many butterflies makes them especially sensitive to habitat disturbance or change. Many species in temperate communities have shown severe declines (see, e.g., reports in Pavlicek-van Beek et al., 1992). In Britain, this decline is considered to have been more extreme than for any other well known group of plants or animals; nearly half of the 59 resident species have experienced major range contractions, and four have recently become extinct (Thomas, 1991; Warren, 1992). The record is even worse in The Netherlands, where 15 of the 75 species have become extinct this century and a further 26 species have an endangered or vulnerable status (Tax, 1989).

The reasons for these types of changes are diverse, but both the contraction and fragmentation of habitat are undoubtedly major factors (Thomas, 1991; Warren, 1992). Many of the remaining populations of endangered species are both of small size and isolated from any other local populations. Butterfly populations are also frequently susceptible to population crashes which may be climatically-induced (Thomas, 1983). Bearing in mind that the effective population size is likely to be substantially smaller than the census number (Brakefield, 1991), genetic processes associated with small numbers are likely to be a major factor influencing genetic diversity in populations of butterflies of concern to conservationists.

Thus, many species of butterfly are likely to be particularly prone to genetic stochasticity and deterioration which is likely to be exacerbated by demographic and environmental stochasticity. Laboratory population cage experiments to investigate empirically the effects of population demography and bottlenecks on genetic diversity and fitness have, to date, mostly involved species of insects, especially *Drosophila* and *Musca*. These species have a more open population structure than most butterflies and other organisms of major concern to conservationists which are characterized by low dispersal and discontinuous distributions. A few similar experiments have been attempted with other animal species including mice (Brewer et al., 1990). Some of the experimental work with diptera has suggested that even severe repeated bottlenecking of populations may not necessarily lead to (substantial) lowering of evolutionary potential (López-Fanjul and Villaverde, 1989; Bryant and Meffert, 1993); theory indicates that a transformation of non-additive components of genetic variance into additive components may contribute to such results (Goodnight, 1987, 1988). These considerations led us to develop an experimental system based on a non-migratory species of butterfly, *Bicyclus anynana*, with a short generation time.

Methodology of butterfly population cage experiments

B. anynana is a small satyrine butterfly which occurs in seasonal environments in sub-Saharan Africa. Brakefield and Reitsma (1991) studied the population biology of several species of *Bicyclus*, including *B. anynana*, over a transition from wet to dry seasons at a forest-edge site in Malawi. The adults are rather sedentary, long-lived, feed only on fallen fruit, and lay eggs on grasses including species of *Oplismenus*. The population structure consists of an unstable mosaic of rather small concentrations of butterflies. Movements of adult butterflies between local habitat patches were detected. Rates of movement may vary with seasonal changes in biotope.

In the laboratory, net population cages ($34 \times 64 \times 47$ cm) comfortably house over 300 adults or 500 larvae. Current experiments suggest that, in these conditions, the effective population size is of the order of one-third to one-half of the absolute number of adults (van Oosterhout et al., unpublished). Adults are fed on mashed banana. Most females mate only once. Mating is allowed to occur for about 1 week following peak adult eclosion before eggs are collected on grass plants or cuttings rooted in water introduced into the cage. Alternatively, butterflies *in copula* or individual gravid females can be removed from mating cages and eggs obtained by holding the females in net-covered small plastic pots containing a grass cutting. Females lay 10–20 eggs *per* day up to a total of over 300 eggs. Families of larvae can be reared on young maize plants to pupation in small net sleeves for offspring-parent analyses of heritability. The generation time at $26°$ C in the experimental climate room with 12L:12D and ca. 70% RH is about 6–7 weeks. A captive stock established in 1988 from approximately 80 gravid females collected at a single site (Nkhata Bay, Malawi) has been maintained at about 500 adults with some overlap between generations. High levels of heterozygosity remain in the stock (Saccheri and Bruford, 1993).

We were concerned to be able to analyze different modes of genetic variation and types of traits in our bottleneck experiments:

Molecular polymorphism. Saccheri and Bruford (1993) have developed fingerprinting techniques for the detection of hypervariable minisatellite DNA markers. Allozyme variation at six polymorphic loci is also being surveyed using a cellulose acetate system. These data will be discussed elsewhere (Saccheri, unpublished data).

Visible polymorphism. The stock is polymorphic for a gene determining pupal color; a recessive allele at a frequency of about 5% changes the normal green color to yellow.

Morphological variation. Heritabilities and additive genetic variances have been obtained for quantitative variation in wing size and a number of wing pattern elements including the size of a particular eyespot and of a pale band (Holloway et al., 1992; Windig, 1993). Measurements are made quickly and accurately on wings removed from frozen bodies (retained for molecular analysis) using a computerized image analysis system. The full multivariate data matrix for all traits can be simplified by applying principal component analysis to produce linear and orthogonal combinations of the individual traits. The different components can be readily interpreted in terms of the traits and can themselves be treated as separate genetic characters.

Variation in insecticide tolerance. We were also concerned with measuring a trait of potential involvement in the ability of a population to withstand a novel environmental stress or challenge. Individual final instar larvae vary in their susceptibility to topical applications of the synthetic pyrethroid insecticide, deltamethrin. A dose-response curve was determined for controls to select an LD_{50} application rate to survey the experimental lines. The proportion of samples of larvae which survive to produce healthy adults is scored.

Variation in egg hatching. Eggs laid on grass cuttings either by populations or individual females can be easily counted. Viable eggs hatch after 4 days when the number of first instar larvae is counted.

Design of the bottleneck experiments

Bottleneck lines were established from single, three, and 10 pairs of fertile founders. Four or six (for the single-pair bottleneck treatment) replicates were used. Each replicate was established using the progeny from a random selection of the requisite number of fertile pairings taken from mating cages of stock butterflies. Four control lines of 300 butterflies (two established from about 75 fertile pairings) were also reared. All treatment lines were allowed to freely increase in size up to a carrying capacity of 300 adults (for single-pair lines this took two to three generations). The five types of traits described above were measured in generation 2 or 3 (some also at additional times). Heritabilities were estimated from about 12 families reared *per* replicate line with five offspring of each sex *per* family.

Most lines were discontinued at generation 4. The single-pair lines were reared on to generation 8, when they were rebottlenecked with several replicates *per* original line. Two of these replicates were allowed to flush up to 300 adults again before repeat measurements were made of each trait (including heritabilities; not discussed below). A similar procedure was also performed on replicated lines established by single pair crosses between three of the original bottleneck lines, and on new single-pair bottleneck lines and a control line all established at the same time as the rebottlenecking.

Results

Polymorphism

Four of the six single-pair lines became fixed for the green pupal color allele. The other two lines had homozygote frequencies indicating that one of the original four alleles at this locus was yellow. The two replicates of one of these latter lines each lost this allele at the second bottleneck. These results and those of the molecular markers (Saccheri, unpublished data) are broadly consistent with the expected effects of genetic drift. It will be interesting to examine how the changes in heterozygosity within and among treatments are correlated with observed changes in fitness-related parameters.

Morphological variation and heritabilities

Fig. 1 illustrates how the single-pair lines produced substantial morphometric differentiation among replicates, while there was little differentiation for the controls. The pattern among 10-pair lines was
indistinguishable from the controls, while the three-pair lines yielded an
intermediate amount of differentiation. There is also evidence from an
analysis of variance in principal component values that the single-pair
(but not three-pair) lines tend to show a reduced phenotypic variation
within lines.

Fig. 2 shows that the single-pair lines after their flush period were
associated with reduced heritabilities for the principal components
describing most of the morphological variation relative to the control
and 10-pair lines. Again, the three-pair lines appear to be rather
intermediate. In general, the additive genetic variances follow a similar
pattern (not shown).

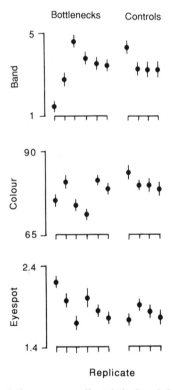

Fig. 1. Phenotypic differentiation among replicated single-pair bottleneck lines (left side) and
unbottlenecked controls (right side) of the butterfly *Bicyclus anynana*. Means with standard
error bars are shown for each of three pattern characters on the ventral hindwing of males:
an index for the area of a pale band across the wings; a measure of the average color in gray
values of a standard area of the wing; the area in square mm of the largest eyespot.

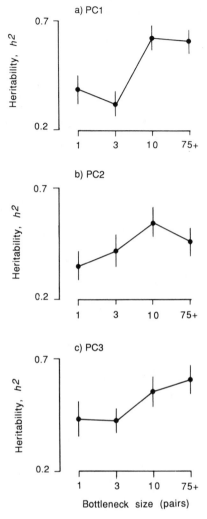

Fig. 2. Mean heritabilities with standard errors attached for each treatment in the bottleneck experiments. Values are for each of the first three principal components: PC1 (explaining 31% of total variance), PC2 (18%) and PC3 (13%) which are primarily measures of eyespot size, color and band with wing size, respectively (see also Fig. 1).

Insecticide tolerance and genetic drift

A summary of the data for mortality of larvae after standard application of insecticide is given in Tab. 1. Two or three of the six original single-pair lines show evidence of reduced tolerance after the first bottleneck-flush cycle. All except one of these lines were highly susceptible (in each of the two replicates tested) to the insecticide after a second

Table 1. Percentage survival in pooled samples from the single-pair bottleneck lines after standard application of the insecticide, deltamethrin: lines A to J were bottlenecked at generations 0 and 8, and lines 4 to 9 were bottlenecked once when established at the time of generation 8. Approximate sample size *per* line is indicated for each group (n.d. = no data)

Generation	F3	F7	F10
A	3% (N = 60)	23% (N = 30)	9% (N = 60–80)
C	n.d.	73%	56%
F	17%	53%	3%
G	0%	30%	0%
H	n.d.	57%	3%
J	40%	47%	7%
4	—	—	43% (N = 40)
7	—	—	42%
8	—	—	3%
9	—	—	45%

severe bottleneck and flush. One of the four new single-pair bottlenecks was also susceptible. These results can be explained by the existence of genetic variation for tolerance in combination with loss of tolerance alleles in many bottleneck lines due to genetic drift. However, we have no independent evidence for such genetic variation.

Egg hatching, inbreeding depression and the purging hypothesis

A dramatic decline in the proportion of eggs which hatch occurs following single-pair bottlenecks. Fig. 3 shows data from population

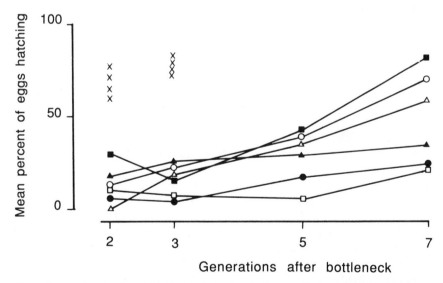

Fig. 3. Population fertility of the single-pair bottleneck lines (symbols connected by lines) and the unbottlenecked controls (×). Changes through time are shown for the bottlenecked lines; each symbol represents an individual replicate.

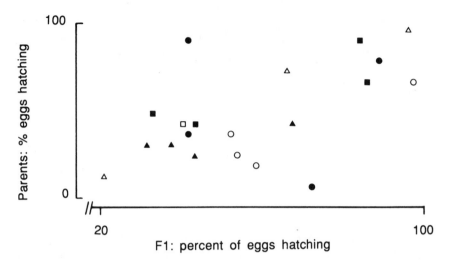

Fig. 4. The relationship between the fertility of females taken from generation 8 of the single-pair bottleneck lines and fertility in full-sib crosses of their offspring (means for 1–5 females). Each symbol indicates one of the six replicates (symbols as in Fig. 3). Heritability ($\pm$ standard error) estimated from the pooled data is 0.69 ± 0.23 ($P < 0.01$).

counts for the six original single-pair lines. There is a 50% or greater reduction relative to controls. Once again, the three-pair lines were intermediate while the 10-pair lines are indistinguishable from the controls. These results are consistent with an extreme inbreeding depression and the existence of many genes with deleterious recessive alleles in our stock population.

This interpretation is supported by the observed "fitness rebound" in some single-pair lines in later generations (Fig. 3). Three of the six replicates recovered much of their egg viability by the seventh generation. These results parallel similar observations in Bryant et al.'s (1990) experiments with houseflies, and are consistent with selection favoring genotypes less affected by inbreeding depression and thus with a purging of some deleterious alleles.

Females *in copula* were taken from the eighth generation of the original single-pair lines to describe individual variation in egg hatching and attempt to demonstrate heritability by rearing and scoring their F1 offspring (after mating of the full sibs). The data for the parent butterflies in Fig. 4 show that substantial phenotypic variation (still) occurred within each line. Furthermore, the pooled offspring-parent regression provides some evidence of a genetic component and thus of the potential for "fitness rebound" (Fig. 4).

Many individual females taken from the eighth generation of each of the single-pair lines were sterile, failing to produce any viable eggs (pooled data: 82 of 129 females). The overall proportion was higher than for females collected at the same time from the stock (1 of 9). A similar difference (but where all scored females were collected *in copula* and each laid eggs) was found two generations after the second bottleneck of each original line (pooled data: 77 sterile of 179 females) relative to the unbottlenecked controls (4 of 32). Thus, any fitness rebound had possibly not led to a full recovery in fertility of matings, at least in part, because some deleterious alleles had become fixed.

The repeat bottleneck experiment examined the purging hypothesis by comparing the loss of fertility in the different treatments (Fig. 5). On the basis of initial analysis of the untransformed data for all fertile F2 females, the second-bottleneck of the pure lines produces a smaller loss of egg hatching (mean with SE: $21.9 \pm 3.2\%$, n = 137) than in either the new single-generation bottlenecks ($43.6 \pm 2.8\%$, n = 68) or in the single-pair crosses of original bottleneck lines ($38.1 \pm 4.3\%$, n = 59; F = 10.96, df = 2 & 261, P < 0.001). The lower average drop in fertility after the repeat bottlenecks of the pure lines is consistent with some previous purging of deleterious alleles. The behavior of the crosses suggests that there were differences among the original bottleneck lines with respect to the identity of the deleterious alleles.

The high inbreeding depression suggests a large genetic load and that the effective population size of the field population in Malawi is very large, with a history lacking founder or bottleneck effects.

Discussion

These experiments show how laboratory bottlenecked-populations of a species of butterfly are susceptible to effects of genetic drift and inbreeding. Inbreeding depression effects on fertility are especially strong. However, these may not be (fully) representative of those expected in wild populations of species with naturally low effective population sizes and low rates of gene flow, or in species with a prolonged recent history of small fragmented populations. Such conditions are likely to be associated with the type of purging of deleterious alleles phenomenon for which we have some evidence in our bottlenecked populations. The effects of genetic drift leading to loss of genetic diversity have been detected in the bottleneck lines for each of the surveyed modes of genetic variation. The demonstration of an increased susceptibility to insecticide is of particular interest in illustrating the potential of such effects to lower the ability of populations to adapt to environmental change or to tolerate any perturbation or stress (see also Krebs and Loeschcke, this volume).

The declines in heritabilities of morphometric traits suggest that the *B. anynana* genome responds to a single bottleneck event in a more classical manner (Falconer, 1989) than as described in some experi-

176

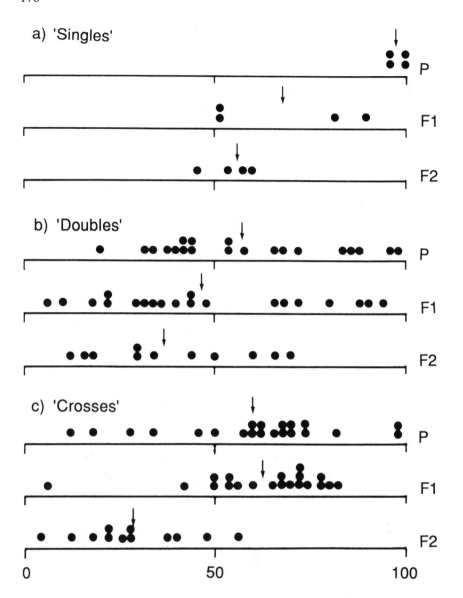

Percent of eggs hatching per new line or cross

Fig. 5. Variation in fertility pooled across replicates for each of the three treatments in the rebottlenecking phase of the experiment: (a) "singles" = four new single-pair bottleneck lines; (b) "doubles" = single-pair rebottlenecks of the six original single-pair bottleneck lines (giving a total of two bottlenecks); (c) "crosses" = lines founded by a single pair with the parents taken from different original single-pair lines. Each P value shows the fertility of the original female parent; each F1 value is the mean egg hatch for several F1 full-sib pairings and each F2 value is the mean for many individual females from the second generation after the bottlenecks (no sterile females are included; only about one-half of F1's were continued on to F2's). Arrows show overall mean values in each generation of each treatment.

ments with *Musca* or *Drosophila* which reported increases in additive genetic variances following single or repeated bottlenecks (López-Fanjul and Villaverde, 1989; Bryant and Meffert, 1993; although additive genetic variance for fertility may have increased in our lines). Recent theoretical treatments have also shown how bottlenecks may lead to a transformation of non-additive into additive components of genetic variance (Goodnight, 1987, 1988), an affect which could partially "compensate" for the loss of heterozygosity and allelic variation due to drift. However, there is no evidence for this phenomenon in our experiments; indeed, the declines we have measured in additive genetic variances are greater than one would expect given our estimates of effective population size associated with the bottleneck episode (Saccheri, unpublished data).

Given that genetic diversity should be maximized, some generally accepted management policies for populations can be listed (Brakefield, 1991):

(1) maintain population size minimizing the incidence and duration of any bottlenecks;
(2) maximize the proportion of breeding adults;
(3) preserve relevant habitat and environmental heterogeneity;
(4) minimize the rate and extent of any (man-induced) environmental change;
(5) retain natural patterns of gene flow.

The experiments reported here are concerned with the development of more specific and, in practice, probably more useful guidelines, and illustrate how quantitative experiments can be designed to investigate questions of genetic management. There is an urgent need for well designed experiments in a wide range of taxa. Our preliminary results support one specific recommendation, namely, that reintroductions of butterfly species, or of similar organisms, which use 10 or more gravid females and which are followed by a rapid flush are likely to be more persistent than those based on fewer founders. Although rare alleles will be lost with a founder size of 10 pairs and this may be of practical importance for the persistence of natural populations, we were unable to detect any detrimental effect resulting from genetic impoverishment due to such a founder/bottleneck event: there was no measurable loss of morphological or physiological variation and no inbreeding depression (in a constant environment with a limited array of stresses).

Some suggestions can be made with respect to the types of experiments which will be useful in focussing attention on the unanswered questions such as those outlined in earlier sections. Experiments could include ones designed to investigate:

(1) the effect of bottleneck or founder events on realized heritabilities as measured by applying artificial selection;

(2) the interactions between adaptive potential and both genetic diversity and inbreeding history by employing arrays of realistic environmental challenges;

(3) the relationships between the (genetic) identity of the founders and the rates of inbreeding depression and recovery ("fitness rebound");

(4) the effects of periods of population "flush" following bottlenecks on genetic variances and the frequencies of deleterious recessives;

Numerous man-made introductions of butterflies have been documented in European countries (Oates and Warren, 1990; Warren, 1992; various papers in Pavlicek-van Beek et al., 1992). Some of these, including examples which involved very few founders, have become established in the short term, but it is not known what overall proportion was unsuccessful in establishment, or whether those established in the short term will persist in the longer term. Our laboratory results suggest that introductions involving at least 10 gravid females from an outcrossed captive stock or a large natural population will be more likely to be successful than when very few founders are used, but will be as likely to persist as those involving many more founders. However, although survival or fitness differences may not occur or be measurable in laboratory environments, the results may be very different in nature. Thus, field-conducted replicate experiments involving different numbers and sources of founders are needed to properly substantiate such laboratory-based guidelines.

Acknowledgements
We thank the many people who in one way or another have contributed to these experiments. We are especially grateful to: Richard Nichols, Rosie Heywood, Fanja Kesbeke, Vicky Silverton, Els Schlatmann, Bert de Winter, Mike Bruford, Paul Jepson, Bob Wayne, and Jack Windig.

References

Brakefield, P. M. (1991) Genetics and the conservation of invertebrates. *In:* Spellerberg, I. F., Goldsmith, F. B. and Morris, M. G. (eds), *The scientific management of temperate communities for conservation*. Blackwell Scientific Publications, Oxford, pp. 45–79.

Brakefield, P. M. and Reitsma, N. (1991) Phenotypic plasticity, seasonal climate and the population biology of *Bicyclus* butterflies (Satyridae) in Malawi. *Ecol. Entomology* 16: 291–303.

Brewer, B. A., Lacy, R. C., Foster, M. L. and Alaks, G. (1990) Inbreeding depression in insular and central populations of *Peromyscus* mice. *J. Heredity* 81: 257–266.

Bryant, E. H. and Meffert, L. M. (1993) The effect of serial founder-flush cycles on quantitative genetic variation in the housefly. *Heredity* 70: 122–129.

Bryant, E. H., Meffert, L. M. and McCommas, S. A. (1990) Fitness rebound in serially bottlenecked populations of the house fly. *Am. Nat.* 136: 542–549.

Falconer, D. S. (1989) *Introduction to quantitative genetics*. 3rd edn. Longman, London.

Holloway, G. J., Brakefield, P. M. and Kofman, S. (1992) The genetics of wing pattern elements in the polyphenic butterfly, *Bicyclus anynana*. *Heredity* 70: 179–186.

Goodnight, C. J. (1987) On the effect of founder events on epistatic genetic variance. *Evolution* 41: 80–91.

Goodnight, C. J. (1988) Epistasis and the effect of founder effects on the additive genetic variance. *Evolution* 42: 441–454.

Lande, R. and Barrowclough, G. F. (1987) Effective population size, genetic variation, and their use in population management. *In:* Soulé, M. E. (ed.), *Viable populations for conservation*. Cambridge University Press, Cambridge, pp. 87–123.

López-Fanjul, C. and Villaverde A. (1989) Inbreeding increases genetic variance for viability in *Drosophila melanogaster*. *Evolution* 43: 1800–1804.

Nunney, L. and Campbell, K. A. (1993) Assessing minimum viable population size: demography meets population genetics. *Trends Ecol. Evol.* 8: 234–239.

Oates, M. R., and Warren, M. S. (1990) *A review of butterfly introductions in Britain and Ireland*. Research and survey in conservation series. Nature Conservancy Council, Peterborough.

Pavlicek-van Beek, T., Ovaa, A. H. and van der Made, J. G. (1992) *Future of butterflies in Europe*. Agricultural University, Wageningen.

Saccheri, I. J. and Bruford, M. W. (1993) DNA fingerprinting in a butterfly, *Bicyclus anynana* (Satyridae). *J. Heredity* 84: 195–200.

Tax, M. H. (1989) *Atlas van de Nederlandse dagvlinders*. Vlinderstichting, Wageningen and Natuurmonumenten, 's-Graveland.

Thomas, J. A. (1983) The ecology and conservation of *Lysandra bellargus* (Lepidoptera: Lycaenidae) in Britain. *J. appl. Ecology* 20: 59–83.

Thomas, J. A. (1991) Rare species conservation: Case studies of European butterflies. *In:* Spellerberg, I. F., Goldsmith, F. B. and Morris, M. G. (eds), *The scientific management of temperate communities for conservation*. Blackwell Scientific Publications, Oxford, pp. 149–197.

Warren, M. S. (1992) The conservation of British butterflies. *In:* Dennis, R. L. H. (ed.), *The ecology of butterflies in Britain*. Oxford Science Publications, Oxford, pp. 246–274.

Windig, J. (1993) *The genetic background of plasticity in wing pattern of Bicyclus butterflies*. Ph.D. Thesis, University of Leiden.

Part IV

Molecular approaches to conservation

Conservation Genetics
ed. by V. Loeschcke, J. Tomiuk & S. K. Jain
© 1994 Birkhäuser Verlag Basel/Switzerland

Introductory remarks

Biodiversity is hierarchically organized. The biotic community of an ecosystem consists of various species that are characterized by numerous phylogenetic relationships. Each evolutionary lineage has its own species diversity. Intraspecies variability among individuals exists in phenotype and genetic constitution. Knowledge about the origin and the maintenance of any kind of biological variability is therefore necessary for the successful conservation of species.

The classical approach in conservation is to study the attributes of a specific population of a species and to determine the factors that influence the species' present or future adaptability to changing conditions. Besides ecological conditions, genetic variability is considered to be important for the survival of a population. New genetic variation obviously originates from mutation. However, the mechanism which maintains genetic variability in populations has raised a dispute among the so-called selectionists and neutralists. The neutralist's view is that most molecular genetic variation is maintained by a balance between mutation and random genetic drift. The continuous input of new mutations is diluted by the random loss of genes due to finite population size (Kimura, 1983). The neutral model predicts an equilibrium of genetic variability in a population with finite population size depending on the mutation rate. On the contrary, the selectionist assumes that natural selection maintains most genetic variability even at the molecular level (Ayala, 1984). An unequivocal proof for any of the two hypotheses, however, is still missing. A few cases indicate that genetic polymorphism is maintained by selection. For example, hemoglobin polymorphism is maintained in malaria epidemic regions in humans, MHC polymorphism is related to resistance against various diseases (see, e.g., Hill et al., 1991), and genetic variability is important for parasite resistance (Hamilton, 1994). These examples demonstrate the importance of the preservation of certain common and even rare alleles that may be necessary for the persistence of a species in the long term (Allendorf, 1986). Focusing only on the sampling of known rare alleles, however, may lead to a reduction of the total genetic variability (Hedrick et al., 1986) and thereby to the loss of potentially important allelic variation. The dilemma that we face is to find a procedure for identifying and maintaining evolutionary important alleles and simultaneously preserving relatively high levels of genetic variability.

We do not yet know the exact number of species on earth. Only rough estimates about the total number of species exist (May, 1988). But before we even can settle this question, biologists must advise how to preserve biodiversity and how to define priorities. Former arguments for specific conservation strategies were mostly anthropomorphic – the beauty and the beast. Vane-Wright et al. (1991) started to tackle this problem of the "agony of choice" by attempting to define objective criteria for weighing evolutionary distinctiveness. These criteria may be applied to identifying nature reserves that contain a high level of biodiversity, including a phylogenetically diverse composition of species. Their suggested procedure has since been modified in various directions.

When complexes of closely related and morphologically similar species exist in the same area, their classification may be difficult. Furthermore, hybridization and gene flow among closely related species living in different niches but the same area might cause problems in breeding and future reintroduction programs (Templeton, 1986). The hybridization of individuals with differently coadapted genes or chromosomal complexes can, for example, result in a decrease in fertility or viability. Released hybrid individuals may not be adapted to any of the local environments and may therefore negatively affect a reintroduction programme. Additionally, the conservation of species complexes consisting of closely related species, the knowledge about species diversity is fundamental to determining the size and attributes of a reserve. Molecular techniques are very powerful for this purpose and provide additional information that complements morphological studies.

Finally, the study of genetic variability within and among species shows clearly that we also have to consider the ecology of species which can be of most importance for the survival of species in natural habitats. Consequently, we have to consider the interactions among species within ecosystems. We can combine systematics and ecology to maintain phylogenetic biodiversity probes and to identify areas that are likely to respond positively to conservation efforts (Brooks et al., 1992). Keystone species should be identified as well as biotic complexes which are necessary for the conservation of endangered species; this implies the conservation of species critical for the survival of the target species.

References

Allendorf, F. W. (1986) Genetic drift and the loss of alleles. *Zoo Biol.* 5: 181–190.
Ayala, F. J. (1984) Molecular polymorphism: how much is there and why is there so much? *Develop. Genet.* 4: 379–391.
Brooks, D. R., Mayden, R. L. and McLennan, D. A. (1992) Phylogeny and biodiversity: conserving our evolutionary legacy. *Trends Ecol. Evol.* 7: 55–59.
Hamilton, W. D. (1994) Parasites and frequency-dependent selection. *J. Heredity*, (in press).
Hedrick, P. W., Brussard, P. F., Allendorf, F. W., Beardmore, J. A. and Orzack, S. (1986) Protein variation, fitness and captive populations. *Zoo Biol.* 5: 91–99.

Hill, A. V. S., Allsop, C. E. M., Kwiatkowski, D., Antsley, N. M., Twumasi, P., Rowe, P. A., Bennet, S., Brewster, D., McMichael, A. J. and Greenwood, B. M. (1991) Common west African HLA antigens are associated with protection from severe malaria. *Nature* 352: 595–600.

Kimura, M. (1983) *The neutral theory of molecular evolution*. Cambridge University Press, Cambridge.

May, R. M. (1988) How many species are there on earth? *Science* 241: 1441–1449.

Templeton, A. R. (1986) Coadaptation and outbreeding depression. *In:* Soulé, M. E. (ed.), *Conservation biology: the science of scarcity and diversity*. Sinauer, Sunderland, MA, pp. 105–121.

Vane-Wright, R. I., Humphries, C. J. and Williams, P. H. (1991) What to protect? – Systematics and the agony of choice. *Biol. Cons.* 5: 235–254.

Conservation Genetics
ed. by V. Loeschcke, J. Tomiuk & S. K. Jain
© 1994 Birkhäuser Verlag Basel/Switzerland

Rare alleles, MHC and captive breeding

P. W. Hedrick and P. S. Miller

Department of Zoology, Arizona State University, Tempe, AZ 85287, USA

Summary. In recent years, more detailed genetic information has become available for individuals of endangered species in captive breeding programs. There have been suggestions that this information be used to identify rare alleles, particularly those at the MHC, that can be subsequently selected for captive breeding programs. First, we summarize the current information on the MHC relevant to conservation genetics, so that such a possible breeding program is seen in a proper perspective. For example, very few specific alleles at the MHC have been identified as selectively advantageous, even though there has been substantial effort to find such alleles in humans and a few other organisms. Further, many of the balancing selection models suggested for MHC variation are based on heterozygotes in general having a higher fitness than homozygotes and not on specific selectively advantageous alleles.

Because there is no detailed data on MHC variability in captive populations, we used transferrin data in Przewalski's horses to evaluate a breeding program to select for rare alleles. In this species, one individual, 1060, has been identified to have the transferrin allele J. We determine the effect on founder contribution of multiply mating 1060 to increase the number of copies of this allele. Since there were 485 individuals in the population at this time, this extra mating had little detrimental effect on the distribution of founder contributions and the number of founder equivalents. We then selected 65, an ancestor of 1060, which had a high likelihood of being the individual that passed on the J allele in the lineage of 1060. We examined the effect of increasing the number of copies of alleles of 65 at a time when the population had only 22 other individuals. In this case, even though the founder contributions were changed more, there was also little effect on the founder contributions and the number of founder equivalents. Overall, it appears that selection that results in a limited change in the number of copies of rare alleles may not always have an overall detrimental effect. However, because other pedigrees may have very different properties, it is essential to perform a detailed pedigree analysis of any such selective breeding program to determine its effect before such a selection program is implemented.

Introduction

Captive breeding of endangered species has several related genetics goals, i.e., to avoid inbreeding depression, avoid adaptation to captivity, and retain genetic variation for future adaptation (e.g., Ralls and Ballou, 1986; Hedrick and Miller, 1992). All of these goals are thought to apply to the genome in general and are not related to specific genes. On the other hand, it has been suggested that genetic variation at some genes be treated differently because alleles at these genes may indicate unique founder contributions or may have important selective effects (e.g., Allendorf, 1986; Fuerst and Maruyama, 1986; Hughes, 1991).

In the proceedings of the Workshop on Genetic Management of Captive Populations (Ralls and Ballou, 1986), Hedrick et al. (1986)

warned that breeding schemes designed to maintain rare alleles would be operationally impossible as well as counterproductive for the maintenance of genetic variation in general. In particular, selecting for specific alleles may result in a faster loss of variation at most of the other loci in the genome. In addition, they pointed out that it would, in principle, be difficult to determine what rare alleles are advantageous, neutral, or detrimental. However, Allendorf (1986) suggested in the same volume in his discussion of rare alleles that "the number of alleles remaining is important for the long-term response to selection and survival of populations and species." Hughes (1991), in a more specific proposal, states that "the most clear-cut examples of loci under overdominant selection are the MHC loci" and, as a result, he supports a breeding program to save any MHC "allele declining in frequency by drift...by selective breeding." To further understand the problems related to rare alleles, we will first discuss the MHC system considered to include many alleles of adaptive importance. We will then use the pedigree of Przewalski's horse to examine the impact on genetic diversity of a hypothetical breeding scheme designed to increase the number of copies of a rare allele.

Major histocompatibility complex

The major histocompatibility complex (MHC) is one genetic region that has been suggested to have alleles of particular selective significance and, in fact, even avowed neutralists have suggested that the pattern of genetic variation of alleles at MHC loci are not consistent with neutrality (Nei and Hughes, 1991). As a result, Hughes (1991) has suggested "a radical reorientation of all captive breeding programs toward the goal of maintaining diversity at MHC loci and preventing loss of MHC alleles." This proposal lead to several responses critical of Hughes' proposal (Gilpin and Wills, 1991; Miller and Hedrick, 1991; Vrijenhoek and Leberg, 1991).

General background

The MHC in humans (called the HLA) has long been studied because of the medical importance of this region in understanding tissue transplantation, autoimmune diseases, and the immune response against viruses, bacteria, and other parasites. As a result, while the MHC is present in all vertebrates, much of the detailed understanding of this group of genes has come from research on the HLA or the mouse MHC (called the H-2). The HLA region consists of about 4000 kilobases of DNA, covering about 2.5 map units and containing at least 82 genes

(Trowsdale et al., 1991), located on the short arm of chromosome 6. The region includes the class I genes (the products of these genes are important in attacking viral-infected cells), the class II genes (the products of these genes are involved in initiating the attack against bacteria and other extracellular parasites), and the class III or complement region. There are also a number of other genes in the MHC that have related functions, making the whole region rich in genes that have a part in the immune response. In fact, one could make the case that the MHC is a supergene (e.g., Darlington and Mather, 1949; Hedrick et al., 1978; Trowsdale, 1993). The MHC in a number of organisms that have been examined is similar to the HLA in many general respects but in other organisms the MHC is quite different, with fewer genes and less polymorphism (Klein and Klein, 1991; Trowsdale, 1993).

The three-dimensional structures of both the HLA class I (Bjorkman et al., 1987) and class II (Brown et al., 1993) antigens are known and this information gives extraordinary insight into the function of the molecule coded for by these genes. The main function of MHC molecules is to recognize short peptides and to present them to T-cells to initiate the immune response (e.g., Grey et al., 1989). Most important for our consideration, these peptides may be from various pathogens such as viruses, bacteria, or parasites, and their presentation may lead to an immune response that results in the destruction of the pathogen. Two subunits or domains of the MHC class I and II molecules form a pocket that contains the peptide being presented to the T-cell. By knowing the position of individual amino acids in the structure, specific amino acids can be categorized as potentially interacting with the peptide, the T-cell receptor, both the peptide and the T-cell receptor, or having none of these functions.

Extent and character of genetic variation

A distinctive feature of the MHC is the high level of polymorphism exhibited by many of the class I and class II loci (see reviews by Hedrick et al., 1991; Nei and Hughes, 1991). These variants were traditionally identified serotypically, but now all new variants are known by their DNA sequence with, as of 1991, for example, for the class I loci HLA-A and HLA-B, 41 and 61 designated alleles, respectively (Bodmer et al., 1992). Further, the distribution of alleles is significantly more uniform than expected from neutrality theory for many populations and many HLA loci (e.g., Hedrick and Thomson, 1983; Klitz et al., 1986). After examining the other potential evolutionary explanations, Hedrick and Thomson (1983) concluded that some form of balancing selection was the explanation most consistent with the pattern of polymorphism observed.

Table 1. Average heterozygosity *per* amino acid position of the HLA-A molecule for sites in four different functional categories for the average of the Caucasian, Asian, and African samples (Hedrick et al., 1991) and for the Havasupai (Hedrick and Markow, 1994). Both the Hardy–Weinberg heterozygosity, HW(*H*), and the observed heterozygosity, O(*H*), are given for the Havasupai because this sample is not in Hardy–Weinberg proportions (Markow et al., 1993). The number in parentheses is the number of sites in each category

Function	Average	Havasupai	
		HW(*H*)	O(*H*)
Peptide (29)	0.264	0.181	0.191
TcR (18)	0.171	0.135	0.142
Peptide-TcR (7)	0.217	0.130	0.141
Other (312)	0.036	0.028	0.030

Amino acid and DNA sequence data are known for a number of HLA serotypic alleles (e.g., Parham et al., 1988). The amino acid variation is concentrated in the residues that interact with the peptides and the T-cell receptor (TcR) (Tab. 1) For example, the average *per* amino acid site heterozygosity in the 29 sites potentially interacting with peptides is 0.264 for HLA-A. For most populations there are between 12 and 20 identified serotypic alleles at HLA-A, but all Native American populations have many fewer alleles. The Havasupai, who live in an isolated side canyon of the Grand Canyon and number less than 700, have only four HLA-A alleles (Markow et al., 1993). However, if we calculate the heterozygosity for the different functional categories, the values are only slightly lower than the world average (Tab. 1). In other words, in this population in which many alleles appear to have been lost, there is still high variation at the putative functionally important sites.

Modes of selection

The most easily identified source of selection on the MHC is that associated with autoimmune diseases in humans, such as ankylosing spondylitis, juvenile diabetes, rheumatoid arthritis, etc. (e.g., Tiwari and Terasaki, 1985; Thomson, 1988). From these studies, it is apparent that a number of HLA alleles have detrimental effects and, presumably, there is a group of MHC alleles in all organisms that also have a negative influence on fitness. Obviously, an effort to increase the frequency of these disease-causing alleles could well result in a lowered, rather than an increased, fitness of a population. However, even in humans it is not known whether the net effect on fitness of any particular allele and its associated genetic background is positive or negative.

Histocompatibility alleles have been shown to differ in their ability to create an immune response to a variety of infectious agents (van Eden et al., 1983), suggesting that the epidemic diseases of the human past may have played a central role in determining the HLA haplotype and allelic frequencies observed in human populations today. In fact, Black (1992) recently suggested that a major reason for the deaths of an estimated 56 million Native Americans, perhaps as much as 90% of the population, to infectious diseases after the Americas were settled by Europeans, was because of a lack of genetic resistance to these introduced pathogens due to low levels of HLA variation.

The result of such immune differences may, in the simplest model, be represented by a symmetrical constant fitness model in which all heterozygous haplotypes have a higher fitness than all homozygous haplotypes (e.g., Black and Salzano, 1981; Nei and Hughes, 1991). In this model, particular alleles may not have special importance but the overall level of heterozygosity is the significant factor. A frequency-dependent selection model would also maintain polymorphism, but in this case rare alleles are thought to have high fitness. Bodmer (1972) suggested that a new, rare allele would allow greater protection against pathogens than more common alleles to which pathogens may have evolved resistance.

In spite of the general acceptance that MHC variation plays an important role in resistance to disease, there have been surprisingly few direct demonstrations that particular MHC alleles confer resistance to specific diseases. One of the best documented associations is that found between resistance to Marek's disease, a viral leukemia fatal in chickens, and the haplotype B^{21} (Briles et al., 1977; Gavora et al., 1986). Recently, an association between resistance to malaria in humans and both the HLA antigen Bw53 and the HLA haplotype DRB1*1302, has been documented in west Africa (Hill et al., 1991). Whether the heterozygote advantage model, the frequency-dependent model, or a specific pathogen resistance model (e.g., Hedrick et al., 1987) is most consistent with these instances of MHC resistance is not clear.

Although resistance to various pathogens is thought to be the basic mode of selection at MHC alleles, it appears that there is also selection at MHC related to other factors, perhaps of a secondary nature. In these instances, the extent of selection also appears strong, making it of potential significance in conservation genetics considerations. For example, in humans, approximately 30% of the couples having two or more spontaneous abortions do not have a demonstrable basis, such as chromosomal or anatomical abnormality, for this fetal loss (Thomas et al., 1985). A number of studies indicate that such couples share antigens for HLA loci to a much greater degree than do control couples, suggesting that there is some kind of maternal-fetal selection favoring couples that differ in their HLA genotypes. The immunological hypothesis to explain this phenomenon suggests that the presence of an immune response occurring when the mother and fetus differ at the HLA loci is necessary for proper implantation and fetal growth (e.g., Gill, 1983; Hedrick and Thomson, 1988).

Second, although there is no evidence for nonrandom mating with respect to HLA types, experimental studies in mice have demonstrated mating preferences with regard to H-2 types (e.g., Yamazaki et al., 1976, 1978). In these studies, males generally mate preferentially with females of a different MHC type. Yamazaki et al. (1988) reported experiments that suggest that this preference is the result of familial imprinting. Two recent reports from other laboratories also indicate that female mice may prefer to mate with males different from them at the H-2 loci (Egid and Brown, 1989; Potts et al., 1991) and a theoretical model (Hedrick, 1992) suggests that this selection is consistent with the deficiency of homozygotes observed by Potts et al..

It appears, from this overall perspective, that the alleles in the MHC would be obvious candidates for a breeding program that seeks to increase the frequency of rare alleles. There is probably stronger evidence for the presence of alleles that have a selective advantage at the MHC than at virtually any other system in vertebrates. We will now consider a hypothetical breeding system that could be used to increase the frequency of MHC alleles and evaluate its effects on the genetic variation in the rest of the genome.

Captive breeding for rare alleles

Methods

Let us assume that a rare allele is identified in a particular individual in a pedigreed population and, as a consequence, a breeding program is subsequently designed to increase the number of copies of that rare allele in the population (Haig et al., 1989, compared a number of alternative breeding schemes for pedigreed populations, but did consider one having this specific goal). In practice, this allele could be at the MHC or at another gene at which it is thought that rare alleles may be of selective importance. The procedure we will use to evaluate such a breeding program is the following. First, we will determine the genetic characteristics of the population at the point immediately after the individual with the rare allele is born, including the expected contribution of the various founders and the number of founder equivalents which is defined as

$$f_e = \frac{1}{\sum p_i^2},$$
(1)

where p_i is the proportion of genes in the current population descended from founder i (Lacy, 1989). If all founders contribute equally to the population, then f_e is equal to the number of founders.

We will then augment the population by breeding the individual with the rare allele two, four, or eight times to determine the effect of maintaining the allele with an expected single copy or increasing the expected number of copies of the allele to two or four (the individual is assumed to be heterozygous for the rare allele). The mates to the individual with the rare allele are assumed to be the next two, four, or eight individuals of the opposite sex to be parents in the actual pedigree. At this point, the population will again be evaluated for the same genetic characteristics to determine the extent of genetic change resulting from this breeding scheme. These populations will be compared to each other and to the actual pedigree after the next two, four, or eight births.

Example from Przewalski's horse

Because we do not, at this point, have detailed MHC variation on a captive pedigreed population, we will use, as an example, data on transferrins in Przewalski's horse (*Equus przewalskii*). The transferrin locus has five variants in Przewalski's horses, one of which, the J variant, has been found in only two individuals (number 1060 and her offspring, 1576) (Bowling, unpublished). At the time of the birth of 1060, May 30, 1982, there were 485 living horses in the population. The complete pedigree of Przewalski's horse is known and all descend from 13 founders, 10 of which were captured around the turn of the century to form the initial breeding group (for a recent review, see Boyd and Houpt, 1994). One of the founders was a domestic horse mare, called DOM. Although the two species have different chromosome numbers, they can successfully hybridize.

Fig. 1 gives the pedigree for 1060 using the marriage node format (e.g., Geyer and Thompson, 1988), showing that seven of the 13 founders are her ancestors. For an average locus, she is expected to have 1.548 unique alleles out of maximum of 2. Founder 11 contributed the highest percentage of unique alleles to 1060, 23.9%, while 39 and 40 contributed the next largest proportions, 20.2% and 15.4%, respectively (the pedigree analysis software package, Pedpack, was used for these calculations; Thomas, 1987; Geyer and Thompson, 1988; Geyer et al., 1989).

The proportionate contribution (p_i values) of the 13 founders to the total living population at the time of the birth of 1060 is given at the left of Tab. 2 (for this and all subsequent calculations, software employing the additive relationship matrix approach of Ballou (1983) was used).

194

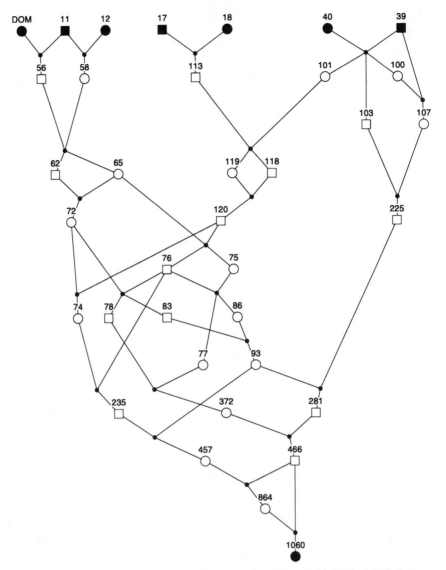

Fig. 1. The pedigree of Przewalski's horses showing the ancestors of individual 1060 (bottom of pedigree) which was found to have a unique transferrin allele. The seven founders that contributed to 1060 are indicated by solid symbols. Squares and circles indicate males and females, respectively; each mating is an unmarked node. DOM is a domestic mare that has contributed to the captive population.

By far the largest contribution is from 212, which contributes more than 20% of the genes while founder 11 contributes the second largest proportion, nearly 12% of the total. The next four columns give the percentage contribution when the next two, four, or eight individuals born in the actual population were included. Because there were 485

Table 2. Proportionate contributions of the 13 Przewalski's horse founders to the world pedigree as of May 30, 1982 (birth of 1060), the contributions following the addition of the next two, four, and eight individuals to the original pedigree, the contributions following the addition of two, four, and eight matings using 1060 as a parent, and the number of founder equivalents (f_e) in each pedigree

Founder	Original pedigree	No. individuals added after birth of 1060			No matings added with 1060		
		2	4	8	2	4	8
1	0.0398	0.0398	0.0397	0.0400	0.0398	0.0396	0.0397
5	0.0199	0.0199	0.0198	0.0200	0.0199	0.0198	0.0198
52	0.0199	0.0199	0.0198	0.0200	0.0199	0.0198	0.0198
11	0.1195	0.1197	0.1197	0.1190	0.1197	0.1200	0.1201
12	0.0597	0.0598	0.0598	0.0595	0.0599	0.0600	0.0600
DOM	0.0597	0.0598	0.0598	0.0595	0.0599	0.0600	0.0600
17	0.0593	0.0593	0.0593	0.0591	0.0593	0.0593	0.0593
18	0.0593	0.0593	0.0593	0.0591	0.0593	0.0593	0.0593
39	0.1072	0.1072	0.1074	0.1075	0.1074	0.1077	0.1080
40	0.0960	0.0959	0.0960	0.0962	0.0960	0.1111	0.0964
211	0.0678	0.0678	0.0678	0.0682	0.0677	0.0676	0.0676
212	0.2035	0.2034	0.2035	0.2046	0.2031	0.2027	0.2027
231	0.0884	0.0882	0.0881	0.0874	0.0882	0.0878	0.0871
f_e	9.493	9.495	9.490	9.472	9.499	9.502	9.504

individuals in the population at this point in time, these additions have little effect on the percentage contribution and the number of founder equivalents (at bottom of Tab. 2).

The last four columns give the values when 1060 is bred to the male parents of the next two, four, or eight individuals. In other words, the contribution from the male side is the same as the pedigreed population, but the contribution from the female side is only from 1060. Again, there is little effect of adding individuals to the population, even though in this case all of the new individuals have the same mother. As expected, the contribution from 11 increases, but only slightly from 11.947% to 12.006%. Again, the founder equivalent number shows little change as the selected individuals are added. Fig. 2 gives these effects as the proportionate change from the initial values for either the next eight individuals or the eight selected matings. Notice that the vertical scale encompasses a quite small range and the largest shift in proportions is for 231 in the selected matings of less than 0.15%.

To determine if this type of breeding program would have a larger effect if the population were smaller, we picked an individual to selectively breed that existed earlier in the population and that has a high likelihood of actually carrying the J transferrin allele. Looking again at Fig. 1, 65, born May 20, 1920, is an obvious candidate because the J allele is most likely to have come from 11 and 65 has 11 as an ancestor (grandfather). In addition, 65 also passes her genes on to the lineages

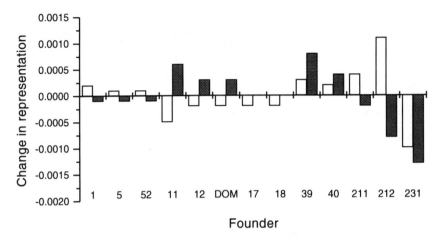

Fig. 2. The proportionate change in founder representation after the next eight individuals after 1060 are included (open bars) and after the progeny of the next eight selected matings are included (shaded bars).

Table 3. Proportionate contributions of the 10 Przewalski's horse founders to the world pedigree as of May 20, 1920 (birth of 65), the contributions following the addition of the next two, four, and eight individuals to the original pedigree, the contributions following the addition of two, four, and eight matings using 65 as a parent, and the number of founder equivalents (f_e) in each pedigree

Founder	Original pedigree	No. individuals added after birth of 65			No. matings added with 65		
		2	4	8	2	4	8
1	0.1739	0.1600	0.1667	0.1774	0.1600	0.1574	0.1532
5	0.1413	0.1300	0.1389	0.1452	0.1300	0.1296	0.1290
52	0.0326	0.0300	0.0278	0.0323	0.0300	0.0278	0.0242
11	0.1522	0.1600	0.1667	0.1452	0.1700	0.1852	0.1936
12	0.0707	0.0700	0.0741	0.0645	0.0750	0.0787	0.0847
DOM	0.0815	0.0900	0.0926	0.0807	0.0950	0.1065	0.1089
17	0.0326	0.0400	0.0370	0.0403	0.0400	0.0370	0.0403
18	0.0326	0.0400	0.0370	0.0403	0.0400	0.0370	0.0403
39	0.1522	0.1500	0.1389	0.1492	0.1400	0.1296	0.1210
40	0.1304	0.1300	0.1204	0.1250	0.1200	0.1111	0.1048
f_e	7.791	8.026	7.924	7.907	8.055	7.932	7.932

leading to 1060 through 72, 75, and 76. At the time of the birth of 65, there were only 22 other living individuals that were not founders in the population. In other words, any selective breeding of 65 should potentially have more impact in this group of 23 than selective breeding of 1060 in a population of 485.

The percentage contribution from the founders after 65 was born is given in the first column of Tab. 3. Notice that only 10 founders had contributed to the population at this time. Five of the founders, 1, 5, 11,

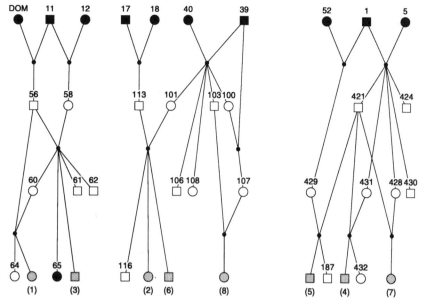

Fig. 3. The pedigree of the population after the birth of 65 and following the addition of the next eight individuals born in the population (lightly shaded symbols with numbers in parentheses below indicating the order of their birth). The symbols are the same as in Fig. 1.

39, and 40, contributed slightly over 75% of the ancestry to the population. Fig. 3 gives the pedigree which includes the next eight individuals born after 65 (these individuals are indicated with shaded circles at the bottom of the pedigree with their birth order in parentheses). The next three columns of Tab. 3 give the data for the population as these individuals are added. These additions generally increase the proportionate contribution of 11, 12, 39, and DOM, while that of 1, 5, 17, and 18 are generally decreased. The number of founder equivalents remains much the same as these individuals are added.

Fig. 4 gives the pedigree when the next eight individuals added are the result of selective breeding with 65. The last three columns of Tab. 3 give the percentage contribution as these individuals are added to the population. In this case, the proportionate contribution is generally increased for 11, 12, 17, 18, and DOM, while it is generally decreased for 1, 5, 39, 40, and 52. Again, the level of founder equivalents is only slightly changed and actually increases from 7.791 when 65 was born to 7.932 when the eight progeny of 65 were added to the pedigree. Figure 5 gives the actual percentage change in contributions from the founders in the population, comparing the times before and after eight individuals are added. In this case, the percentage change is much larger than when individuals were added after 1060. The largest change is the increase for 11, 4.14%, when there were eight selected matings.

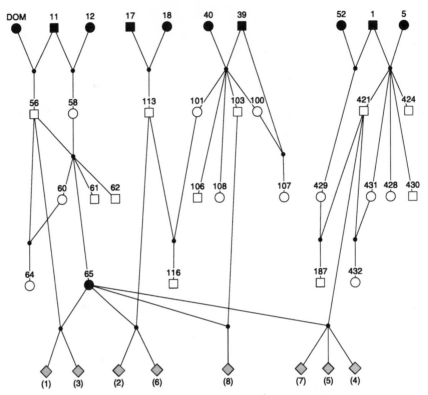

Fig. 4. The pedigree of the population after the birth of 65 and following the addition of the eight progeny of the selected matings (lightly shaded symbols with numbers in parentheses below indicating the order of their birth). The symbols are the same as in Fig. 1.

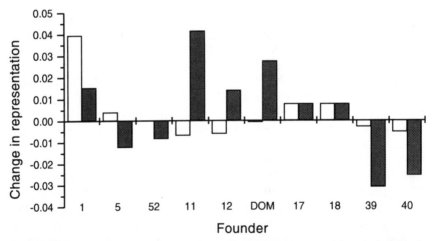

Fig. 5. The proportionate change in founder representation after the next eight individuals after 65 are included (open bars) and after the progeny of the next eight selected mating are included (shaded bars).

Discussion

As we know more about the genetics of captive endangered organisms, we will obviously find unusual and rare alleles at a number of loci. What should be done with these alleles, if anything? Certainly these alleles and variable genetic markers in general can be used to identify paternity (and even maternity) (see references in Hedrick and Miller, 1992), identify populations (e.g., Schreiber and Tichy, 1992), and perhaps to suggest relationships between founders of unknown ancestry (Templeton et al., 1987; Geyer et al., 1993). But should attempts be made to specifically maintain them through programs of selective breeding? From our discussion of MHC variants and our examination of the expected results from such a breeding program in Przewalski's horses, we have some general findings and feel that we can make some preliminary recommendations.

Increasing MHC alleles

First, it is not clear that all MHC alleles are advantageous. For example, some of these rare alleles may actual be rare because there has been selection against them as appears for some HLA variants that cause autoimmune diseases. Efforts to find a specific selective advantage for a HLA allele have not been particularly successful (for a discussion of this problem, see Hill, 1991) and the two best documented selectively favored alleles were among the most frequent alleles at these loci in these populations (Hill et al., 1991).

Second, many of the models that are suggested to explain or influence the variation at MHC loci (e.g., heterozygote advantage, maternal-fetal interaction, and female mate selection) do not necessarily discriminate among alleles and really are models that select for heterozygosity *per se*. The general breeding protocols that are used to avoid inbreeding and to equalize founder representation (e.g., Haig et al., 1989) should result in maintenance of heterozygosity at MHC loci just as they are designed to do at all loci.

Third, it is not clear that genetic variation at MHC is the only strategy that is employed by vertebrates to combat pathogens. For example, there are a number of other seemingly unrelated loci that are thought to be important in combating pathogens (e.g., Wakelin and Blackwell, 1988). In addition, a number of vertebrates have low numbers of MHC genes or have little variability at their MHC genes. There are suggestions that some of these species, such as the cheetah (Evermann et al., 1988; O'Brien and Evermann, 1988) and cottontop tamarin (Watkins et al., 1988) or human populations such as Native Americans (Black, 1992) are quite susceptible to various pathogens. Although this cause-effect situation may be true, more research would be useful to confirm such specific connections. There are, however, some populations having low MHC variation that appear to be either increasing in number (Ellegren et al., 1993) or do not appear to have problems with pathogen susceptibility (Trowsdale et al., 1989; Slade, 1992).

Finally, when MHC variation for a given locus is known for a given species there will probably be several rare alleles, not just one. As a result, the appropriate selection program would be one that increases the frequency of several rare alleles simultaneously. Although we have not examined such a program, it seems likely that the impact on overall genetic variation would also depend upon the structure of the pedigree as did the program we evaluated here. However, when enough loci are examined so that we would expect that nearly every individual has a rare allele, then such a selection scheme may not even be possible.

Selective breeding of individuals with rare alleles

The example that we used above to augment the number of copies of a rare allele in the pedigree of Przewalski's horse showed that even when the expected number of copies was increased to four (by breeding the heterozygous individual with the rare allele eight times), there was little impact on the genetic constitution of the population. This was surprising to us, but, upon reflection, is explainable. When we increased the copy number of the rare allele from individual 1060, the number in the population was large (485 living individuals). As a result, an addition of eight individuals is unlikely to have a substantial effect. In any case, the mates displayed a variety of founder contributions so the progeny would reflect this diversity even though the contribution from the other parent was limited to one individual.

When we augmented the copy number of the allele in individual 65 (there were only 23 individuals, excluding founders, in the population at this time), we were even more surprised that this breeding protocol did not distort founder representation and reduce the number of founder equivalents. At this point in the pedigree, there were only 10 founders contributing so if all contributed equally, then the expected contribution from each is 10%. Individual 65 had contributions from only three founders, 11, 12, and DOM (all her grandparents). Two of the ancestors of 65 had contributions to the population of less than 10%, 12 with 7.065%, and DOM with 8.152%. Therefore, increasing the number of progeny from 65 actually increased the contribution from these individuals, making the breeding program consistent in some respects with one that equalizes overall founder contribution. Further, two other individuals, 39 and 40, which were initially overrepresented with contributions of 15.22% and 13.04%, respectively, had their contributions reduced to 12.10% and 10.48%, closer to 10%, by the selective breeding program.

Notice, however, that the contribution of 11, the founder thought to have had the J allele, was initially an overrepresented founder at 15.2%, and this representation was increased even more, to 19.4%, by the selective breeding program. In other words, the effect of the program depended on the structure of this particular pedigree in which the founder thought to have the rare allele was in a lineage with two underrepresented founders. How often a program to increase the number of copies of a rare allele would share some objectives with a program to equalize founder contributions because of a pedigree structure such as this is not clear.

Perhaps an association with rare alleles and low founder contribution may occur primarily in pedigrees with a small number of founders or individuals. If we assume that a pedigree has structure due to these factors, then the frequency of a rare allele may be related to the frequency of other alleles in that individual, i.e., the founder contributions in that individual. However, since we do not know how general this effect may be, it seems imperative that a thorough pedigree analysis should be carried out to determine the influence of a particular selection program before it is implemented. We have initiated further investigation of the effects of breeding schemes to increase rare alleles (Miller and Hedrick, 1994) and find that the same general conclusions are true for other individuals that were likely carriers of the transferrin J allele in the Przewalski's horse population.

Acknowledgements
This research was supported by NSF BSR-8923007. We appreciate Ann Bowling sharing her unpublished data on the transferrin allele.

References

Allendorf, F. W. (1986) Genetic drift and the loss of alleles versus heterozygosity. *Zoo Biol.* 5: 181–190.

Ballou, J. D. (1983) Calculating inbreeding coefficients from pedigrees. *In*: Schoenwald-Cox, C. M., Chambers, S. M., MacBryde, F. and Thomas, L. (eds), *Genetics and conservation: A reference for managing wild animal and plant populations.* Benjamin/Cummins, Menlo Park, CA, pp. 509–520.

Bjorkman, P. J., Saper, M. A., Samraoui, B., Bennett, W. S., Strominger, J. L. and Wiley, D. C. (1987) The foreign antigen binding site and T cell recognition regions of class I histocompatibility antigens. *Nature* 329: 512–518.

Black, F. L. (1992) Why did they die? *Science* 258: 1739–1741.

Black, F. L. and Salzano, F. M. (1981) Evidence for heterosis in the HLA system. *Amer. J. Hum. Genet.* 33: 894–899.

Bodmer, W. (1972) Evolutionary significance of the HL-A system. *Nature* 237: 139–145.

Bodmer, J. G., Marsh, S. G. E., Albert, E. D., Bodmer, W. F., Dupont, B., Erlich, E. R., Mach, B., Mayr, W. R., Parham, P., Sasauzuki, T., Schreuder, G. M., Strominger, J. L., Svejgaard, A. and Terasaki, P. I. (1992) Nomenclature for factors of the HLA system, 1991. *Tissue Antigens* 39: 161–173.

202

Boyd, L. and Houpt, K. (1994) *Przewalski's horses: The History and biology of an endangered species*. SUNY Press, Albany, NY.

Briles, W. E., Stone, H. A. and Cole, R. K. (1977) Marek's disease: Effects of B histocompatibility alloalleles in resistant and susceptible chicken lines. *Science* 195: 193–195.

Brown, J. H., Jardetsky, T. S., Gorga, J. C., Stern, L. J., Urban, R. G., Strominger, J. L. and Wiley, D. C. (1993) Three-dimensional structure of the human class II histocompatibility antigen HLA-DR1. *Nature* 364: 33–39.

Darlington, C. D. and Mather, K. (1949) *The Elements of Genetics*. Allen and Unwin, London.

Egid, K. and Brown, J. L. (1989) The major histocompatibility complex and female mating preferences in mice. *Anim. Behav.* 38: 548–550.

Ellegren, H., Hartman, G., Johansson, M. and Andersson, L. (1993) A rapidly expanding population of beavers is virtually monomorphic at MHC and DNA fingerprinting loci. *Proc. Nat. Acad. Sci. USA* 90: 8150–8153.

Evermann, J. F., Heeney, J. L., Roelke, M. E., McKeirnan, A. J. and O'Brien, S. J. (1988) Biological and pathological consequences of feline infectious peritonitis virus infection in the cheetah. *Arch. Virol.* 102: 155–171.

Fuerst, P. A. and Maruyama, T. (1986) Considerations on the conservation of alleles and of genic heterozygosity in small managed populations. *Zoo Biol.* 5: 171–179.

Gavora, J. S., Simenson, M., Spencer, J. L., Fairfull, R. W. and Gowe, R. S. (1986) Changes in the frequency of major histocompatibility haplotypes in chickens under selection for both high egg production and resistance to Marek's disease. *J. Anim. Breed. Genet.* 103: 218–226.

Geyer, C. J., Ryder, O. A., Chemick, L. G. and Thompson, E. A. (1993) Analysis of relatedness in the California condors from DNA fingerprints. *Mol. Biol. Evol.* 10: 571–589.

Geyer, C. J. and Thompson, E. A. (1988) Gene survival in the Asian wild horse (*Equus przewalskii*): I. Dependence of gene survival in the Calgary breeding group. *Zoo Biol.* 7: 313–327.

Geyer, C. J., Thompson, E. A. and Ryder, O. A. (1989) Gene survival in the Asian wild horse (*Equus przewalskii*): II. Gene survival in the whole population, in subgroups, and through history. *Zoo Biol.* 8: 313–329.

Gilpin, M. and Wills, C. (1991) MHC and captive breeding: A rebuttal. *Cons. Biol.* 5: 554–555.

Grey, H. M., Sette, A. and Buus, S. (1989) How T cells see antigen. *Sci. Amer.* (November), 56–64.

Haig, S. M., Ballou, J. D. and Derrickson, S. R. (1989) Management options for preserving genetic diversity: Reintroduction of Guam rails to the wild. *Cons. Biol.* 4: 290–300.

Hedrick, P. W. (1992) Female choice and variation in the major histocompatibility complex. *Genetics* 132: 575–581.

Hedrick, P. W., Brussard, P. F., Allendorf, F. W., Beardmore, J. A. and Orzack, S. (1986) Protein variation, fitness and captive propagation. *Zoo Biol.* 5: 91–99.

Hedrick, P. W., Jain, S. K. and Holden, L. (1978) Multilocus systems in evolution. *Evol. Biol.* 11: 101–184.

Hedrick, P. W., Klitz, W., Robinson, W. P., Kuhner, M. K. and Thomson, G. (1991) Evolutionary histories of HLA. *In:* R. K. Clark, A. G. and Whittam, T. S. (eds), *Evolution at the molecular level*. Sinauer, Sunderland, MA, pp. 248–271.

Hedrick, P. W. and Markow, T. (1994) High amino acid heterozygosity for HLA genes in an isolated group, the Havasupai Manuscript.

Hedrick, P. W. and Miller, P. S. (1992) Conservation genetics: Techniques and fundamentals. *Ecol. Appl.* 2: 30–46.

Hedrick, P. W. and Thomson, G. (1983) Evidence for balancing selection at HLA. *Genetics* 104: 449–456.

Hedrick, P. W., Thomson, G. and Klitz, W. (1987) Evolutionary genetics and HLA: Another classic example. *Biol. J. Lin. Soc.* 31: 311–331.

Hedrick, P. W., Whittam, T. S. and Parham, P. (1991) Heterozygosity at individual amino acid sites: Extremely high levels for *HLA-A* and *-B* genes. *Proc. Nat. Acad. Sci. USA* 88: 5897–5901.

Hill, A. V. S. (1991) HLA associations with malaria in Africa: some implications for MHC evolution. *In:* Klein, J. and Klein, D. (eds), *Molecular evolution of the major histocompatibility complex*. Springer-Verlag, New York, pp. 403–420.

Hill, A. V. S., Allsop, C. E. M., Kwiatkowski, D., Antsley, N. M., Twumasi, P., Rowe, P. A., Bennet, S., Brewster, D., McMichel, A. J. and Greenwood, B. M. (1991) Common west African HLA antigens are associated with protection from severe malaria. *Nature* 352: 595–600.

Hughes, A. (1991) MHC polymorphism and the design of captive breeding programs. *Cons. Biol.* 5: 249–251.

Klein, J. and Klein, D. (1991) *Molecular evolution of the major histocompatibility complex.* Springer-Verlag, New York.

Klitz, W., Thomson, G. and Baur, M. P. (1986) Contrasting evolutionary histories among tightly linked HLA loci. *Amer. J. Hum. Genet.* 39: 340–349.

Lacy, R. C. (1989) Analysis of founder representation in pedigrees: founder equivalents and founder genome equivalents. *Zoo Biol.* 8: 111–123.

Markow, T., Hedrick, P. W., Zuerlein, K., Danilovs, J., Martin, J., Vyvial, T. and Armstrong, C. (1993) HLA polymorphism in the Havasupai: Evidence for balancing selection. *Amer. J. Hum. Genet.* 53: 943–952.

Miller, P. S. and Hedrick, P. W. (1991) MHC polymorphism and the design of captive breeding programs: Simple solutions are not the answer. *Cons. Biol.* 5: 556–558.

Miller, P. S. and Hedrick, P. W. (1993) Selective breeding for rare alleles in pedigreed populations. (in preparation).

Nei, M. and Hughes, A. L. (1991) Polymorphism and evolution of the major histocompatibility complex in mammals. *In:* Selander, R. K., Clark, A. G. and Whittam, T. S. (eds), *Evolution at the Molecular Level.* Sinauer, Sunderland, MA, pp. 222–247.

O'Brien, S. J. and Evermann, J. F. (1988) Interactive influence of infectious disease and genetic diversity in natural populations. *Trends Ecol. Evol.* 3: 254–259.

Parham, P., Lomen, C. E., Lawlor, D. A., Ways, J. P., Holmes, N., Coppin, H. L., Salter, R. D., Wan, A. M. and Ennis, P. D. (1988) Nature of polymorphism in HLA-A, -B, and -C molecules. *Proc. Nat. Acad. Sci.* USA 85: 4005–4009.

Potts, W. K., Manning, C. J. and Wakeland, E. K. (1991) Mating patterns in seminatural populations of mice influenced by MHC genotypes. *Nature* 352: 619–621.

Ralls, K. and Ballou, J. D. (1986) Proceedings of the workshop on genetic management of captive populations. *Zoo Biol.* 5: 81–238.

Schreiber, A. and Tichy, H. (1992) MHC polymorphisms and the conservation of endangered species. *Symp. Zool. Soc. Lond.* 64: 103–121.

Slade, R. W. (1992) Limited MHC polymorphism in the southern elephant seal: Implications for MHC evolution and marine mammal population biology. *Proc. Roy. Soc. Lond. B* 249: 163–171.

Templeton, A. R., Davis, S. K. and Read, B. (1987) Genetic variability in a captive herd of Speke's gazelle (*Gazella spekei*). *Zoo Biol.* 6: 305–313.

Thomas, A. (1987) *Pedpack: User's manual. Technical Report 99*, Dept. Statistics, Univ. Washington, Seattle, WA.

Thomas, M. L., Harger, J. H., Wagener, D. K., Rabin, B. S. and Gill, T. J. (1985) HLA sharing and spontaneous abortion in humans. *Amer. J. Obstrec. Gynecol.* 151: 1053–1058.

Thomson, G. (1988) HLA disease associations: Models for insulin dependent diabetes mellitus and the study of complex human genetic disorders. *Annu. Rev. Genet.* 22: 31–50.

Tiwari, J. L. and Terasaki, T. I. (1985) *HLA and disease associations.* Springer-Verlag, New York.

Trowsdale, J. (1993) Genomic structure and function in the MHC. *Trends Genet.* 9: 117–122.

Trowsdale, J., Groves, V. and Arnason, U. (1989) Limited MHC polymorphism in whales. *Immunogenetics* 29: 19–24.

Trowsdale, J., Ragoussis, J. and Campbell, R. D. (1991) Map of the human major histocompatibility complex. *Immunol. Today* 12: 443–446.

Van Eden, W., Devries, R. R. P. and Van Rood, J. J. (1983) The genetic approach to infectious disease with special emphasis on the MHC. *Dis. Markers* 1: 221–242.

Vrijenhoek, R. C. and Leberg, P. L. (1991) Let's not throw the baby out with the bathwater: A comment on management for MHC diversity in captive populations. *Cons. Biol.* 5: 252–254.

Wakelin, D. and Blackwell, J. M. (1988) *Genetics of resistance to bacterial and parasitic infection.* Taylor and Francis, London.

Watkins, D. I., Hodi, F. S. and Letvin, N. L. (1988) A primate species with limited major histocompatibility complex class I polymorphism. *Proc. Nat. Acad. Sci.* USA 85: 7714–7718.

Yamazaki, K., Beauchamp, G. K. and Kupniewski, D. (1988) Familial imprinting determines H-2 selective mating preferences. *Science* 240: 1331–1332.

Yamazaki, K., Boyse, E. A., Mike, V., Thaler, H. T., Matheison, B. J., Abbott, J., Boyse, J., Zayas, Z. A. and Thomas, L. (1976) Control of mating preferences in mice by genes in the major histocompatibility complex. *J. Expt. Med.* 144: 1324–1335.

Yamazaki, K., Yamaguchi, M., Andrews, P. W., Peake, B. and Boyse, E. A. (1978) Mating preferences of F_2 segregants of crosses between MHC-congenic mouse strains. *Immunogenetics* 6: 253–259.

Conservation Genetics
ed. by V. Loeschcke, J. Tomiuk & S. K. Jain
© 1994 Birkhäuser Verlag Basel/Switzerland

Andean tapaculos of the genus *Scytalopus* (Aves, Rhinocryptidae): A study of speciation using DNA sequence data

P. Arctander[1] and J. Fjeldså[2]

[1]*Department of Population Biology, University of Copenhagen, Universitetsparken 15, DK-2100 Copenhagen, Denmark*
[2]*Zoological Museum, University of Copenhagen, Universitetsparken 15, DK-2100 Copenhagen, Denmark*

Summary. Tapaculos of the genus *Scytalopus* are secretive birds which tunnel like mice through dense understory of humid forest in the Andes, Central America, and south-eastern Brazil. Their agoraphobic habits make *Scytalopus* species highly sensitive to habitat disconti-nuities, so they are well suited for analyzing diversification patterns in montane forest biota. This study uses DNA sequence data to test hypotheses about past speciation events. The DNA data support that allopatric and parapatric populations with different songs represent different species. The high degree of phylogenetic resolution obtained by DNA-data permits a better description of geographical patterns of endemism. The data suggests that the commonly observed biogeographic pattern, where related species have long linear distribu-tions along the Andes in different altitudinal zones, arose by divergence in disjunct isolates rather than by parapatric divergence. The approach seems well suited for identifying areas that have a special role for the diversification process. The paper finally discusses how detailed phylogenetic studies can be used to test interpretations of biogeographic patterns of high relevance for pinpointing top priority areas for conservation.

Introduction

Tapaculos of the genus *Scytalopus* are sooty-gray colored birds, often with some brown mottling, inhabiting dense understory of humid forest and scrub in the Andes, in Central America, and in southeastern Brazil. They belong to the family Rhinocryptidae, which contains 26–28 spe-cies in 12 genera; six of the genera being monotypic, but *Scytalopus* comprising 12 species and a large number of subspecies (Meyer de Schauensee, 1966; Sibley and Monroe, 1990).

Scytalopus is one of the taxonomically most difficult groups of birds. Many species have several subspecies, and species that replace each other in different habitats or altitudinal zones show a very subtle degree of morphological differentiation. A high incidence of asymmetric adop-tion of adult plumage (e.g., immature feathers retained on one side of the body only) and partial albinism indicate that their external appear-ance is subject to little adaptive selective pressure. On the other hand, many populations which are barely distinguishable morphologically have markedly different songs, or use similar notes in different combinations or in different situations (Fjeldså and Krabbe, 1990, pp. 423–444; Vielliard, 1990; Krabbe and Schulenberg, 1994). Such populations often show strikingly sharp altitudinal replacement. The vocal variations in altitudinal range of each population are correlated with the local presence or absence of other species, suggesting strong interference competition, and, in general, lack of introgression. Since the songs of suboscine birds are generally thought to be genetically determined (e.g., Kroodsma, 1984), the vocal differences could suggest that natural taxonomic units and number of species are higher than described in the current essentially typological classification (Zimmer, 1939).

Recent collections, especially by Krabbe and Schulenberg (1994), have provided a large series of well documented *Scytalopus* specimens accompanied by tissue samples and, in most cases, also recordings of the song and other vocalizations of the individual in question. Krabbe and Schulenberg (in press) have, e.g., re-described the Ecuadorian populations in great detail, combining morphological, bioacoustic, geographical, and ecological data. The Peruvian and Bolivian *Scytalopus* populations are less well documented.

In this paper, we re-examine the relationships of *Scytalopus* populations of Ecuador and the eastern Andean slope of Peru using DNA sequence data. Our principal aims are to improve the understanding of taxic diversity, defining patterns of endemism, and to interpret a common biogeographic pattern among Andean birds, where related species have long linear distributions along the Andes in different altitudinal zones (see Fig. 1).

We will:

- compare genetic distances among allopatric populations that look alike but have more or less different vocalizations, within populations, and among sympatric populations with different songs.
- investigate the natural unit of *Scytalopus*: Are Zimmer's (1939) species natural units, or just ecological equivalents replacing each other in a specific habitat or altitudinal zone in different parts of the Andes?
- investigate the mode of speciation of *Scytalopus*: Is the excessive density of limited-range birds found in certain parts of the Andes caused by sequential differentiation along altitudinal gradients or random divergence of small local populations (parapatric divergence), or by the standard avian model of divergence of widely disjunct (allopatric) populations?
- comment on the phylogenetic affiliations of *Melanopareia*: Is it a tapaculo (and suitable as outgroup for the phylogenetic analysis of *Scytalopus*), a spinetail (Furnariidae) or an antbird (Formicariidae), as indicated in various earlier taxonomic treatments?

– finally, discuss possible implications for conservation, of identification of natural evolutionary units.

A review of the *Scytalopus* populations studied

We use a traditional naming of taxa (Zimmer, 1939; Meyer de Schauensee, 1966; Fjeldså and Krabbe, 1990), and refer to yet unnamed local populations by the name of the collecting locality (*viz.*, "Ampay Bird," "El Oro Bird," "Esmeraldas Bird," 'Millpo Bird" and "Zapote Najda Bird") as defined in Tab. 1. Collecting sites for the samples analyzed are shown in Fig. 1, together with geographical range of the taxa involved. Below follows a brief review of characteristics, habitat, altitude, etc. of the different "species."

Scytalopus unicolor (map Fig. 1A) comprises populations in which adult males are uniform in color. Northern populations with blackish males, giving distinctive low-pitched frog-like whistles (the parapatric taxa *latrans* and *subcinereus*) and dark gray southern populations whose songs are short to very long series of 12–16 notes per second (*parvirostris* group) are sympatric in northern Peru, indicating that at least two different species are involved.

Scytalopus femoralis (map Fig. 1B) comprises fairly large, long-tailed and stout-billed birds of humid premontane forest, distributed from Venezuela to southern Bolivia. The blackish *S. femoralis atratus* (and similar forms) is narrowly bound to primary forest and is characterized by white crown-patch and monotonous series of sharp notes. Recent studies show that it is sympatric with *S. f. micropterus* from Colombia to extreme northern Peru and *S. f. femoralis* further south in Peru (Krabbe and Schulenberg, in press). The two latter taxa give an endless series of well spaced resonant notes which are single (*femoralis*) or distinctively double (*micropterus*). Yet other vocalizations are given by populations in northern Colombia and in Bolivia.

Scytalopus vicinior (map Fig. 1B) is medium-sized, fairly long-tailed and gray, and gives a rapid series of notes of increasing amplitude; it inhabits humid forest at 1250–2000 (2350) m on the Pacific slope of Colombia and northwestern Ecuador. Very similar but smaller birds, usually with only 10 rectrices, inhabit mature, wet foothill forest in western Ecuador (map Fig. 1C). The *Esmeraldas Bird* represents the southern extreme of a population inhabiting the Colombian Choco region. The *El Oro Bird* represents a small isolated population in southwestern Ecuador. Both have songs composed of long series of rapid notes (double notes in the El Oro Bird).

Scytalopus latebricola (map Fig. 1C) refers to a complex group of fairly small to large brown-rumped birds in the northern part of the Andes. *S. l. spillmani* is slightly heavier than *S. vicinior*, and gives long,

208

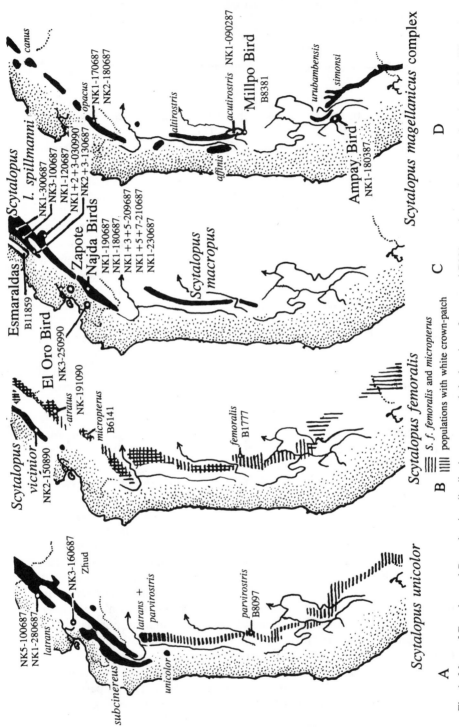

Fig. 1. Maps of Ecuador and Peru showing distribution ranges and the location of compared DNA specimens. A condensed review of the differentiation of taxa is given in the text.

Table 1. Specimens analyzed. For all the DNA samples, a traditional study skin is also available, and in most cases even a tape recording. Most tissue samples are in the Zoological Museum, University of Copenhagen (ZMUC) and Louisiana State University Museum of Zoology (LSU). Skins are kept in ZMUC and in the Academy of Natural Sciences of Philadelphia (ANSP). Tape recordings made by Krabbe are in the Bioacoustic Laboratory, University of Aarhus. All samples used for analyzing the variation within populations with similar song patterns were collected by Krabbe who confirms that the specimens compared had similar songs.

Scytalopus unicolor latrans (ZMUC NK5-100687 and NK1-280687) Ecuador: Imbabura, 15 km SE Apuela, 2800 m; (ZMUC NK3-160687) Ecuador: Cañar, 11 km N Zhud, 2850 m (an isolated population, see Fig. 1).

Scytalopus unicolor parvirostris (tissue LSU B8097, skin specimen in ANSP 128572) Peru: Pasco, Playa Pampa, 2300 m.

Scytalopus femoralis femoralis (tissue LSU B1777, skin specimen ANSP-106110) Peru: Pasco, Santa Cruz, 2050 m.

Scytalopus femoralis micropterus (tissue LSU B6141, skin specimen in ANSP) Ecuador: Morono Santiago, Cutucú mountains, 1700 m.

Scytalopus femoralis atratus (ZMUC NK1-191090) Ecuador: Napo: south slope of Guacamayos Mountains, 1200 m.

Scytalopus, unnamed (called *Esmeraldas Bird* in the following) (tissue LSU B11859, skin specimen ANSP 180149) Ecuador: Esmeraldas, El Placer, 650 m.

Scytalopus, unnamed (called *El Oro Bird* in the following) (ZMUC NK3-250990) Ecuador: El Oro, 9 km W Piñas, 900 m.

Scytalopus vicinior (ZMUC NK2-150890) Ecuador: Pichincha, near el Obelisque on the Mindo road, 1700 m.

Scytalopus latebricola spillmanni (ZMUC NK1-300687) Ecuador: Sucumbios, 10 rd km E Sta Barbara, 2500 m; (ZMUC NK2 and NK3-130687) Ecuador: Pichincha, Lloa-Mindo road, 2550 m; (ZMUC NK3-100687 and NK1-120687) Ecuador: Imbabura, 15 km SE Apeula, 2800 m; (ZMUC NK1, NK2 and NK3-030990) Ecuador: Napo, Guacamayos Ridge, 2100 m.

Scytalopus, unnamed (called *Zapote Najda Birds* in the following) (ZMUC NK3-170387, NK1-180687, NK1-190687, NK1, NK3 and NK5-200687, NK1, NK5 and NK7-210687, NK1-230687) Ecuador: Morono-Santiago, Zapote Najda mts., 2250 m.

Scytalopus magellanicus unnamed subspecies (called *Ampay Bird* in the following) (ZMUC NK1-180387) Peru: Apurímac, north side of Nevado Ampay, at 3500 m.

Scytalopus magellanicus unnamed subspecies (called *Millpo Bird* in the following) (tissue LSU B8381; skin ANSP 128624) Peru: Pasco, Millpo, 3600 m.

Scytalopus magellanicus acutirostris (ZMUC NK1-090287) Peru: Huánuco, Carpish mountains, at 2760 m, above the road tunnel.

Scytalopus magellanicus opacus (ZMUC NK1-170687 and NK2-180687) Ecuador: Azuay/Morono-Santiago border, Zapote-Najda mountains, 3450 m.

Melanopareia maximiliani (ZMUC NK1-180487) Bolivia: Mizque in Cochabamba (outgroup).

---

fast trills. It inhabits tall forest and bamboo at 1900–3200 m in northern and central Ecuador. The *Zapote Najda Birds* represent a population previously included in *S. l. spillmanni*, but having a slower song which descends at first. It inhabits large stands of bamboo, in particular, at 2200–3200 m along the continental divide in southern Ecuador.

Scytalopus macropus (map Fig. 1C) is a large, blackish gray form inhabiting dense, mossy, cloud- and elfin forest in northern Peru. Its song is very distinctive – a monotonous series of low-pitched notes.

Scytalopus magellanicus (map Fig. 1D) comprises several populations of small, fine-billed birds, most of them with densely barred rump and flanks, replacing each other along the Andes in ecotone habitat at the treeline (3200–4200 m in the Tropical Andes Region). Among the 15 recognized populations, most use various combinations of well spaced single or vibrant notes, but some have strikingly different songs. The *Ampay* and *Millpo Birds*, both probably of very restricted distribution, resemble *S. m. simonsi* (silvery gray superciliary), and have distinctive songs of well spaced *tras* notes and well spaced vibrating *chirps*, respectively (Fjeldså and Krabbe, 1990, pp. 440–441), but similar call notes. *S. m. opacus* is nearly a uniform gray and its song an endless, fast trill with a stuttering start; it is distributed throughout Ecuador, and replaced by a very similar form, *S. m. canus* (with similar song) in the western and central Andean ranges of Colombia. The name *S. m. acutirostris* has been applied to three to four different populations, but the type probably represents a population inhabiting dense cloud forest shrubbery at 2400–3450 m on both sides of the gap formed in the Cordillera Central by the Upper Huallaga River, in the Carpish Mountains in Huánuco and in the Panao Mountains in Pasco (Krabbe and Schulenberg, 1994; this form presented as an unnamed species in Fjeldså and Krabbe, 1990, pp. 427–428). It is virtually indistinguishable from *S. u. parvirostris*, but the vocalizations are qualitatively more like Peruvian *S. magellanicus* populations, a monotonous series of well spaced *keek* calls and a song of short trills *keek-keekrrrr*. It is replaced near the treeline by the *Millpo Bird* in Pasco and by *S. magellanicus altirostris* in Huánuco.

Materials and Methods

We analyzed 33 tissue samples, comprising individuals of the same local populations, of different local populations with a consistent song type, and of populations with different songs replacing each other over a short distance (Tab. 1). All Ecuadorian *Scytalopus* populations with distinctive song types were covered, except two populations of *S. unicolor* with weakly different songs. Some specimens from the eastern Andean slope of Peru are also included. In order to interpret the modes of differentiation, we analyzed samples mainly from the two areas with maximum *Scytalopus* diversity (Fig. 1), in northwestern Ecuador and in mountains flanking the Upper Huallaga Valley in Huánuco/Pasco, Peru. This valley forms a deep gap with an arid subtropical climate on the bottom interrupting the humid temperate forest. Unfortunately, we lack samples of *S. macropus* (which inhabits both sides of the Huallaga gap) and *S. magellanicus altirostris* (north of the gap).

For the 33 *Scytalopus* samples, 285 base pairs (bp) were sequences of the mitochondrial cytochrome *b* (*cyt b*) gene. Nineteen sequences from five local populations were identical. To determine the systematic position of the species used as outgroup, *Melanopareia maximiliani*, 293 bp *cyt b* was compared with the corresponding sequence of a Formicaridae (*Grallaria andicola*), a Furnariidae (*Asthenes dorbignyi*), and a Phytotomidae (*Phytotoma rutila*) of the flycatcher/cotinga assembly.

DNA was extracted from blood samples stored in APS buffer according to Arctander (1988). The primers HL14841:5'-CAACATCTCAG-CATGATGAAA-3' (Kocher et al., 1989) and HH-15149:5'AAACT-GCAGCCCCTCAGAATG-3' (shorter sequence than given by Kocher et al., 1989) were used for PCR amplifications as well as dideoxy termination sequencing (standard protocols; e.g., Innis et al., 1990). Sequences are deposited in GenBank with accession numbers: U06158–78.

For the analysis of sequence data, we used the programs EDEE 1,09d (E. Cabot, unpublished program), PAUP 3.0 (Swofford, 1991) and CS 3. = (H. Siegismund, unpublished program). Phylogenetic trees were constructed using PAUP (heuristic search option and random addition of sequences). To get an impression of the support of certain branching orders four-taxon tests were performed. The three possible alternative trees for four taxa, one of them an outgroup, are evaluated by testing whether the number of phylogenetically informative sites supporting each tree are significantly different (χ^2 with two degrees of freedom).

Results

Pairwise comparisons of the number of base substitutions, transitions (ts), and transversions (tv) between all specimens (except the 19 identical sequences) are given in Tab. 2. *Melanopareia* differs from 20 different *Scytalopus* sequences in 25–36 ts (average 30.8), and in 15–21 tv (average 17.2). This gives a ts/tv ratio of 1.8. The ts differences between specimens which differ in no or only one tv suggest a ts/tv rate of 20 (Edwards et al., 1991). The much lower ts/tv ratio in intergeneric comparisons demonstrates a strong degree of multiple hits saturation of transitions (in third codon positions), as these substitutions may shift back and forth an unknown number of times.

Variation within populations with similar song patterns

Two *S. unicolor latrans* from Apuela (northwestern Ecuador) differ from each other in one third position transition. For *S. l. spillmanni*, identical sequences were found in two birds from Pichincha and two from Imbabura, while one out of three from Napo differed from the two others by 1 ts. The three populations differed from each other by 1–6 ts and 0–1 tv. Ten *Zapote Najda Birds* had identical sequences, except for

Table 2. Matrix showing base substitutions (transitions and transversions) in pairwise comparisons of all specimens with different base sequences. The ratio of transitions to transversions for the whole data-set is 4.3. Comparing only specimens with similar songs, the ratio is 9.5, while all pairwise comparisons amongst *Scytalopus* gives 5.3, and pairwise comparisons of *Scytalopus* and *Melanopareia* (outgroup) gives 1.8, indicating multiple hits and substitutional saturation for transitions

Upper triangle: transversions — Lower triangle: transitions

	1	2	3	4	5	6	7	8	9	10	11	12	13	14	15	16	17	18	19	20	21
1. *Millpo Bird* B8381		5	10	3	6	6	3	2	5	3	3	3	3	3	3	10	3	2	2	2	17
2. *Ampay Bird* NK1-180387	27		9	6	9	9	6	5	4	6	6	6	6	6	8	9	8	7	7	7	16
3. *S. m. opacus* NK1-170687	34	21		9	12	16	9	10	7	9	9	10	10	11	11	8	11	10	5	10	19
4. *S. f. femoralis* B1777	42	32	33		3	7	2	3	2	2	0	2	2	2	2	8	2	1	1	1	16
5. *S. f. micropterus* B6141	34	26	31	8		10	5	6	5	3	3	5	5	5	5	10	5	4	4	4	16
6. *S. f. atratus* NK1-191090	29	34	33	34	30		7	6	9	7	7	5	5	5	7	15	7	6	13	6	21
7. *Esmeraldas Bird* B11859	34	32	31	22	21	29		1	6	2	2	2	3	3	2	10	2	1	1	1	18
8. *S. u. parvirostris* B8097	34	32	37	38	30	28	30		3	2	2	3	3	3	2	2	3	2	2	2	19
9. *Zhud latrans* NK3-160687	32	35	20	15	15	29	25	34		17	18	27	24	22	30	19	22	24	23	13	16
10. *S. u. latrans* NK1-280687	38	19	31	6	10	10	24	38	17		0	26	22	22	32	25	20	24	23	23	16
11. *S. u. latrans* NK5-100687	41	29	30	9	9	31	25	39	18	2		26	22	24	33	25	21	25	24	30	16
12. *Zapote Najda* NK1-210687	36	35	33	27	24	30	25	32	27	26	26		19	18	31	26	18	17	16	31	18
13. *Zapote Najda* NK5-210687	34	32	30	23	21	28	24	30	24	22	22	0		19	30	22	15	18	17	29	17
14. *S. l. spillm.* NK1-030990	33	32	26	27	23	25	27	31	22	22	24	19	19		32	23	6	3	2	27	18
15. *El Oro Bird* NK3-250990	35	32	26	27	27	33	28	34	30	32	33	31	30	32		29	30	34	33	26	18
16. *S. vicinior* NK2-150890	32	18	7	29	29	23	30	33	19	25	25	26	22	23	29		20	24	23	11	17
17. *S. l. spillm.* NK1-300687	33	29	26	24	24	22	25	33	22	20	21	18	15	6	30	20		34	33	30	18
18. *S. l. spillm.* NK3-130687	36	33	28	28	28	25	26	32	24	24	25	17	18	3	34	24	34		0	31	17
19. *S. l. spillm.* NK3-100687	35	32	27	27	27	24	25	31	23	23	24	16	17	2	33	23	33	0		30	17
20. *S. m. acutir.* NK1-090287	31	16	11	25	24	29	25	36	13	23	30	31	29	27	26	11	30	31	30		15
21. *Melanopareia maxi.*	29	36	33	34	30	30	34	32	30	30	30	26	25	28	35	28	30	31	30	34	

1 ts in one individual. Two *S. m. opacus* specimens from the Zapote Najda Mountains had identical sequences. Surprisingly, the *S. u. latrans* specimen from Zhud in Cañar, from an isolated population on the Pacific Slope, differed from those from Imbabura in 17–18 ts and 2 tv (see below).

Differences between sympatric populations (= definite biospecies)

The following taxa, representing different song populations, are found within a restricted area on different altitudes or in different habitats. We regard these taxa as essentially sympatric. They are: in northwest Ecuador, *S. u. latrans*, *S. l. spillmanni* (Pichincha specimen), *S. vicinior* and the *El Placer Bird*; in Napo, *S. f. atratus* and *S. l. spillmanni*; in southern Ecuador, *Zapote Najda Birds* and *S. m. opacus*; in Pasco, *S. u. parvirostris*, *S. f. femoralis* and the *Millpo Bird*. These birds differed from each other by 23–42 ts (average 28.33), and 1–13 tv (average 4.69), ts/tv 6.04:1 (Tab. 1).

Differences amongst allopatric populations with different songs

Pairwise comparisons include *s. u. latrans versus S. u. parvirostris*: *S. f. femoralis versus micropterus*; treeline forms, *S. m. opactus* and the *Millpo* and *Ampay Birds*; *Esmeraldas vs. El Oro Birds*; two *Zapote Najda Bird* sequences *vs.* four *S. l. spillmanni* sequences. Except for *S. u. latrans/parvirostris* and *S. m. opacus*, these populations are almost indistinguishable morphologically.

Comparison of *S. f. femoralis* and *micropterus* gave only 8 ts and 3 tv. The remaining comparisons gave base sequence divergences of 16–34 ts (average 24.5) and 0–10 tv (average 2.9), ts/tv = 8.5.

Phylogenetic relationships

Tentative phylogenies are shown in Fig. 2. Fig. 2A places *Melanopareia* with *Scytalopus*. Despite rather low bootstrap values, this suggests that it is a tapaculo. The next two nodes have swapped position compared to the result of DNA × DNA hybridization data by Sibley and Ahlquist (1990, Figs 190–205 and 371–372). However, these mid-tertiary nodes are densely packed (Sibley and Ahlquist, *cit. op.*), and a full resolution cannot be expected without a very long base-sequence.

Melanopareia was used as an outgroup in the phylogenetic analysis of *Scytalopus*. A heuristic search where transversions were weighted 20 times over transitions, gave one single most parsimonious tree (Fig. 2B).

214

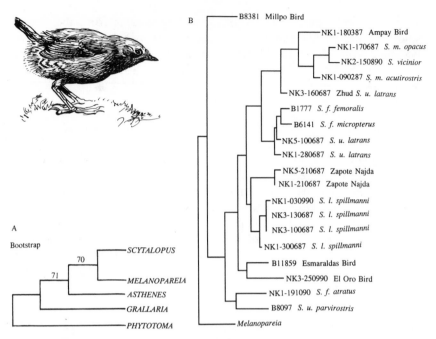

Fig. 2. Phylogenetic relationships of tapaculos. The single most parsimonious trees, generated by the computer program PAUP (Swofford, 1991) using heuristic search and random addition of sequences. In A, bootstrap values (out of 100 replications) are given above the branches. Transversions in third codon position are weighted 20 times over transitions, because this is the transition/transversion ratio found in pairwise comparisons of taxa which differ in no or one single transversion (see also Edwards et al., 1991). The weight is applied because transitions in third codon position clearly are into saturation when comparing specimens with dissimilar songs (see text and Tab. 2). Omitting transitions in third codon position altogether only results in minor changes near the base of the tree, joining B8097 with B11859, and NK1-191090 in a monophyletic group with NK3-250990 as sister group.

Almost identical trees were obtained if transitions in 3rd positions were omitted. However, if all substitutions are used with no weights applied, a quite different tree topology will emerge. Because of the strong indications of multiple-hits saturation of transitions (see above), only the transversions appear to give reliable phylogenetic information. Since a full phylogenetic resolution certainly requires longer sequences, we will not discuss the tree topology in full, but will only consider some specific questions:

What are the natural units of S. unicolor and S. femoralis?

Birds traditionally placed in *S. unicolor* and *S. femoralis* fall into distinctive song groups. A high sequence difference between *S. u. latrans* and *parvirostris* and between *S. f. femoralis*, *micropterus*, and *atratus*

(Tab. 1) was therefore expected. We were nevertheless surprised to find one representative of *S. unicolor*, namely, *latrans* from Apuela, falling together with *S. f. femoralis* and *S. f. micropterus* in the phylogeny; and *S. unicolor parvirostris* falling together with *S. f. atratus* (Fig. 2B). The relationship of *S. u. latrans* with *S. f. femoralis* and *S. f. micropterus* is supported in all six comparisons made using four-taxon tests ($p < 0.01–0.0001$). The relationship between *S. f. femoralis* and *micropterus* was supported in any comparison made ($p < 0.001$). However, the relationship of *S. u. parvirostris* to *S. f. atratus* is not significant in any other comparison tried.

The *S. u. latrans* specimen from the marginal population in *Zhud* groups together with populations of the *S. magellanicus* complex (Fig. 2B). Four-taxon tests show that this bird is closer to *S. m. opacus* or *acutirostris*, *S. vicinior* or the *Ampay Bird* than to other *S. u. latrans* specimens or *S. femoralis* ($p < 0.05–0.001$).

What is the relationship of S. vicinior?

Although *S. vicinior* is confusingly similar to populations (*S. l. spillmanni* and the *Esmeraldas Bird*) inhabiting adjacent zones of tropical and lower temperate forest in northwestern Ecuador and western Colombia (Fig. 1), it clusters together with the quite different-looking treeline form *S. m. opacus* (Fig. 2B). This relationship is supported in four-taxon tests (see above) involving comparisons with *S. u. latrans* ($p < 0.001$), the *Zhud Bird* ($p < 0.05$), *S. m. acutirostris* ($p < 0.05$), the *Ampay Bird* ($p < 0.05$), *S. l. spillmanni* ($p < 0.001$) and *El Oro* ($p < 0.001$) and *Esmeraldas Birds* ($p < 0.001$).

What is the relationship of S. magellanicus acutirostris?

This taxon, morphologically confusingly similar to *S. u. parvirostris* (which replaces it below 2400 m) but with a distinctive song, was presented as an "unnamed species" in Fjeldså and Krabbe (1990, pp. 427–428) (see Krabbe and Schulenberg, in press). The DNA data places it together with members of the *S. magellanicus* complex, which is represented in the adjacent treeline habitat by *S. m. altirostris* (Fig. 2B). Four-taxon tests place it close to the *Ampay Bird* in comparison with the *Millpo Bird*, with *S. u. parvirostris*, and with *S. f. femoralis* ($p < 0.01–0.001$).

What are the natural relationships of the yet unnamed populations?

The Millpo Bird, although closely similar to the *Ampay Bird* and *S. m. simonsi* by morphology, and having structurally fairly similar vocalizations, forms a very deep branch in Fig. 2B. In four-taxon tests, none of the seven combinations tested gave a significant association with another specimen.

Zapote Najda Birds were previously included in *S. l. spillmanni*, but differ from it by 15–19 ts and 0–2 tv (Tab. 2). Four-taxon tests support a sister-group relationship between the two in comparisons with *S. u. latrans* and *S. m. opacus, Esmeraldas* and *El Oro Birds* ($p < 0.01–0.001$) (but not in a comparison with *S. vicinior*).

Isolated populations recently found in tropical foothill forest in western Ecuador (*Esmeraldas* and *El Oro Birds*) have considerably different DNA sequences (27 ts 2 tv, Tab. 2) but appear to be sister taxa (Fig. 2B). This relationship was supported in comparison with *S. vicinior* ($p < 0.001$) and *S. opacus* ($p < 0.01$), but not in comparisons with other Ecuadorian populations.

Parapatric or allopatric divergence?

If the divergence was parapatric, then some populations which live adjacent to each other in areas with an excessive species richness (see Fig. 1) would be sister taxa. Suitable cases for testing this assumption exist for two areas. In northwestern Ecuador, *latrans* (Imbabura specimens), *spillmanni* (Pichincha specimen), *vicinior* and the *Esmeraldas Bird* were collected essentially on the same slope, where they replace each other in semi-humid temperate shrubbery, lower temperate-zone forest with bamboo, subtropical forest, and in tall tropical forest, respectively. In Pasco, the *Millpo Bird, parvirostris*, and *atratus* replace each other in a similar way from the treeline to subtropical forest (with *S. macropus* and *S. m. acutirostris* also present).

Furthermore, parapatric populations of the *Zapote Najda Bird* and *S. m. opacus* were sampled in southern Ecuador, and parapatric populations of *S. f. atratus* and *S. l. spillmanni* were sampled in Napo.

Four-taxon tests reject a close relationship between parapatric populations in comparison with the relationships suggested in Fig. 2B. Vertically adjacent populations are genetically highly distinctive (average 28.70 ts, 5.78 tv), and each of them is closer to a population in another part of the Andes or to a population separated from it on the same slope by one or two other species. *S. u. latrans* (Ecuador) was genetically closer to the Peruvian *S. f. femoralis* than to the Ecuadorian *S. femoralis micropterus* (Tab. 1). In the case of *S. l. spillmanni*, the southernmost sample (NK1-030990, from south of the Río Pastaza barrier), showed the strongest divergence from the *Zapote Najda Birds* (Fig. 2B and Tab. 1).

Discussion

Mode of speciation

The time scale for tapaculos speciation is indicated by divergence data obtained from DNA hybridization. Sibley and Ahlquist (1990) compared two *Conopophaga* species, *Pteroptochus*, *Lioscelis*, and two *Scytalopus*, *S. latebricola* and *S. femoralis* (information about which populations were involved is lacking; G. C. Sibley, *in litt.*). $T_{50}H$ values ranging from 10.7 to 1.9 place the family's early radiation to the Oligocene/Miocene transition, and suggest that the *Scytalopus* radiation took place in the Pliocene and Pleistocene. During this period, the group would have been affected by intensive orogenesis in the Andes, tectonic subsidence between the Andean and Brazilian ranges, and by great climatic changes. As the climate became cool and arid repeatedly (Hooghiemstra, 1991), the treeline was suppressed and possibly the montane shrubbery and forest remained permanently evergreen only in certain Andean slopes.

Endler (1982) proposes that the clinal variation resulting from differential selection on ecological gradients may lead to parapatric speciation without a geographical gap. In the Andes, sharp ecotones exist, for example, between subtropical forest and mossy montane cloud forest, and at the treeline. In many parts of the Andes, *Polylepis* woodlands form a distinctive community which is separated by a habitat discontinuity from the cloud-forest zone. Other sharp habitat gradients and habitat mosaics exist where deep valleys intersect the Andean slopes. Gentry (1982) suggests random and essentially parapatric divergence of small local populations (of plants) within such habitat mosaics. The more landslides, the more species.

Areas with a high degree of habitat mosaics do indeed have many species, but this does not necessarily mean that the divergence of new species took place within that mosaic. As long as some migration is possible between the local populations, the alternative exists that tiny local populations inhabiting a habitat mosaic would form a metapopulation, and that only populations separated by a distinctive geographical gap diverged to become different species. The metapopulations may evolve rapidly as a result of random divergence of the local part populations (Levin, 1993).

Interpretations of whether the speciation took place within an ecotonal area or by divergence of widely allopatric populations needs to be based on studies of phylogenetic relationships of adjacent and remote populations.

The four-taxon test clearly excludes close relationships among populations which are segregated in different-quality habitat and on different altitudes within areas with excessive species richness. The very consistent difference between all taxa inhabiting adjacent parts of a species-rich area indicates that the initial divergence was not across ecotones within that area. However, a full interpretation requires samples from vertical transects in many other parts of the Andes. Northern Ecuador has two taxa which are closely related: *S. vicinior* and *S. m. opacus*. However, the geographical and ecological gap between them (Fig. 1) is considerable, and is occupied by two other species. In absence of DNA from *S. m. canus* (probably a close relative of *S. m. opacus*; Fjeldså and Krabbe, 1990, p. 443) from the treeline habitat of southwestern Colombia, above *S. vicinior*, we cannot exclude that the nearest relatives of *S. vicinior* lives in the nearest treeline habitat. Even so, there would be two remotely related species inserted on the habitat gradient between the sister taxa. In this case, the speciation scenario could have started by colonization by a pre-*canus/opacus* stock of the ridgecrest habitat in the Western Cordillera in Colombia, which is cut off from the Central Andes. Because of the low altitude of this ridge, this probably happened in a cold climatic period. However, the population could only survive here through subsequent warm periods by adapting to a subtropical climate. During a later cold period, the top ridges were colonized by the more distantly related *S. u. latrans* and *S. latebricola*, whereby *S. vicinior* was forever isolated from the *canus/opacus* stock.

The only other case of disjunct altitudinal segregation known among Andean birds is in *Chamaepetes goudotii* (Cracidae). This bird normally inhabits steep subtropical hills, and is replaced in the temperate zone by *Penelope montagnii*. The isolated *C. goudotii rufiventris* lives above *P. montagnii* (Remsen and Cardiff, 1990). We suspect that a full phylogenetic resolution of other species-rich Andean groups would reveal additional cases.

Unfortunately, we lacked DNA samples of two taxa from the Huánuco/Pasco area in Central Peru. *S. macropus* is very distinctive, morphologically and vocally. However, we cannot exclude a close relationship between *S. m. acutirostris* and *altirostris*, which replace each other in the lower temperate zone and treeline zone in Carpish.

Although the phylogenetic resolution of *Scytalopus* populations of Ecuador and Peru is still incomplete, the most parsimonious conclusion from molecular, morphological, bioacoustic, and habitat data would be that divergence was allopatric, in small disjunct isolates in different parts of the Andes (Fjeldså, 1992), and that the pattern of geographically overlapping populations segregated in different altitudinal zones was secondary. The same conclusion was drawn from a well-designed study of DNA divergence in three species of the Andean rodent genus *Akodon* (Patton and Smith, 1992). Using samples from seven vertical transects in southeastern Peru, the authors rejected the gradient model of differentiation and associated the divergence with specific sectors of the Andean slope, separated by deep valleys (see also Patton, 1990).

Classification

When comparing sympatric species (confirmed biospecies), the ts values are nearly as high as in the *Melanopareia/Scytalopus* comparison (Tab. 1), suggesting that they have diverged to reach the multiple hit zone for transitions. Slightly lower values were found for allopatric populations with different songs. DNA sequence differences cannot determine species status in its biological sense. However, giving a measure of the "time since the common ancestor," the DNA-based genetic distances provide a helpful clue. The numbers of substitutions revealed in this study between "song populations" are in most cases at the level of, or higher than, what is usually found among species of the same genera (Arctander, 1991; Smith et al., 1991; Rickman and Price, 1992). We hereby support Krabbe and Schulenberg (in press) in assigning species rank to *Scytalopus* populations with distinctly different songs. Only the difference between *S. f. femoralis* and *micropterus* (8 ts 3 tv), whose songs are only slightly different, is below the usual difference between definite biospecies.

The isolated *Scytalopus* population in Zhud, assigned to *S. u. latrans* on basis of morphology and song, is puzzling. One possible explanation would be past integradation leading to the transfer of a mitochondrial genome from a pre-*opacus* female into a population of *latrans* morphotype. The two inhabit adjacent habitats. The sharp local replacements of populations with different songs seem to indicate lack of introgression, but we cannot exclude rare or local hybridization. Because of this, a more comprehensive future analysis should include nuclear as well as mitochondrial genes.

S. unicolor and *S. femoralis*, as defined by Zimmer (1939), are not natural units. The northern *S. u. latrans* is closely related to *S. f. femoralis/micropterus*, while the southern *S. u. parvirostris* and forms with a white crown-patch (including *S. f. atratus*) represent deep but yet unresolved branches of the phylogeny.

The *Millpo*, *Esmeraldas* and *El Oro* Birds represent deep branches which cannot be resolved with the 270 bp sequence studied by us. All these must be regarded as legitimate species. Birds inhabiting bamboo thickets in the lower temperate zone in southern Ecuador (*Zapote Najda Birds*) are related to *S. latebricola spillmanni*. However, the degree of genetic divergence, together with parapatric contact in the Río Paute area, supports species rank.

Several forms of the *S. magellancius* complex differ only by degree and by different combinations of rather similar song elements. However, the branches shown in Fig. 2B are equally deep as many sympatric species in other groups. The deep node of the *Millpo Bird* was highly surprising, considering its similarity to the *Ampay Bird* and certain other forms of the *S. magellanicus* complex. Some populations placed by

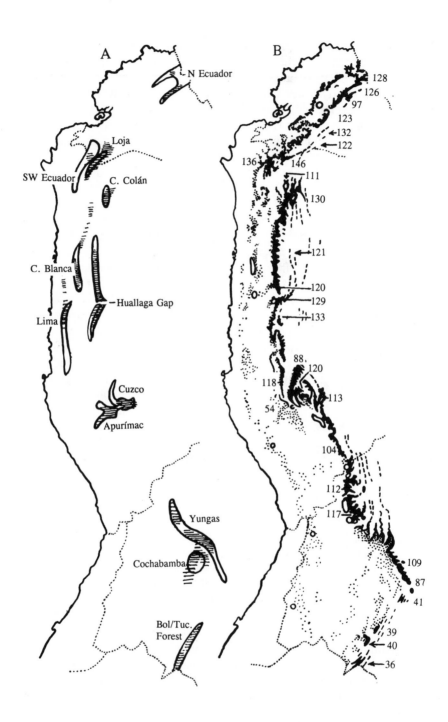

Zimmer (1939) in *S. magellanicus* have strikingly aberrant songs, suggesting that this is a polyphyletic assemblage of forms which has in common only an adaptation to the tangled dwarf forest and ericaceous shrub in the treeline zone. One possible interpretation could be that the primitive *Scytalopus* forms were small treeline forms with a light supercilium or forecrown (as found today in some populations inhabiting Andean treeline habitat, Patagonian forest, and some mountains in Central America). Larger forms of lower-altitude forest then represent secondary specializations. *S. m. acutirostris* is a distinctive species related to the treeline taxa, but displaced to cloud-forest on a lower altitude. We suspect that the 15 populations traditionally placed in *S. magellanicus* (all described in Fjeldså and Krabbe, 1990) represent 8–12 different species in at least three different clades.

Evidently, a full phylogenetic resolution of *Scytalopus* would unravel a very complex geographical pattern of endemicism, which was completely obscured in the traditional classification. In this case, the true diversity was not overlooked because of the traditional biological species concept as such, but rather because of the lack of bioacoustic and ecological data that could be correlated with the specimens.

Implications for making conservation priorities

In groups with slight morphological variation, molecular data provide a more natural picture of the species diversity than the traditional morphological approach based on skins. Such data, together with better geographical and bioacoustic data, will provide a better basis for defining patterns of endemicism in the Andes. Definition of endemic bird areas represents one of the most efficient tools today for rapid identification of top priority areas for biodiversity conservation in the tropics (ICBP, 1992). However, the geographical resolution and the ranking of the priorities can still be much improved by defining the

Fig. 3. To the left, summary map of peak densities for neoendemic and relict bird populations for birds of montane evergreen forest and shrub (from Fjeldså, in press). Encircled: exclusive areas of endemism, the line demarcating the core areas beyond which less than half of the representative species extend, each taxon included being part of a young (Pleistocene?) vicariance pattern. Shaded: core areas for taxa with a relict distribution, judging from significant geographical gaps separating distinctive subspecies, or a significant geographical gap towards the sister taxon. Some of the relict lineages may be of Miocene age. To the right the distribution of humid montane (temperate) forest (stippled: cloudforest patches and evergreen shrub in seasonally arid zones; dashed lines: ridgecrests in the subtropical zone). Figures show the variation in total richness of montane forest birds for 28 well-studied areas along the eastern Andes slope. The endemicity is not correlated with species richness, but a good correlation exists between patterns of neoendemics and older species with relict distributions (see Fjeldså in press for statistical testing).

natural units of groups representing strong recent radiations, such as *Scytalopus*.

We believe that good interpretations of the evolutionary histories behind patterns of species richness and endemicity may have even greater implication for making priorities for conservation. We will briefly outline here a possible hypothesis about the evolution of complex biota in tropical montane forests, and explain the potential conservation perspective. We will also outline how we think that DNA divergence data can be used to test it.

In the Andes, the topographic relief would appear to provide plentiful opportunities for isolation, e.g., by random drift after jump dispersal. Treeline birds show a high degree of replacement where deep (arid) valleys intersect the humid Andean slope (Graves, 1985, 1988). However, the ranges of *S. u. latrans*, *S. macropus* and *S. m. acutirostris* extend across such gaps (Fig. 1), and a detailed mapping of all Andean montane-forest birds (Fjeldså and Krabbe, 1990; Fjeldså, 1994) demonstrates that a large number of restricted-range species are established right across a habitat gap, indicating that the distributions are not only determined by habitat gaps, but also by special ecological conditions on both sides of a gap. Although *Scytalopus* and several other cloudforest birds are often regarded as agoraphobic, it is evident that at least the species inhabiting the treeline shrubbery sometimes manage to cross habitat gaps. Poulsen (in press) has shown, on a very local scale, that they may leave the vegetation cover during dark and rainy weather.

Because of the climatic influence from the humid lowlands, the eastern Andean slope would probably have had a permanent band of forest and shrub habitat throughout the Pleistocene. However, the forest quality may have varied, with epiphyte-laden evergreen conditions maintained continuously only in certain places. It is our assumption that the evolution of new species happened mainly when once-widespread species survived periods of cold and arid climate as widely separated relict populations in such places. If biological diversification was driven by periodic rapid divergence of disjunct isolates of once-widespread taxa in ecologically stable places, then older species which may have little genetic potential for coping with shifting ecological conditions would be expected to survive as relict populations, mainly in places with many neoendemics. Fjeldså (in press) demonstrated that this correlation exists in the Tropical Andes Region (see Fig. 3A for a simplified illustration). The pattern is not correlated with the variation in total species richness (Fig. 3B).

The *Scytalopus* species follow the general endemicity pattern fairly well (compare Figs 1 and 3). The peaks are found along the eastern Andean slope where systems of valleys and projecting ridges form "funnels" directing the moist air into high massifs (Fig. 3B), leading to permanent accumulation of clouds and condensation. Endemicity peaks outside the most raindrenched parts of the Andes have fog zones caused by persistent climatic fronts or atmospheric inversions.

Implications of such a model are considerable because of ecological effects of montane forests. By lack of vertical rainfall, forests of mountain ridges with persistent fog and clouds are capable of "combing" moisture out of the atmosphere (Bruijnzeel, 1990). There is some evidence that small forest patches, once established, maintain a stable local climate that make them self-sustaining even in fairly dry regions (Kerfoot, 1968; Pocs, 1974). Because of this effect, destruction of forest cover may well disrupt a local water cycle, requiring a very long regeneration time. Places that remained stable while the global climate fluctuated would seem to be optimal places for securing stable core areas through shifting global climates. According to this rationale, identification of places with aggregates of birds representing natural lineages of strong Pleistocene radiations has perspectives far beyond identifying playground for conservationists! If politicians do not find the species of these areas important *per se*, they should at least regard the endemicity as an indicator of other factors of high potential life-support value.

An extension of the approach presented in this paper could have great potential for robust testing of the speciation model suggested above, and for identifying places of particular life-support value.

A powerful test of the model of divergence of relict populations in places which remained ecologically stable would be to study population phylogenies using DNA sequence data. We assume that birds living in the center of stability would represent a deep phylogenetic branch compared with peripheral derived subpopulations of the same species. Results presented in this paper are suggestive (compare Figs 2B and 3 for *S. l. spillmanni* and *Zapote Najda Birds*), but the approach must be extended, e.g., by samples from the Río Paute Valley where the two above-mentioned taxa meet. A complete resolution of geographical patterns of closest relationships between *Scytalopus* populations would be a vast undertaking. Detailed studies could be limited to a few species groups, and a detailed study of all *Scytalopus* forms inhabiting the narrow treeline zone would probably be particularly rewarding.

Acknowledgements
First of all we wish to thank N. Krabbe for providing the majority of tissue samples used in this study, and for many good discussions over population structure and species limits in *Scytalopus*. T. S. Schulenberg provided other tissue samples from the LSU collection. We are grateful to J. Lovett for stimulating discussions and linguistic revision of the manuscript. The study was made on grants from the Carlsberg Foundation (to P.A.) and the Danish Natural Science Research Council (grant J.nr. 119033 to J.F.).

224

References

Arctander, P. (1988) Comparative studies of avian DNA by restriction fragment length polymorphism analysis: convenient procedures based on blood samples from live birds. *J. Orn.* 129: 205–216.

Arctander, P. (1991) Avian systematics by sequence analysis of mtDNA. *Acta XX Congr. Int. Orn.*: 619–628.

Bruijnzeel, L. A. (1990) *Hydrology of moist tropical forests and effects of conversion: a state of knowledge review.* UNESCO Int. Hydrol. Progr., Free University Amsterdam.

Edwards, S. V., Arctander, P. and Wilson, A. (1991) Mitochondrial resolution of a deep branch in the genealogical tree for perching birds. *Proc. Royal Soc. London. B.* 243: 99–107.

Endler, J. A. (1982) Pleistocene forest refuges: Fact or fancy?. *In:* G. T. Prance (ed.), *Biological diversification in the tropics.* Columbia Univ. Press, N.Y., pp. 641–657.

Fjeldså, J. (1992) Biogeographic patterns and evolution of the avifauna of relict high-altitude woodlands of the Andes. *Steenstrupia* 18: 9–62.

Fjeldså, J. (1994) Geographical patterns of neoendemic and relict species of Andean forest birds: the significance of ecological stability areas. *Mem. New York. Bot. Gard.*, in press.

Fjeldså, J. and Krabbe, N. (1990) *Birds of the high Andes.* Apollo Books, Svendborg, and Zoological Museum, Copenhagen.

Gentry, A. H. (1982) Neotropical floristical diversity: phytogeographical connections between Central and South America, Pleistocene climatic fluctuations, or an accident of the Andean orogeny? *Ann. Missouri Bot. Gard.* 69: 557–593.

Graves, G. R. (1985) Elevated correlates of speciation and intraspecific geographical variation in plumage in Andean forest birds. *Auk* 102: 556–579.

Graves, G. R. (1988) Linearity of geographical range and its possible effect on the population structure of Andean birds. *Auk* 105: 47–52.

Hooghiemstra, H. (1991) Long continental pollen record from a tropical intermontane basin; Late Pliocene and Pleistocene history from a 540-meter core. *Episodes* 14: 107–115.

ICBP (1992) *Putting biodiversity on the map: Priority areas for global conservation.* International Council for Bird Preservation, Cambridge.

Innis, M. A., Gelfand, D. H., Sninsky, J. J. and White, T. J. (1990) *PCR protocols.* Academic Press, San Diego.

Kerfoot, O. (1968) Mist precipitation on vegetation. *For. Abstr.* 29: 8–20.

Kocher, T. D., Thomas, W. K., Meyer, A., Edwards, S. V., Pääbo, S., Villablanca, F. X. and Wilson, A. C. (1989) Dynamics of mitochondrial DNA evolution in animals: amplification and sequencing with conserved primers. *Proc. Natl. Acad. Sci. USA* 86: 6196–6200.

Krabbe, N. and Schulenberg, T. S. (1994) The *Scytalopus* tapaculos (Rhinocryptidae) of Ecuador, with descriptions of three new taxa and notes on extralimital forms. Manuscript.

Kroodsma, D. E. (1984) Songs of the alder flycatcher (*Empidonax alnorum*) and willow flycatcher (*Empidonax traillii*) are innate. *Auk* 101: 13–24.

Levin, D. A. (1993) Local speciation in plants: the rule not the exception. *Syst. Bot.* 18: 197–208.

Meyer de Schauensee, R. (1966) *The species of birds of South America and their distribution.* Livingston Publ. Comp., Narberth, Penn.

Patton, J. L. (1990) Vicariance versus gradient models of diversification: the small mammal fauna of eastern Andean slopes of Peru. *In:* G. Peters and R. Hutterer (eds), *Vertebrates in the tropics.* Mus. Alex König, Bonn, pp. 355–371.

Patton, J. L. and Smith, M. F. (1992) mtDNA phylogeny of Andean mice: a test of diversification across ecological gradients. *Evolution* 46: 174–183.

Pocs, T. (1974) Bioclimatic studies in the Uluguru mountains (Tanzania, East Africa). I. *Acta Bot. Acad. Sci. Hung.* 20: 115–135.

Poulsen, B. O. (1993) Change in mobility among crepuscular ground-living birds in an Ecuadorian cloud forest during overcast and rainy weather. *Orn. Neotrop.* 4: 103–105.

Remsen, J. R., Jr. and Cardiff, S. W. (1990) Patterns of elevational and latitudinal distribution, including a "niche switch," in some guans (Cracidae) of the Andes. *Condor* 92: 970–981.

Rickman, A. D. and Price, T. (1992) Evolution of ecological differences in the Old World leaf warblers. *Nature* 355: 817–821.

Sibley, C. G. and Ahlquist, J. E. (1990) *Phylogeny and classification of birds. A study in molecular evolution.* Yale Univ. Press, New Haven – London.

Sibley, C. G. and Monroe, B. L., Jr. (1990) *Distribution and taxonomy of birds of the world.* Yale Univ. Press, New Haven – London.

Smith, E. F. G., Arctander, P., Fjeldså, J. and Amir, O. G. (1991) A new species of shrike (Laniidae: *Laniarius*) from Somalia, verified by DNA sequence data from the only known individual. *Ibis* 133: 227–235.

Swofford, D. L. (1991) *PAUP: Phylogenetic Analysis Using Parsimony. Version 3.1.* Computer program distributed by the Illinois Natural History Survey, Campaign, Illinois.

Vielliard, J. M. (1990) Estudo bioacústico das aves do Brasil: o gênero *Scytalopus. Ararajuba* 1: 5–18.

Zimmer, J. T. (1939) Studies of Peruvian birds. No. XXXII. The genus *Scytalopus. Amer. Mus. Novitates* 1044.

Conservation Genetics
ed. by V. Loeschcke, J. Tomiuk & S. K. Jain
© 1994 Birkhäuser Verlag Basel/Switzerland

Genetic distances and the setting of conservation priorities

R. H. Crozier and R. M. Kusmierski

Department of Genetics and Human Variation, La Trobe University, Bundoora, Victoria 3083, Australia

Summary. We develop the Genetic Diversity (GD) approach (Crozier, 1992) to assess the likely biodiversity of reserves using indicator groups. This approach seeks to maximise the preservation of genetic information, and hence requires distances based on divergence and not substitutions. A product formula enables estimation of the GD of the complete set of species and of all subsets implied by possible reserve systems. The bootstrap is suggested as a means of estimating the confidence limits of GD-based priorities. In an example using mitochondrial DNA data from bower birds (*Ptilonorhynchidae*) the order of priorities appeared to be more robust than might have been expected. Faith's (1992) Phylogenetic Diversity (PD) measure is discussed; although PD can return incorrect estimates of genetic biodiversity, it seems likely that it will usually return the same result as GD. When branch lengths are available, use of GD or PD is recommended. In the absence of branch length information, the cladistic dispersion method of Williams and Humphries (1993) is recommended.

Introduction

The goal of conservation biology must be to end the loss of the Earth's biodiversity. Such a goal will require dramatic changes in human society and in our rate of population growth. Such changes are not likely to come about soon, and until they do, our concerns must be to encourage policies which reduce the loss of biodiversity, even if we cannot yet stop this loss.

One of the current approaches to preserving biodiversity is the establishment of reserves. Because of the competing uses for land, not all potential reserves can be established. It is therefore important to assess various possible reserves with the aim of maximising the maintenance of biodiversity. There are various means of assessing potential reserves. One of the ways in which potential reserves can be assessed is in terms of the species present on each which would be preserved if the reserves were to be established. We agree with Vane-Wright et al. (1991) that the use of species diversity (number of species) is an imperfect guide because it does not weight species by their evolutionary distinctness. As Wilson (1992, pp. 73–74) notes, species will vary in their conservation value according to their evolutionary distinctiveness. Wilson (1992, p. 161) proposes that sequence diversity is the final measure of species diversity. Vane-Wright et al. (1991; see also Williams

et al., 1991), Crozier (1992), and Faith (1992) proposed phylogeneti-
cally-based measures of species distinctness. We will here review and
extend Crozier's measures, and illustrate the application of a distance-
based measure to a set of mtDNA sequence data.

A cladogram-based measure

Vane-Wright et al. (1991) proposed a simple measure based on clado-
grams (rooted-trees with no distance information for the branches).
Under this scheme, a cladogram is deduced for the species concerned
and each species is given a conservation weighting inversely proportional to the number of nodes linking it and the root of the tree. This
method presents a number of problems, such as in being so dependent
on the correct placement of the root of the tree. Williams and
Humphries (1993) therefore recommend a *cladistic dispersion* measure:

$$CD = STL + e^{-s.d.(ND_{ij})/\overline{ND_{ij}}}, \tag{1}$$

where, STL = length of tree conserved by a subset of species measured
as the number of species preserved plus the number of (internal) nodes
retained. ND_{ij} = number of nodes in the path linking taxa i and j, plus
2 (the number of terminal nodes, the two species defining the path).

Using the cladistic dispersion measure, the conservation worth of a
single taxon would be the decrease in CD resulting from its loss. The
worth of subsets of taxa can be evaluated using Eq. (1) for the portions
of the complete tree each subset would preserve.

Cladistic-dispersion solves the problems of using rooted trees for
conservation assessment, but two problems remain:

(1) No account is taken of the actual degree of genetical differentiation between taxa (rapid speciation can lead to taxa which are mini-
mally distinct, yet these will be weighted as much as slowly speciating
species with large amounts of differentiation).

(2) Concentration on species rather than populations renders the
analysis prone to instability due to taxonomic redefinition of taxa.

The second objection can be partially dealt with by using populations
rather than species as the units of choice. However, this answer is not
definitive, because species may differ in the number of populations in
the analysis and hence the results will be biased.

The genetic distance method

The genetic distance method (Crozier, 1992) meets both objections
raised to the cladistic methods of the preceding section. Basically, in

agreement with the notion that biodiversity can be defined in terms of genetical information content (Crozier, 1992; Wilson, 1992, pp. 161–162), the set of taxa preserved should maximise the genetic diversity in the system.

The genetic diversity (GD) approach leads naturally to definitions assessing the conservation value of a species within its larger taxon, or of a subset of species from this larger taxon. The values of such subsets enable comparisons between potential reserves in terms of the proportion of the total biodiversity each would maintain.

As noted previously (Crozier, 1992), one definition of the conservation value of a taxon is the probability of it having a unique allele according to the genetic distance measure used. This measure is difficult to calculate, so that an alternative is desirable (Crozier, 1992). The effect of preserving only some species could be examined by defining the value of the group in terms of the probability that it contains more than one allele (Crozier, 1992). This measure, GD, is relatively easily calculated using Eq. (2) for n species:

$$GD = P(>1) = 1 - \prod_{k=1}^{2n-3} (1 - b_k), \qquad (2)$$

where $P(>1)$ = probability of the group having more than one allele; b_k = length of branch k.

The genetic distance that is required is the probability that one allele is replaced by another, not the more usual one giving the probable number of substitutions. We discuss this further below.

The value of a single taxon within the group can then be defined simply as the proportional decrease in the conservation value of the group if this taxon is removed. The value of localities as potential reserves can then be assessed using the same approach and deleting groups of species from the tree.

The appropriate taxonomic unit to use seems to be the population, provided, of course, that the information is available to do this. Using populations minimises the effects of taxonomic decisions. Because the number of branch points between taxa is unimportant, inclusion of closely-related populations will not bias the analysis as it would using cladogram-based measures.

The principle followed here, of maximising the preservation of biodiversity defined as genetic information, is paralleled by the concept of complementarity (Williams et al., 1991). Under the complementarity principle, the aim is to preserve the highest possible overall evolutionary diversity according to various measures based on species distinctness weights.

The appropriate genetic distance

It is important to separate the methods used to construct a tree from what comes next. For tree-building using genetic distances, additivity of the distances is desirable. But the appropriate distances important in then assessing conservation priorities from a given tree will not in general be additive.

The above point can be illustrated using the case of sequence data. For tree-building, the appropriate distance to use would be the substitution distance. On the final tree, the branch lengths estimate the number of substitutions per site between nodes (taxa or internal nodes). However, the substitution distance compensates for multiple substitutions at the same site. Hence the substitution distance, while correctly estimating elapsed time since divergence (to the extent that there is a molecular clock), overestimates the actual difference between two nodes. The departure of the number of different sites from the number of substitutions increases with substitution rate and with time. The required quantity is the percent difference between the nodes, because this then fits the meaning of genetic biodiversity we have followed.

As an example of the properties of the divergence distance, consider Fig. 1 showing a dendrogram of three species. If the distances were substitution distances, then the overall distance expected between species A and B would $= 0.01 + 0.02 = 0.03$. However, if the correct difference distance is used, then the expected distance between A and $B = 0.01 + 0.02(0.99) = 0.0298$.

Substitution distances will generally have been derived from percent difference distances using various well-known methods (e.g., Li and Graur, 1991, pp. 50–54). One approach would be to simply run the methods in reverse for each branch length, rederiving the divergence distance. However, is this necessary? The example just given shows that the differences between the two distances will in general be small and naturally will be in the same direction for every branch, because each distance is a monotonic function of the other. It therefore seems unlikely that significant errors will eventuate if substitution distances are used instead of divergence distances, especially if the values are small.

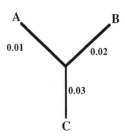

Fig. 1. A three-species tree with genetically-based distances as indicated. The probability of this tree having more than one allele according to these distances and Eq. (1) $= 1 - (1 - 0.01)(1 - 0.02)(1 - 0.03) = 0.0589$.

Similar distance measures

Faith (1992) suggested using the length of tree preserved as a measure of the worth of a taxon or set of taxa (his PD, or Phylogenetic Diversity, measure). While this measure does not lead directly to the preservation of genetic biodiversity, it does approximate Eq. (2) if the genetic distance values are small (Crozier, 1992).

Subsequently, Faith (1993) has suggested the following measure, based on his PD method, instead of Eq. (2):

$$1 - e^{-\sum b_k} \tag{3}$$

For the three-species case, estimating which pair of species to preserve, Eq. (3) will give the same answer as Eq. (2). However, the two equations give different answers in various cases involving larger numbers of species, as for the earlier PD measure (Faith, 1992). Thus, considering the network in Fig. 2 and asking when set {A, B, D} should be preferred over set {A, B, C}, we have the condition from Eq. (2):

$$1 - (1 - a)(1 - b)(1 - f)(1 - g)(1 - d)$$
$$> 1 - (1 - a)(1 - b)(1 - f)(1 - c),$$

which reduces to:

$$g + d > c + gd.$$

The equivalent condition from Eq. (3) is:

$$1 - e^{-(a + b + f + g + d)} > 1 - e^{-(a + b + f + c)},$$

which simplifies to:

$$g + d > c.$$

For small distance values the difference between the two results will be negligible, but the chance of Eq. (3) giving a different result increases as

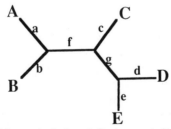

Fig. 2. A five-species tree with genetically-based distances as indicated.

distance values increase. Which measure is correct? According to the logic we have developed here and earlier (Crozier, 1992), Eq. (2) gives the right answer, but there is scope for further analytical work, especially in terms of the predictive power of the two approaches with respect to further aspects of the organisms' biology.

Pamilo (1990) proposed a method of constructing phylograms using the information content of each species. This method could be readily modified to estimate conservation value along the general lines suggested by Crozier (1992) and Faith (1992), by determining the percentage loss of the information content of the whole dendrogram that would be occasioned by the loss of particular sets of species.

Weitzman (1993) proposes a method combining economics theory with evolutionary distinctiveness to assess conservation value. Unlike Crozier (1992) and Faith (1992), Weitzman (1993) does not use a phylogenetic tree for assessing evolutionary distinctiveness but uses the matrix of distances from which a tree could be derived, and stresses more the preservation of particular species than the assessment of potential reserves.

Use of genetically-based phylogenies to assess reserve networks

Eq. (2) can be used to estimate the genetic biodiversity for the complete set of species and for subsets which would be preserved by possible alternative sets of reserves (Crozier, 1992). The various possibilities that fit within political realities can then be compared with regard to the preservation of biodiversity and appropriate policies formulated. Various methods now exist for statistically evaluating phylogenetic schemes (e.g., Felsenstein, 1985, 1988), and these can be applied to the reserve problem. For example, sequence or other data can be bootstrapped (Felsenstein, 1985) to yield confidence intervals for the relative value of sets of reserves. Without such evaluations, it is likely that policy formulations would lack sufficient force to avoid being labelled as merely conjectural.

Towards an example

Birds of the family *Ptilonorhynchidae* are characterised by the construction by the males of many species of elaborate architectural bowers which function as part of the courtship display (Borgia, 1986). Species differ in whether the males are polygynous and build a bower of one of a number of kinds, or are monogynous and build no bower. Bower birds are thus of high intrinsic conservation interest themselves and are therefore perhaps a poor example of a poor 'indicator' group, but they

Table 1. A set of reserves and their characteristics capable of preserving the bowerbird species considered in this analysis

1. Atherton Tablelands
   Far North East Queensland montane rainforest
2. Border Ranges
   Subtropical montane rainforest, north New South Wales (NSW), south Queensland
3. Budawang Ranges
   Temperate rainforest, south NSW
4. Pilliga Scrub
   Semi arid woodland and *Acacia* scrub, north central NSW
5. Barkly Tablelands
   Monsoonal vine scrub and humid woodland, Northern Territory
6. New Guinea Highlands
   Montane rainforest

do provide data of the sort which is becoming readily available (Kusmierski et al., 1993).

Tab. 1 lists a set of hypothetical reserves covering the range of habitats for bower birds in our data set. This list was developed purely for illustrative purposes for use in this example, and is not intended as a proposal for a set of national parks.

We used 870 bp of sequence from the mitochondrial cytochrome b gene to infer the phylogeny of a number of bower bird species differing markedly in behavior (bower type). One hundred bootstrap samples were prepared from this data set using the program SEQBOOT from the PHYLIP 3.4 package (Felsenstein, 1990), and genetic distances derived for each sample according to the Kimura two-parameter method using DNADIST in PHYLIP 3.4. Phylogenies were derived from these distance matrices according to the Neighbor-Joining method (Saitou and Nei, 1987) using NEIGHBOR in the PHYLIP package. The NJ tree from the complete data set, with bootstrap values derived from the bootstrap samples, together with the occurrence of the various bowerbird taxa in the hypothetical reserves, is shown in Fig. 3.

Using 100 bootstrapped samples shows that the phylogeny, while comparatively well-supported, is not completely stable. Applying Eq. (2) to this tree results in a priority listing of systems of reserves (Tab. 2). Systems of reserves that preserve only one species of the group have undefined biodiversity preserved: they have all that is left.

Examining 50 of these bootstrap samples showed that, in 40 of these, reserve 1 preserved more biodiversity than reserve 2, and in 10 the reverse was true. Examining Tab. 2 shows that the priorities acorded to the two reserves differed only at the third decimal place. The relative stability of the ranking is in fact surprising and encouraging!

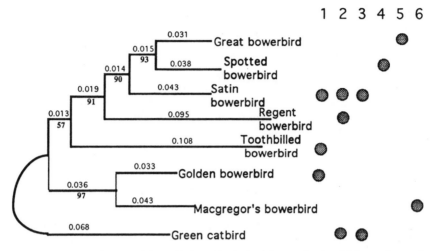

Reserves

1 2 3 4 5 6

Fig. 3. Phylogeny of a set of bowerbirds as determined from mitochondrial cytochrome b gene sequences (see text). Genetic distances are shown, and the number of times the group, implied by a branch, occurred in 100 bootstrap replicates are shown. The presence of a species in any of the six reserves listed in Tab. 1 is shown by a solid dot. The root of the tree is on the branch to the Green Catbird, as determined by outgroup analysis (Kusmierski et al., 1993).

Table 2. Genetic diversity values for all possible systems of reserves according to the dendrogram in Fig. 3

Reserve system	Genetic diversity
ALL	0.444
1	0.240
2	0.239
3	0.159
4	?
5	?
6	?
1 & 2	0.359
1 & 3	0.301
1 & 4	0.280
1 & 5	0.275
1 & 6	0.273
2 & 3	0.239
2 & 4	0.279
2 & 5	0.274
2 & 6	0.298
3 & 4	0.204
3 & 5	0.198
3 & 6	0.225
4 & 5	0.068
4 & 6	0.165
5 & 6	0.159

Prospects

A Macintosh computer program is being developed to reduce the large amount of trivial yet onerous calculation required. The intent is to allow choice of measure (GD or PD), selection of several systems of reserves, and bootstrap determination of the confidence limits of reserve priorities taking input from appropriate PHYLIP programs (normalised against the biodiversity of the whole set of species for each bootstrap sample). Such a computer program would allow sufficient bootstrap replicates to be assessed to give reasonable confidence in the results (Hedges, 1992). A preliminary program, calculating GD and PD, is available from RHC in binhexed form *via* e-mail.

Discussion

We stress that we do not support the genetic distance method to the exclusion of all others, but rather propose it as a further source of management information along with considerations of ecological role and cultural value and, of course, the inescapable economic constraints of the present political system. Although the method we have discussed could be used to assess species in particularly significant groups for preservation, we agree with Vane-Wright et al. (1991) and Williams et al. (1991) that the major use for phylogenetic methods will be to identify ecological communities of apparent high biodiversity importance, as evidenced by indicator groups. As such the phylogenetic method will supplement others useful for identifying likely communities of high conservation value, such as landform analysis (Pressey et al., 1993). (Williams and Gaston (1993) point out that it will often be difficult to select indicator groups appropriate globally, and suggest using the number of families as an indicator of biodiversity. Such an approach would greatly reduce the effort required for estimating biodiversity but, as Williams and Gaston (1993) acknowledge, choice of any higher taxon rank is difficult. The main problem is the difference in tendency to split in different systematic traditions, e.g., birds and ants constitute groups believed to have similar numbers of species and to be of similar ages (Hölldobler and Wilson, 1990, p. 23; Sibley and Ahlquist, 1990, p. 273), yet all ants are placed into one family whereas birds are divided into many.)

Crozier (1992) presented two measures concerned with genetic biodiversity: an estimate of species uniqueness (defined as the per site probability that the species has a unique allele, and provided to counterpoint the original Vane-Wright et al. (1991) proposal) and the genetic diversity measure developed further here. In the same issue, Faith (1992) presented his PD measure. Faith (1993) supports the use of

236

distance measures for determining conservation priority, but criticises aspects of Crozier's (1992) paper. Crozier (1992) proposed scaling distances using a constant; we agree with Faith (1993) that the constant is unnecessary. Faith (1993) advocates use of GD (or PD) alone, because the species uniqueness measure is less convenient to calculate; Crozier (1992, p. 14) already took this approach and we have developed it further here. Faith's (1993) 'correction' of Crozier's specific species uniqueness formula stems from Faith's use of a two-allele model, whereas an infinite alleles model (e.g., Hedrick, 1985, pp. 334–337) is more appropriate, was the one used, and, as can be shown, is closer to the finite-allele answer even for the restrictive four-state model generally applicable to nucleic acid sequences. Faith's (1993) remaining criticisms of Crozier (1992) all seem to stem from his preference for substitution distances over divergence distances whereas, as we have shown, divergence distances are more appropriate.

Other population-genetic aspects relevant to conservation management could be included in the approach developed here. For example, a non-dichotomous tree approach (reviewed by Maynard Smith, 1989) could be used, say by replacing the tree with a matrix representation. Incorporation of differing levels of gene diversity within populations might be approached by continuing the phylogenetic analysis down into the population level using a phylogeny of alleles. It is likely that each of these suggestions would be hard to implement separately, and much more so to implement together. It is also unclear at present whether the effort would be cost-effective in terms of significantly improving the precision of our estimates, especially given the need to estimate empirical sufficiency!

Our recommendations are therefore to use the GD or, perhaps, the PD method if distance information is available when using phylogenetic information as an input to conservation decisions. If no estimates of branch lengths are available, then we should use CD, which should also be used with some statistical method to estimate sufficiency, such as the bootstrap.

Acknowledgements
We thank the Australian Research Grants Scheme and the Ian Potter Foundation for grants to RHC for work on bowerbirds and evolutionary genetics generally, Dr P. H. Williams for discussions and timely reprints, and Dr D. P. Faith for sending us a copy of his 1993 paper before publication and for discussions.

References

Borgia, G. (1986) Sexual selection in bower birds. *Sci. Amer.* 254: 91–101.
Crozier, R. H. (1992) Genetic diversity and the agony of choice. *Biol. Cons.* 61: 11–15.
Faith, D. P. (1992) Conservation evaluation and phylogenetic diversity. *Biol. Cons.* 61: 1–10.
Faith, D. P. (1993) Genetic diversity and taxonomic priorities for conservation. *Biol. Cons.* (in press).

Felsenstein, J. (1985) Confidence limits on phylogenies: an approach using the bootstrap. *Evolution* 39: 783–791.

Felsenstein, J. (1988) Phylogenies from molecular sequences: Inference and reliability. *Annu. Rev. Genet.* 22: 521–565.

Felsenstein, J. (1990) *PHYLIP manual version 3.3*. University Herbarium Berkeley, CA.

Hedges, S. B. (1992) The number of replications needed for accurate estimation of the bootstrap P value in phylogenetic studies. *Mol. Biol. Evol.* 9: 366–369.

Hedrick, P. W. (1985) *Genetics of populations*. Jones & Bartlett, Boston, MA.

Hölldobler, B. and Wilson, E. O. (1990) *The ants*. Harvard Univ. Press, Cambridge, MA.

Kusmierski, R. M., Borgia, G., Crozier, R. H. and Chan, B. (1993) Molecular information on bowerbird phylogeny and the evolution of exaggerated male characteristics. *J. Evol. Biol.* 6: 737–752.

Li, W-H. and Graur, D. (1991) *Fundamentals of molecular evolution*. Sinauer, Sunderland, MA.

Maynard Smith, J. (1989) Trees, bundles or nets? *Trends Ecol. Evol.* 4: 302–304.

Pamilo, P. (1990) Statistical tests of phenograms based on genetic distances. *Evolution* 44: 689–697.

Pressey, R. L., Humphries, C. J., Margules, C. R., Vane-Wright, R. I. and Williams, P. H. (1993) Beyond opportunism: key principles for systematic reserve selection. *Trends Ecol. Evol.* 8: 124–128.

Saitou, N. and Nei, M. (1987) The neighbor-joining method: A new method for reconstructing phylogenies. *Mol. Biol. Evol.* 4: 406–425.

Sibley, C. G. and Ahlquist, J. E. (1990) *Phylogeny and classification of birds*. Yale University Press, New Haven.

Vane-Wright, R. I., Humphries, C. J. and Williams, P. H. (1991) What to protect? – Systematics and the agony of choice. *Biol. Cons.* 55: 235–254.

Weitzman, M. L. (1993) What to preserve? An application of diversity theory to crane conservation. *Quart. J. Econ.*, Feb. 157–183.

Williams, P. H. and Gatson, K. J. (1993) Measuring more of biodiversity: can higher-taxon richness predict wholesale species richness? *Biol. Cons.* (in press).

Williams, P. H. and Humphries, C. J. (1993) Biodiversity, taxonomic relatedness and endemism in conservation. *In:* Forey, P. L., Humphries, C. J. and Vane-Wright, R. I. (eds), *Systematics and conservation evaluation*. Oxford Univ. Press, Oxford, UK.

Williams, P. H., Humphries, C. J. and Vane-Wright, R. I. (1991) Measuring biodiversity: taxonomic relatedness for conservation priorities. *Aust. Syst. Bot.* 4: 665–679.

Wilson, E. O. (1992) *The diversity of life*. Harvard Univ. Press, Cambridge, MA.

Conservation Genetics
ed. by V. Loeschcke, J. Tomiuk & S. K. Jain
© 1994 Birkhäuser Verlag Basel/Switzerland

Multi-species risk analysis, species evaluation and biodiversity conservation

L. Witting[1], M. A. McCarthy[2] and V. Loeschcke[1]

[1]*Department of Ecology and Genetics, University of Aarhus, Ny Munkegade, DK-8000 Aarhus C, Denmark*
[2]*Faculty of Agriculture and Forestry, University of Melbourne, Parkville, Victoria 3052, Australia*

Summary. The multi-species conservation approaches starting from Vane-Wright et al. (1991) search for a biologically valid basis to evaluate biodiversity. Such evaluations can be used to optimize reserve boundaries so reserves contain a high level of biodiversity. However, these optimization procedures do not minimize the future loss of biodiversity. We provide a method that can be used to minimize the loss of biodiversity. The method integrates an evolutionary species evaluation with an ecological multi-species risk analysis to estimate the expected loss of phylogenetic diversity. A minimization of this loss will optimize the preservation of phylogenetic diversity.

Introduction

In this contribution we describe how to optimize conservation strategies to maximize biodiversity preservation. But let us first briefly relate our methodology to the literature of the field of conservation biology. Most papers on conservation consider only a single population or a single species. In the single species context a general conservation goal is to maximize the viability of the species. This is often done by the use of a method which is referred to as a population viability analysis (PVA) (*e.g.*, Ginzburg et al., 1982; Gilpin and Soulé, 1986; Boyce, 1992; Burgmann et al., 1993). These PVAs can guide the management and the protection of a single species. A typical scenario could be when the effects of several strategies are investigated. Each strategy will alter the parameters in a model and, consequently, the risk to the population can be quantified for each strategy. From these results an acceptable strategy can be chosen.

At the other end of the conservation spectrum we find the multi-species approaches initially suggested by Vane-Wright et al. (1991) and later elaborated by others (*e.g.* Crozier and Kusmierski, this volume; Brooks et al., 1992; Erwin, 1991; Crozier, 1992; Faith, 1992; Pressey et al., 1993). These methods search for a biologically valid basis to evaluate biodiversity. The evaluations are often based on a phylogenetic tree and they can be used to define reserve boundaries so reserves contain a phylogenetically diverse composition of species (*e.g.* Vane-Wright et al., 1991; Pressey et al., 1993). However, these diversity optimization procedures do not consider the risks experienced by the different species. Thus, as the diversity is optimized within the reserves the *loss* of biodiversity *is not minimized* (Witting and Loeschcke, 1993). It is easy to imagine scenarios where a diversity optimization selects areas with a high diversity of relatively non-threatened species, whereas areas that contain more endangered species, but otherwise are low in species diversity, are discarded. We provide procedures that integrate an evolutionary diversity evaluation with an ecological multi-species risk analysis. This type of method was first introduced by Witting and Loeschcke (1994) and it can be used to estimate the expected loss of phylogenetic diversity, $D^{(-)}$. A minimization of $D^{(-)}$ will then optimize the preservation of phylogenetic diversity in agreement with the aim of conservation biology to preserve the present biodiversity.

In the first section, we present a description of biodiversity. This is done to clarify the biodiversity concept and, in particular, to describe the interconnection between biodiversity and a phylogenetic species evaluation. Hereafter, we continue to biodiversity conservation, where we focus on different methods that can be used to estimate the expected loss of phylogenetic diversity. The conservation section is partitioned into three subsections. In the first subsection we describe the estimation of $D^{(-)}$ when the extinction events of species are independent, or when the correlations among the extinction events are known. The correlations among extinction events represent a convenient measure from which to estimate $D^{(-)}$. However, an estimation procedure for the correlations is not easily available. We do not consider this problem in more detail, instead we leave it for future investigations. The correlations among extinction events also contain no information on causality. Therefore, we focus in the next subsection on some methods that can incorporate information on extinction causality. These approaches have the advantage that they can not only be used to estimate $D^{(-)}$, but the methods can also analyze the viability of a community, where they among other things can be used to identify keystone species. These PVA based procedures can be used to estimate background extinctions from specific areas as, for example, nature reserves. However, throughout the world one major threat to species diversity is habitat fragmentation. In this context, not only background extinctions from reserves takes place, but also, and probably more important, an immediate threat exists that the habitat fragmentation will destroy the habitat required by the different species. Our third subsection is an attempt to create a method that can be used to minimize the loss of biodiversity when data are sparse and habitat fragmentation takes place.

Biodiversity

Biodiversity is described by the total amount of biotic information of a biological system. Biotic information is contained within or arises from three basic and mutually exclusive domains. These domains are the materially-transmitted, the non-materially transmitted, and the non-transmitted information. Following the concept of Williams (1992), a major proportion of the materially transmitted information can be ascribed to the codical domain of genes. The codical domain reflects the information (messages) coded into the genes. Non-materially transmitted information can, for example, be neurally coded and transmitted from parents to offspring *via* learning, whereas non-transmitted information is generated by each individual through plastic responses. These three domains form the basis from which biotic information at higher levels arises. High level information can, for example, be information associated with patterns of species interactions. This type of information may not be preserved by *ex situ* conservation. However, if preservation of the basic entities (species and individuals) results in preservation of all information at the basic level, high level information can be regenerated (assuming that the physical surroundings remain intact). Before deciding on an eventual *ex situ* preservation strategy it is important to be aware to what extent a biological system can regenerate. If all information at the basic level is contained within the genes, it is apparent that a system is able to regenerate from the basic entities. However, important information may be contained within the non-materially transmitted and the non-transmitted domains. For example, the pattern in which the species is distributed may contain essential information for the persistence of a biological system. This type of information is likely to be lost *ex situ* and, consequently, a system may become unable to regenerate from *ex situ* preserved entities.

If we can evaluate the information contained within the three basic domains we will have a measure of the diversity of a biological system. Let us, however, restrict ourselves to a description of the information contained within the genetic messages of the codical domain. Wiener (1949) defined the information of a message as

$$D = -\ln(p_i), \tag{1}$$

where p_i is the probability that the message occurs. A gene can be viewed as a sequence of genetic messages which, for example, could code for a protein. For an amino acid in an amino acid site within a protein the probability, p_i, would be the evolutionary determined probability that the given amino acid is present at that particular site. Since there are no means by which these probabilities can be determined it might be assumed that the probability is the same for each message, i.e.,

$p_i = p$. It then follows that the information of a gene is proportional to the number of genetic messages, n, contained within the gene

$$D = -\ln(p^n) \sim n. \tag{2}$$

If the total number of message sites per gene is large and the number of messages *per* site is also not too small, each change in the genetic message is likely to be a unique event. In such a situation we can ignore the possibility that two distant species contain the same genetic allele when more closely related species do not. The diversity in genetic information among a set of species can then be described by an information tree. In such a tree the weight of a path from a common ancestor is proportional to the total number of message sites that differ between the evolutionary lineages. Fig. 1 illustrates such a tree. Here, species A and B differ from species E and F on nine different sites. Species C and D lack one message site when compared to the other species. Species C and D, furthermore, differ from species A and B on eight sites, from species E and F on four sites, and from each other on one site. Species with more genes will most often contain more message sites and, consequently, have a higher level of information than will species with fewer genes. The genetic information contained within a set of species is given as

$$D = \sum_{i=1}^{m} D_i, \tag{3}$$

where D_i is the weight of the ith internode of an information (phylogenetic) tree containing all and only all the species common to the species

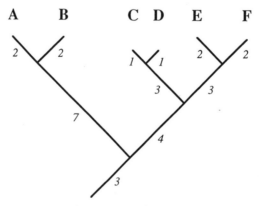

Fig. 1. A tree illustrating the differentiation in genetic information among six species, A, B, C, D, E, F. Species A and B differ from species E and F on nine different message sites. Species C and D lack one message site when compared to the other species. Species C and D differ from species A and B on eight sites, from species E and F on four sites, and from each other on one site. The genetic information, D, contained within the species set is given as $D = 3 + 7 + 4 + 3 + 3 + 2 + 2 + 1 + 1 + 2 + 2 = 30$.

set, and m is the number of internodes common to the information tree. This evaluation is comparable to the growing recognition that phylogenies should play a role in multi-species evaluation (*e.g.*, Vane-Wright et al., 1991; Brooks et al., 1992; Erwin, 1991; Crozier, 1992; Faith, 1992; Pressey et al., 1993).

Expected loss of genetic information

The expected loss of genetic information can be estimated by an integration of an evolutionary species evaluation and an ecological multi-species risk analysis. The evolutionary evaluation will be based on a phylogenetic tree as in Fig. 1. The problem of constructing the tree will not be dealt with here; we simply assume that a tree, with internode weights equal to the properties described in the preceding section, can be constructed. Our method of ecological risk analysis requires estimates of the probability that species will become extinct within some defined time-period. The estimation of extinction probabilities is to some extent described in the literature (*e.g.*, Boyce, 1992; Burgman et al., 1993).

As an introduction to the estimation of the expected loss of genetic information, consider a situation where the extinction events of the different species are independent. This could be the case if the population dynamics of the different species are uncorrelated. The probabilities of multi-species extinction can then be calculated by probability multiplication. Hence, the expected loss of genetic diversity from a set of species is given as

$$D^{(-)} = \sum_{i=1}^{m} D_i \prod_{j=1}^{n_i} p_j, \qquad (4)$$

where m is the number of internodes common to the phylogenetic tree and n_i the number of species terminal to the ith internode. Fig. 2 illustrates this calculation.

Correlated extinction events

Extinction events of species are in most cases likely to be correlated with one another. These correlations can be caused either by species interactions or because the different species respond to fluctuations in the same abiotic environmental factors. The extinction events might then be characterized by an extinction correlation matrix:

$$\begin{bmatrix} p_A & - & - & \cdot & - \\ r_{AB} & p_B & - & \cdot & - \\ r_{AC} & r_{BC} & p_C & \cdot & - \\ \cdot & \cdot & \cdot & \cdot & \cdot \\ r_{A\phi} & r_{B\phi} & r_{C\phi} & \cdot & p_\phi \end{bmatrix}, \qquad (5)$$

244

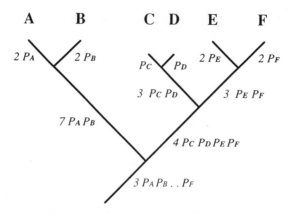

Fig. 2. An illustration of how to calculate the expected loss of genetic information, $D^{(-)}$, when the extinction events of the different species are independent. $D^{(-)} = P_C + P_D + 2(P_A + P_B + P_E + P_F) + 3(P_C P_D + P_E P_F) + 7 P_A P_B + 4 P_C P_D P_E P_F + 3 P_A P_B \cdots P_F$. The tree resembles that of Fig. 1.

where p_A is the probability that species A becomes extinct unconditional of the extinction of other species (this is the overall extinction probability of species A), and r_{AB} is the correlation between the extinction events of species A and B. If we consider these extinction correlations, then they contain sufficient information to determine the probability of all multi-species extinction events and, consequently, to estimate $D^{(-)}$. To see this, consider an extinction as a Bernoulli event, so that the extinction of species A is the event $A = 1$ and the persistence of the species is the event $A = 0$. We can define the correlation between the extinction of two species A and B as

$$r_{AB} = cov_{AB} / \sigma_A \sigma_B, \tag{6}$$

where cov_{AB} is the covariance between events A and B, and σ represents the standard deviation of the event. Covariance is equal to

$$cov_{AB} = E(A \cdot B) - E(A)E(B), \tag{7}$$

where E represent the mean value. So, for Bernoulli random events, we have

$$cov_{AB} = p_{AB} - p_A p_B,$$
$$\sigma_A = \sqrt{p_a(1 - p_A)}. \tag{8}$$

Given the overall probabilities that species A and B will become extinct, and given the correlation between these two events, it is possible to calculate the probability that both will become extinct as,

$$p_{AB} = p_A p_B + r_{AB} \sigma_A \sigma_B. \tag{9}$$

Note that the probability of other outcomes also can be obtained. For example, the probability that A will survive B becomes extinct is

$$p_{A'B} = p_{A'} p_B - r_{AB} \sigma_A \sigma_B, \tag{10}$$

where $r_{A'B} = -r_{AB}$ and $p_{A'} = 1 - p_A$.

The application to a three-species system is obtained by considering partial correlation coefficients. Using conditional probability, we have

$$p_{ABC} = p_{AB|C} p_C,$$ (11)

and from Eq. (9),

$$p_{AB|C} = p_{A|C} p_{B|C} + r_{AB|C} \sigma_{A|C} \sigma_{B|C},$$ (12)

where $p_{A|C}$ is the probability of event A occurring given C, and $r_{AB|C}$ is the correlation between the extinctions of species A and B given the outcome of C. Note that $r_{AB|C}$ is equal to $r_{AB|C'}$ so we can consider the partial correlation between A and B which is equal to (Sokal and Rohlf, 1981)

$$r_{AB.C} = \frac{r_{AB} - r_{AC} r_{BC}}{\sqrt{(1 - r_{AC}^2)(1 - r_{BC}^2)}}.$$ (13)

The conditional probabilities in Eq. (12) are equal to

$$p_{A|C} = p_{AC}/p_C,$$ (14)

with p_{AC} obtained from Eq. (9). This approach can be extended to any number of species. The generalized equations are

$$p_{AB..\phi} = p_{AB|C..\phi} p_{C..\phi},$$

$$p_{AB|C..\phi} = p_{A|C..\phi} p_{B|C..\phi} + r_{AB.C..\phi} \sigma_{A|C..\phi} \sigma_{B|C..\phi},$$

$$r_{AB.C..\phi} = \frac{r_{AB.D..\phi} - r_{AC.D..\phi} r_{BC.D..\phi}}{\sqrt{(1 - r_{AC.D..\phi}^2)(1 - r_{BC.D..\phi}^2)}},$$ (15)

$$p_{A|C..\phi} = p_{AC..\phi}/p_{C..\phi}.$$

Eq. (15) can be used to determine the probabilities of multi-species extinction and, consequently, the expected loss of genetic information can be calculated as

$$D^{(-)} = \sum_{i=1}^{m} D_i p(S_i),$$ (16)

where m is the number of internodes common to the phylogenetic tree and $p(S_i)$ is the probability that all species terminal to the ith internode become extinct (calculated by the use of Eqs (15)).

Causality and community viability

The correlations among extinction events contain no information on the causal connections of the risks experienced by the different species. The optimization of conservation strategies is, however, a search to identify specific factors by which it is possible to minimize $D^{(-)}$ as much as possible. Such factors that simultaneously influence the risk of several species can have two origins. They can be either biotic, where a species due to, for example, competition and predation affects other species, or abiotic, where fluctuations or changes in underlying abiotic factors affects different species.

When conservation programs focus on abiotic factors the risk of multi-species extinction can be estimated by combining single species viability analysis (PVAs) and correlated random numbers. In a single-species viability analysis, environmental fluctuations are simulated by randomly generating a growth rate, or a set of survival and fecundity rates, at each time step. This leads to fluctuations in the abundance of individuals which will contribute to the risk of decline faced by the population. Modeling environmental variation in the multi-species case is analogous to the single-species situation, but we need to account for the appropriate correlations in the demographic parameters. In terms of population viability analysis, this means that the random numbers must account for the correlations between these parameters of the different species. Correlated random numbers can be generated from linear combinations of uncorrelated random numbers (Ripley, 1987; Burgman et al., 1993). Once these numbers have been generated each species can be simulated one by one. Recording all extinction events can then provide a method to estimate the probability that each possible combination of species becomes extinct. From this and an information tree the expected loss of genetic information can be calculated.

Biologically caused correlations among extinction events can be dealt with by using either a multi-species simulation model that incorporates the interactions explicitly, or by PVAs combined with an analytical multi-species risk model as proposed by Witting and Loeschcke (1994). Here, we will only make a specific reference to the latter type of model since it explicitly incorporates all possible extinction pathways by which the extinction risk experienced by one species affects the extinction risks experienced by all other species. The method requires that the parameters of the PVA model of each species are estimated when all species of the community are present and when directly interacting species are absent one by one. Since the multi-species risk model incorporates all possible extinction pathways, the method cannot only be used to estimate $D^{(-)}$, but it can also give detailed information on the ecological interdependence of the risks experienced by the different species. Hence, the method can be used to analyze the biological component of the viability of species within a community, a methodology we refer to as a community viability analysis (CVA). In a multi-species conservation situation a CVA can give at least four different types of relevant information. Starting from the single species point of view, the influence from the extinction of one species on the extinction risk experienced by another species can be given. For a single species the method can also evaluate the degree to which the extinction risk of the species is dependent on the extinction of directly and indirectly interacting species and, consequently, the degree to which the extinction of a species remains independent of the presence or absence of other species. This type of information can be summarized for all species to describe the proportion of species which are beneficially, negatively, and neutrally affected by the extinction of other species. This will give some insight into the risk that a community faces, i.e., of collapsing once the initial extinctions start to occur. In this context it is also possible to identify keystone species, i.e., species that are of crucial importance for the persistence of other species. This can explicitly be given as the probability that the extinction of a species induces a cascade extinction of two, three, or more additional species. For a formularization of the CVA see Witting and Loeschcke (1994).

Habitat alteration

Throughout the world, one major threat to species is habitat alteration/ destruction. In connection with habitat alteration there is both a direct and an indirect extinction effect. The direct effect is the immediate decline in species diversity caused by alteration of the habitat required by the different species, while the indirect effect is the background extinctions occurring among the remaining species. Let $p_{A,j}$ be the probability that species j becomes extinct as an immediate consequence of habitat alteration, and let, for a species that survived the immediate threat, $p_{B,j}$ be the probability that species j becomes extinct from the remaining habitat. The overall loss of genetic information from an area where habitat alteration takes place is then given as

$$D^{(-)} = \sum_{i=1}^{m} D_i \prod_{j=1}^{n_i} (p_{A,j} + (1 - p_{A,j})p_{B,j}). \qquad (17)$$

Population viability analyses deal with the indirect background type of extinction effects where the PVA based on population dynamic information can be used to estimate the risk experienced by the population. One goal of establishing a reserve system is to ensure that the species within the reserves have a high probability of persistence. Hence, for such reserves all $p_{B,j} \to 0$ and, consequently,

$$D^{(-)} \to \sum_{i=1}^{m} D_i \prod_{j=1}^{n_i} p_{A,j}. \qquad (18)$$

It is then evident that when a self-sustainable reserve system is established the loss of genetic information is given by the immediate effects of the habitat alteration taking place outside the reserves. Habitat alteration is then the only factor causing the increased risk experienced by the different species. Species can then be categorized according to the sensitivity of the species to the habitat alteration. A very crude separation would consider only habitat-sensitive and habitat-insensitive species. The persistence of habitat sensitive species will depend on the presence of the original habitat type and, consequently, these species will disappear as the habitat is changed. In contrast, the habitat insensitive species can persist in the altered habitat. A generally applicable and easily collectable form of data could consist of information on whether a species is present or absent from a habitat patch. To ensure that the estimates of extinction risk are comparable among patches, each patch must be of a similar size. In this way a habitat patch is defined according to the scale on which data can be collected. Thus, when information is plentiful, we will operate with many small patches, whereas when information is sparse, we will have only a few large patches. Since we consider extinctions caused directly by the habitat alteration it is reasonable to make the simplifying assumption that the persistence of the different species, both within and among patches, are independent from one another.

Extinctions can take place among the habitat-sensitive species that are unable to persist in the changed habitat. For these species the extinction probability from a given patch will be zero if no habitat of the patch is changed. The extinction probability is likewise one if the total patch is changed. If we assume a linear relationship between the risk of species extinction and the relative amount of the patch being altered, the extinction probability of a habitat sensitive species is estimated as

$$p = A_C / A_0, \tag{19}$$

where A_C is the area of the changed habitat within the patch and A_0 is the initial area of the patch. The extinction probabilities of the habitat insensitive species are by definition zero. The expected loss of genetic information from the overall area is then given as

$$D^{(-)} = \sum_{i=1}^{m} D_i \prod_{j=1}^{n_i} \prod_{l=1}^{r_j} p_{j,l}, \tag{20}$$

where $p_{j,l}$ is the extinction probability of the jth species within the lth patch (given by Eq. (19)), m is the number of internodes common to the information tree containing all species of concern, n_i is the number of species terminal to the ith internode, and r_j is the number of patches where the jth species is present before the habitat change took place. A minimization of $D^{(-)}$ with respect to the distribution of the habitat alteration will then, in connection with the establishment of self-sustainable reserves, save most genetic information from being lost. Note, that when, for Eq. (17), all $p_{A,j} \to 0$ and all $p_{B,j} \to 0$, our conservation goal is achieved.

Conclusion

The method proposed in this chapter is only one of many possible ways to evaluate multi-species conservation strategies. Obviously, real ecosystems will not conform precisely to the idealized model system considered in this paper. However, our message is not that our mathematics is the only valid path for multi-species conservation. Instead, we want to stress that conservation programs confronted with multi-species preservation should base eventual decisions on an integration of a species evaluation system and a multi-species risk analysis. In this context we have formulated some considerations that explicitly focus on the biological components of multi-species conservation.

Acknowledgements
Jon Fjeldså is thanked for helpful discussions, and Jürgen Tomiuk and Subodh Jain for helpful comments on a draft manuscript. This study has been supported by a grant from the Danish National Research Council (#11-9025-1 and #11-9639-1) and a Ph.D. stipend from the Faculty of Natural Science, University of Aarhus.

References

Boyce, M. S. (1992) Population viability analysis. *Annu. Rev. Ecol. Syst.* 23: 481–506.

Brooks, D. R., Mayden, R. L. and McLennan, D. A. (1992) Phylogeny and biodiversity: Conserving our evolutionary legacy. *Trends Ecol. Evol.* 7: 55–59.

Burgman, M. A., Ferson, S. and Akçakaya, H. R. (1993) *Risk assesment in conservation biology.* Chapman & Hall, London.

Crozier, R. H. (1992) Genetic diversity and the agony of choice. *Biol. Cons.* 61: 11–15.

Erwin, T. L. (1991) An evolutionary basis for conservation strategies. *Science* 253: 750–752.

Faith, D. P. (1992) Conservation evaluation and phylogenetic diversity. *Biol. Cons.* 61: 1–10.

Gilpin, M. E. and Soulé, M. E. (1986) Minimum viable populations: processes of species extinction. *In:* Soulé, M. E. (ed.), *Conservation biology: the science of scarcity and diversity.* Sinauer, Sunderland, MA, pp. 19–34.

Ginzburg, L. R., Slobodkin, L. B., Johnson, K. and Bindman, A. G. (1982) Quasiextinction probabilities as a measure of impact on population growth. *Risk Anal.* 2: 171–181.

Pressey, R. L., Humphries, L. J., Margules, C. R., Vane-Wright, R. I. and Williams, P. H. (1993) Beyond opportunism: Key principles for systematic reserve selection. *Trends Ecol. Evol.* 8: 124–128.

Ripley, B. D. (1987) *Stochastic simulation.* Wiley, New York.

Sokal, R. R. and Rohlf, F. J. (1981) *Biometry*, 3rd edition. W. H. Freeman and Company, New York.

Vane-Wright, R. I., Humphries, C. J. and Williams, P. H. (1991) What to protect? – Systematics and the agony of choice. *Biol. Cons.* 55: 235–254.

Wiener, N. (1949) A new concept of communication engineering. *Electronics* 22: 74–77.

Williams, G. C. (1992) *Natural selection. Domains, levels, and challenges.* Oxford University Press, New York.

Witting, L. and Loeschcke, V. (1993) Biodiversity conservation: Reserve optimization or loss minimization? *Trends Ecol. Evol.* 8: 417.

Witting, L. and Loeschcke, V. (1994) The conservation of interacting species. Manuscript.

Part V

Case studies

Conservation Genetics
ed. by V. Loeschcke, J. Tomiuk & S. K. Jain
© 1994 Birkhäuser Verlag Basel/Switzerland

Introductory remarks

Many species share similar basic properties, but each one has some particular characteristics. Knowledge about the distribution of a species, its population size and the conditions affecting its population structure is fundamental for an effective conservation program. Case studies of endangered species are therefore necessary to characterize a species' reaction to changing environmental conditions, and conditions for its successful conservation.

In nature, geographically widespread species are often subdivided into local populations that experience different environments. Differences in the genetic structure among such local populations can be caused by adaptive or random processes, where adaptation to an environment can occur by the reorganization of the genetic population structure or by phenotypic changes of the individuals. The population structure represents an evolutionary state that has evolved in response to a past set of environmental conditions. When this natural equilibrium is disturbed by habitat fragmentation or reduction, then the knowledge about the evolutionary history of a species is important for determining the potential detrimental influence to the species. If, for example, species have adapted to high inbreeding during their evolution, then small population sizes might not really affect its population fitness. On the other hand, species with large population sizes and random mating might be sensitive to such episodic drastic reductions of their population sizes (bottlenecks). Further, the consequences of habitat fragmentation depend on the migration rate; pollinators, for example, can maintain high migration rates in certain plants even among distant habitat fragments. One may envisage, generally speaking, that as natural habitats become increasingly fragmented by human activities, population size bottlenecks could bring about a significant loss of genetic variability and might even increase genetic load of the population.

Except for catastrophes, changes in nature are relatively slow and populations can adapt step-by-step to new conditions through changes of their population structure. In nature, individuals additionally show large ranges of phenotypical adaptation to environmental changes. Genotypes can adaptively modify their phenotype to different environments. The characteristic function of a genotype that describes its ability to change phenotypically as a reaction to environmental conditions is named its norm of reaction (Woltereck, 1909; Schmalhausen, 1949).

However, natural habitats today are affected seriously by human-induced changes; for example, serious incidents of chemical and physical pollution are today frequent catastrophes for natural communities, and are those to which most species might not adapt genetically or phenotypically. To respond successfully, physiological or phenotypic changes that exceed the normal range of individual adaptability might be required. A discussion of adaptive constraints in terms of quantitative genetics and life history theories is to be found in several works (e.g., Loeschcke, 1987; Roff, 1992; Stearns, 1992). Human activities expose individuals to stress which, in the general sense, "is the disturbance of the normal functioning of an individual by environmental factors, as detected by the departure from a steady state" (Hoffmann and Parsons, 1991). When the environmental conditions experienced by individuals are stressful, then the risk of population extinction increases. In a recent study (Kareiva et al., 1993), population and community responses to environmental changes are assessed in genetic and physiological terms. Many scenarios predict rather dramatic shifts in the surviving biota and their gene pools. Species must be considered as threatened when their habitats become fragmented or even partly destroyed. A conservation program may be the regeneration of populations by the release of individuals from different geographic localities or by the release of individuals bred in managed programs (e.g., in fishery and forestry programs). For most rare and endangered plants, habitat protection through the establishment of nature reserves is of high priority; genetic and demographic information may contribute to the design and management of reserves. In any case, the local population structure must be considered and the stress factors which endanger species have to be determined. Sometimes, it then may be possible to protect populations from exposure to stress, but at other times individuals would have to be selected that tolerate the stressful environment.

References

Hoffmann, A. A. and Parsons, P. A. (1991) *Evolutionary genetics and environmental stress.* Oxford University Press, Oxford.

Kareiva, P. M., Kingsolver, J. G. and Hey, R. B. (1993) *Biotic interactions and global changes.* Sinauer, Sunderland, MA.

Loeschcke, V. (1987) *Genetic constraints on adaptive evolution.* Springer, Berlin-New York.

Roff, D. A. (1992) *The evolution of life histories. Theory and analysis.* Chapman & Hall, New York.

Schmalhausen, I. I. (1949) *Factors of evolution.* Blakiston, Philadelphia.

Stearns, S. C. (1992) *Evolution of life histories.* Oxford University Press, Oxford.

Woltereck, R. (1909) Weitere experimentelle Untersuchungen über Artveränderungen, speziell über das Wesen quantitativer Unterschiede bei Daphniden. *Ver. dtsch. zool. Gesell.* 1919: 110–173.

Conservation Genetics
ed. by V. Loeschcke, J. Tomiuk & S. K. Jain
© 1994 Birkhäuser Verlag Basel/Switzerland

On genetic erosion and population extinction in plants: A case study in *Scabiosa columbaria* and *Salvia pratensis*

R. Bijlsma[1], N. J. Ouborg[2] and R. van Treuren[1]

[1]*Department of Genetics, University of Groningen, Kerklaan 30, NL-9751 NN Haren, The Netherlands*
[2]*Department of Plant Population Biology, Netherlands Institute of Ecology, P.O. Box 40, NL-6666 ZG Heteren, The Netherlands*

Summary. Due to human activities, populations of many species have become small and isolated. In this situation they become subject to genetic drift and inbreeding, resulting in loss of genetic variation, an increase in homozygosity, and possibly a decrease in viability (inbreeding depression). This process, here referred to as genetic erosion, may significantly increase the extinction probability of populations or even species, and is therefore currently an important issue in conservation biology. The research presented here aimed to determine the occurrence, extent, and significance of genetic erosion in natural populations of *Scabiosa columbaria* and *Salvia pratensis*, two species that are considered endangered in The Netherlands.

We observed decreased levels of both allozyme variation and phenotypic variation in the smaller populations of both species, indicating the occurrence of substantial genetic drift. Experiments that compared outcrossed progeny to selfed progeny showed a severe reduction in fitness for many fitness components, while on the other hand crosses between different populations revealed substantial heterosis. This proved that in both species there is potentially sufficient genetic load to cause inbreeding depression when populations become small. In contrast to expectation, no correlation between population size and the level of inbreeding depression due to selfing was observed, suggesting that purging of genetic load had not (yet) occurred in the small populations. This most probably indicates that the small populations had become small only recently. To evaluate the possible future consequences of the observed amount of genetic load, data on inbreeding depression were integrated with demographic data and the environmental variance in transition models. These computer simulations showed that most populations, both small and large, have a high probability of becoming extinct in the near future, and also that inbreeding depression did significantly increase this extinction probability of populations, except of those populations that, due to the high environmental stochasticity, already had a very high extinction chance in the absence of inbreeding depression. Implications of the results for possible management measures and future research are discussed.

Introduction

Due to human activities, populations of many organisms have become small, fragmented, and isolated in recent years. Populations that decline in size become increasingly affected by stochastic events (causing fluctuations in population size and demographic parameters) and, as a consequence, the risk of populations to become extinct in the "near" future increases significantly (Soulé, 1987). These stochastic processes that threaten "small" populations with extinction are often subdivided into four types (Shaffer, 1981, 1987): (i) *Demographic stochasticity*: random effects causing variation in survival and reproduction of individuals; (ii) *Environmental stochasticity*: random fluctuation in the physical and biotic environment of a population (e.g., nutrient supply, weather, diseases, predators, competitors, etc.); (iii) *Natural catastrophes*: an extreme form of environmental stochasticity such as droughts, floods, fires, storms, etc.; (iv) *Genetic stochasticity*: random genetic drift, inbreeding, and founder events that cause changes in genetic composition of small populations and change survival and reproduction of invidivuals. Whereas these stochastic factors all increase in importance when populations sizes are low, natural catastrophes and environmental stochasticity seem more important at moderate population sizes than the other two factors (Shaffer, 1987; Menges 1991b). It is also evident that the different factors may interact, as both environmental and genetic stochasticity will change the demographic parameters, and integration of the effects of the various factors is needed for a proper evaluation of their consequences for population extinction (Gilpin and Soulé, 1986; Menges, 1991a,b).

Many rare and endangered species occur in small and isolated populations, often still decreasing in size. It is therefore of utmost importance to assess the impact of the different stochastic processes in order to develop effective management and conservation measures to minimize the risk of extinction of such populations and to ensure long-term persistence.

Genetic component

Until the early 1980s, nature conservation had chiefly concentrated on the demographic and environmental component, but since then the role of population genetics in the management and conservation of species has become a major topic in conservation biology (e.g., Frankel and Soulé, 1981; Schonewald-Cox et al., 1983; Soulé, 1987; Falk and Holsinger, 1991). Based on population genetic theory, we may expect that small and isolated populations will become subject to genetic drift and inbreeding. For one thing, this will lead to loss of genetic variation, not only of deleterious and neutral alleles, but also favorable alleles may be lost. The magnitude of these effects strongly depends on the (effective) population size and increases when population size decreases. The loss of variation may significantly decrease both the short-term and long-term adaptability of populations in variable and changing environments. For another thing, inbreeding will cause increased homozygosity and may be accompanied by inbreeding depression, resulting in a significant decrease in fitness. The entire process, which we will refer to as "genetic erosion," thus reduces both adaptability and fitness and, consequently, population sizes will decline even more. Because of this, population sizes may be trapped in a downward spiral ultimately leading to extinction.

However, the importance of genetic erosion for population extinction is not undisputed. Many plant species, for instance, may be relatively insensitive to the genetic consequences of small population size because they show restricted gene flow by nature and therefore may already be adapted to breeding in small local neighborhoods and to a certain degree of inbreeding, while also many plant species show a high degree of selfing (Bijlsma et al., 1991). It is also argued that genetic problems may be often secundary to the demographic problems of small populations due to environmental stochasticity, and that extinction will therefore occur before the genetic problems have become evident (Lande and Barrowclough, 1987; Menges, 1991a).

Empirical data on the significance of genetic erosion are still very scarce, but are clearly needed to properly evaluate genetic stochasticity in relation to the other stochastic forces. Therefore, we studied two plant species to possibly answer the question of whether genetic erosion occurs in natural populations and if so, whether it has any significance for the extinction probability of populations.

The species

The two plant species investigated were *Scabiosa columbaria* (small scabious) and *Salvia pratensis* (meadow sage or common meadow clary). *Scabiosa columbaria* (*Dipsacaceae*) is found at dry grassy sites on calcareous soils. In The Netherlands it is found along major rivers and in the limestone district. *Salvia pratensis* (*Labiatae*) is found in calcareous grasslands and occurs mostly on dikes along large rivers. Both species are perennial and predominantly outcrossing (for outcrossing rates, see van Treuren et al., 1993a, 1994). In The Netherlands the number of still extant populations of both species has decreased by more than 50% during the last three decades (Bijlsma et al., 1991). Undoubtedly, this decrease, for the greater part, has been due to human activities which have caused destruction and alteration of the habitat (e.g., dike and road (re)construction, sand digging, agricultural management, etc.). Because of this large decrease in the number of populations the species are recognized as endangered in The Netherlands and designated as most vulnerable and vulnerable respectively in the *Red Data List of Dutch Plants* (Weeda et al., 1990), even though several large populations of both still exist. The presence of a sufficient number of populations highly variable in size, from only a few to thousands of individuals, makes the species very suitable for research.

Population size and the amount of genetic variation

According to theory, the first step in the genetic process is that small (and isolated) populations compared to larger populations show decreased levels of genetic variation. Is this actually observed in nature? By means of electrophoresis, we determined the amount of allozyme variation in 12 *S. columbaria* populations, ranging in size from 14 to 100 000 flowering plants, and 14 *S. pratensis* populations, size range 5 to 1500 flowering plants (van Treuren et al., 1991). The results showed a highly significant correlation between population size and both the proportion of polymorphic loci and the mean observed number of alleles (Tab. 1), indicating that, as expected, small populations show decreased levels of genetic variation. The relation between size and gene diversity (mean expected heterozygosity) was, however, not significant. This could reflect the possibility that the small populations only recently have been reduced in size, because bottlenecks do cause a reduction in the number of alleles, especially low frequency alleles, rather than loss of heterozygosity (Barrett and Kohn, 1991).

Not only allozyme variation seems to be affected, but also the amount of phenotypic variation (averaged over a number of morphological, growth, and reproductive characteristics) correlated positively with population size (Tab. 1; Ouborg et al., 1991), although this correlation was significant only in *S. pratensis*. Assuming that the phenotypic variation is at least partly genetically based and, more likely, of greater adaptive significance than allozyme variation, we can infer that genetic drift might have been largely responsible for the loss of variation in the small populations. This is confirmed by the observation that genetic differentiation (expressed as the coefficient of gene differentiation, G_{ST}) was substantially larger among small populations than among large populations for both species (Tab. 2). In contrast to the differentiation observed between populations, not much differentiation was in general observed within populations of both species, except in small populations of *S. columbaria* that showed a spatially discontinuous population structure (van Treuren et al., 1993a, 1994).

Table 1. Correlation coefficients between log population size and variability measures (data from van Treuren et al., 1991 and Ouborg et al., 1991)

	S. columbaria		S. pratensis	
Allozymes				
– Proportion polymorphic loci	r = 0.713	(P = 0.005)	r = 0.619	(P = 0.009)
– Mean observed number of alleles	r = 0.819	(P < 0.001)	r = 0.540	(P = 0.023)
– Gene diversity	r = 0.490	(P = 0.053)	r = 0.309	(P = 0.143)
Total phenotypic variation	r = 0.529	(P = 0.140)	r = 0.915	(P = 0.015)

Table 2. Analysis of gene diversity (Nei, 1987) calculated over all loci for *S. columbaria* and *S. pratensis* either for all populations or for small and large populations separately. The coefficient of gene diversity (G_{ST}) is the proportion of the total gene diversity (H_T) that can be attributed to the average gene diversity between populations $[(H_T - H_S)/H_T]$. N is the number of populations. (Data from van Treuren et al., 1991)

	H_T	H_S	G_{ST}	N
S. columbaria				
– all populations	0.129	0.107	0.175	12
– small populations	0.131	0.100	0.236	5
– large populations	0.124	0.112	0.101	7
S. pratensis				
– all populations	0.136	0.115	0.156	14
– small populations	0.123	0.105	0.181	8
– large populations	0.144	0.127	0.115	6

Given these observations, we conclude that small populations show less genetic variation than large populations, thus indicating the presence of substantial genetic drift in small populations of both species.

Selfing and inbreeding depression

Deleterious effects of inbreeding on fitness have been well documented from artificial selection and husbandry (for examples see Falconer, 1981), and zoo populations (Ralls et al., 1988). On the other hand, many organisms regularly inbreed, e.g., many hermaphroditic plants are predominantly selfing, without apparent ill effects. For inbreeding depression to be important, sufficient genetic load should be present in populations of *S. columbaria* and *S. pratensis*. As both species are predominantly outcrossing (van Treuren et al., 1993a, 1994) we expected this to be the case. This was investigated by comparing the performance of selfed *versus* outcrossed progeny for a number of populations for both species (van Treuren et al., 1993b; Ouborg, 1993). As *S. columbaria* is the most extensively studied species in this respect, we will present these data to illustrate our findings (van Treuren et al., 1993b).

From six populations of varying size, plants were used to make three types of crosses by means of hand pollination: self-crosses, outcrosses within the same population, and outcrosses between different populations, resulting in selfed seed (SELF), outcrossed seed within populations (WPC), and outcrossed seed between populations (BPC). These progenies were grown in a greenhouse and their performance was examined throughout an entire life cycle, during which different fitness components were measured. The difference in pollination treatment did not so much affect some fitness components like percentage of flowering

Table 3. The effect of pollination treatment on different fitness components: mean percentage survival during three life stages, biomass production (mean dry weight of the aboveground biomass produced during 35 days) and mean seed set per flowerhead after a second hand pollination with either outcross or self pollen. All measures are averaged over six populations ($\pm$ S.E.). The level of inbreeding depression (δIBD) is calculated as (WPC – SELF)/WPC, and the level of heterosis δHET is calculated as (BPC – WPC)/BPC. (Data from van Treuren et al., 1993b)

	SELF	WPC	BPC	δIBD	δHET
Survival					
– germination	73.4 $\pm$ 3.3	78.5 $\pm$ 2.6	89.4 $\pm$ 4.5	0.06	0.12
– seedling to adult	26.5 $\pm$ 0.8	26.0 $\pm$ 3.6	33.4 $\pm$ 3.4	−0.02	0.22
– adult stage	32.6 $\pm$ 5.6	74.0 $\pm$ 3.9	85.4 $\pm$ 3.0	0.56	0.13
Total survival	6.6 $\pm$ 1.4	15.1 $\pm$ 2.3	26.3 $\pm$ 4.0	0.56	0.43
Biomass production					
– without competition	1.75 $\pm$ 0.16	2.88 $\pm$ 0.15	3.22 $\pm$ 0.12	0.39	0.11
– with competition	1.22 $\pm$ 0.12	2.43 $\pm$ 0.20	3.02 $\pm$ 0.13	0.50	0.20
Seed set					
– outcross pollination	18.8 $\pm$ 2.3	27.3 $\pm$ 2.1	30.7 $\pm$ 1.8	0.31	0.11
– self pollination	8.1 $\pm$ 2.0	19.6 $\pm$ 2.4	24.7 $\pm$ 2.3	0.59	0.21

plants and the number of flowerheads, but for other characteristics such as adult survival, biomass production, and seed set, severe inbreeding depression was observed, while at the same time also strong heterotic effects were found for the crosses between populations (Tab. 3).

Conspicuous differences between the progeny groups were also observed for root development (Fig. 1): on average very poor root development was observed for SELF plants (71.2% of these plants fell in category 1), while WPC and BPC plants showed mostly extensive root development (51.4% and 59.0% of these plants fell in category 3, respectively). These differences in root development correlated well with the differences found in adult survival and biomass production (Tab. 3). The differences were also reflected in the regenerative ability of the plants after the above ground biomass had been removed: plants with poor root development also had poor regenerative ability. Thus, growth-related characteristics are clearly impaired by one generation of selfing in *S. columbaria*.

These data on survival and reproductive fitness components even more clearly demonstrate the effects of selfing and outcrossing when combined to calculate overall fitness during one complete life cycle from filled seed to filled seed (Fig. 2). The significant decrease in fitness of SELF progeny compared to WPC progeny demonstrates the presence of a substantial genetic load in populations of *S. columbaria*. Crosses between individuals of different populations (BPC) clearly increase fitness of the offspring compared to WPC, indicating that the different populations have been diverged genetically with respect to fitness loci.

Root categories

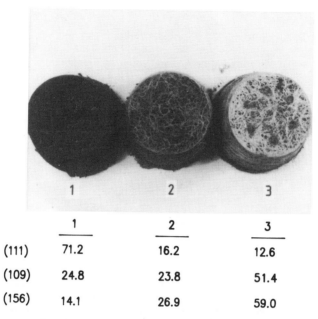

		1	2	3
SELF	(111)	71.2	16.2	12.6
WPC	(109)	24.8	23.8	51.4
BPC	(156)	14.1	26.9	59.0

Fig. 1. The three categories of root development and the percentage of plants per pollination treatment in each category at the end of the experiment. Total sample size per treatment given in parentheses (after van Treuren et al., 1993b).

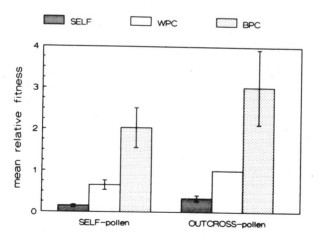

Fig. 2. Mean relative fitness over a whole life cycle of SELF, WPC, and BPC progeny after a second hand pollination by either self-pollen or outcross-pollen. Means are averaged over six *S. columbaria* populations. Within each population the fitness values are expressed relative to the fitness of the WPC-outcross-pollen plants. Vertical bars indicate standard errors (modified after van Treuren et al., 1993b).

262

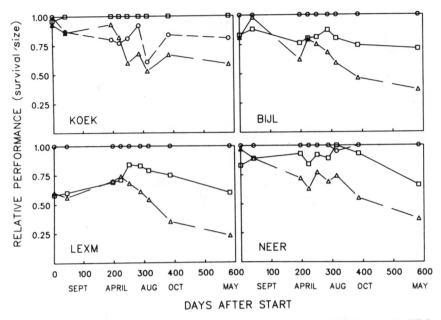

Fig. 3. Relative performance in the field for the three progeny groups SELF ($\triangle - \triangle$), WPC ($\square - \square$), and BPC ($\bigcirc - \bigcirc$) of two large (top figures) and two small (bottom figures) *S. pratensis* populations. For each population performance is expressed relative to the best performing progeny group within each population and plotted against time (from Ouborg, 1993).

Given the observed level of heterosis, artificially increasing gene flow towards the small populations might be an important management measure for *S. columbaria*, although the persistence of the heterosis effects has not yet been investigated.

For *S. pratensis* comparable results have been found: significant inbreeding depression was demonstrated for seed weight, germination, plant size, regenerative capacity, and survival (Ouborg, 1993). In this case fitness differences were not only observed under greenhouse conditions but also in a field experiment, in which the performances of the different progeny groups were tested in the maternal sites (Fig. 3).

Thus, in both species sufficient genetic load seems to be present to cause a loss in fitness when populations become inbred.

Population size and the level of inbreeding depression

Having shown that substantial genetic load is potentially present in the species, the next question should be: does the magnitude of the observed fitness reduction relate to population size? In other words, do small and large populations react differently to selfing and outcrossing?

Inbreeding depression is generally assumed to be caused by an increase in homozygosity at fitness loci, provided that these loci show some degree of dominance (Falconer, 1981). Although the genetic mechanisms underlying inbreeding depression are still not completely understood, it is evident that inbreeding does reduce fitness in many organisms. Of the two existing theories that account for inbreeding depression, the partial dominance hypothesis is the most common explanation (Charlesworth and Charlesworth, 1987). Under this hypothesis, inbreeding depression is the consequence of the presence of (slightly) deleterious recessive mutations in diploid organisms that become homozygous due to drift and inbreeding. An alternative hypothesis is the overdominance hypothesis that assumes that heterozygotes are fitter than homozygotes and inbreeding depression is thought to be the consequence of low heterozygosity. Under this theory, heterozygosity by itself is advantageous, most probably due to a greater developmental homeostasis of heterozygotes (Mitton and Grant, 1984).

Whatever mechanism causes inbreeding depression, populations that have been small and subject to genetic drift and inbreeding for a prolonged period will have become increasingly homozygous. In this process they will have been purged of part of their genetic load due to selection against deleterious homozygotes (Barret and Charlesworth, 1991). Additional inbreeding through one generation of selfing in such small populations then should cause less inbreeding depression (defined here as (WPC − SELF)/WPC) than in large populations that have not experienced inbreeding previously. We can thus expect a positive correlation between population size and the level of inbreeding depression due to one generation of selfing. On the other hand when crosses are made between populations, heterosis (defined here as (BPC − WPC)/BPC) is expected to be larger for the more homozygous small populations than for large populations. This should result in a negative correlation between population size and the magnitude of heterosis.

Although incidentally a weak correlation between population size and the level of inbreeding depression was observed for a few fitness components, we did in general not observe any consistent relationship between size and level of inbreeding depression (δIBD) or the level of heterosis (δHET), for *S. columbaria* (van Treuren et al., 1993b) or *S. pratensis* (Ouborg, 1993). This is easily illustrated by Tab. 4, which shows the level of inbreeding depression and heterosis calculated from the relative fitness values over an entire life cycle for the six *S. columbaria* populations used in the previous section. The data clearly show that δIBD and δHET are quite substantial for all populations, but there is no apparent trend with population size (compare, for example, the small-large pairs Kwartierse Dijk−Wrakelberg and Ruitenberg-Olst).

Table 4. The level of inbreeding depression (δIBD) and the level of heterosis (δHET) observed for six *S. columbaria* populations that differed in size (shown in parentheses). Values were calculated from the relative fitness values for each population separately. For further explanation see text and the legend of Tab. 2 (after van Treuren et al., 1993b)

	2nd Pollination outcross		2nd Pollination self	
	δIBD	δHET	δIBD	δHET
Kwartierse Dijk (35)	0.41	0.17	0.63	0.38
Ruitenberg (90)	0.87	0.67	0.90	0.53
Terwolde (200)	0.63	0.50	0.0	0.91
Bemelerberg (300)	0.75	0.66	0.82	0.75
Olst (50 000)	0.83	0.86	0.78	0.82
Wrakelberg (100 000)	0.55	0.32	0.85	0.26

How do we explain this unexpected lack of correlation?
One can think of several more or less plausible explanations, e.g., the fact that the fitness measurements were, for the greater part, made under greenhouse conditions and not in the field where under local selection pressure differences would possibly have been revealed, or that the effective size of the small populations is still too large to show significant inbreeding effects. However, we think that it is most probably a matter of the history of the populations. In populations that have become small only recently, the effects of genetic drift and inbreeding will accumulate gradually and homozygosity will increase slowly. This increase in homozygosity is expected to proceed earlier and faster for neutral loci (as allozymes in plants are often thought to be) than for fitness loci (Barrett and Kohn, 1991). Hence, purging of genetic load will be delayed and will become evident only a long time after the population becomes small. The finding stated earlier that in small populations allozymes showed decreased levels of genetic variation with respect to the proportion of polymorphic loci and the mean observed number of alleles, whereas no effect was found for gene diversity, corroborates this view.

In conclusion, the results so far indicate that for both species the small populations are in an early phase of the genetic erosion process and have not yet experienced the proposed negative effects of drift and inbreeding. But the high level of genetic load currently observed in all populations of the species suggests that significant reductions in fitness can be expected in future generations. The high level of heterosis observed in inter-population crosses, even for large populations, suggests that the natural populations of both species have become diverged genetically with respect to fitness loci.

Inbreeding depression and population extinction prediction

We concluded that the deleterious effects of genetic erosion may become manifest sooner or later in populations of *S. columbaria* and *S. pratensis*. A logical question then is: what will be the expected effect of genetic erosion on the extinction probablity in future generations? To evaluate this, we integrated demographic data from the natural populations with the data on inbreeding depression in both time invariant and stochastic matrix projection models (Ouborg, 1993). These models, especially stochastic models, are very useful in analyzing and estimating extinction probabilities of populations (Menges, 1991a,b).

For both *S. columbaria* and *S. pratensis* two large and two small populations were monitored demographically for 3 years, and the probabilities of transition between distinct life history stages were estimated. In general, transition parameters were extremely variable within each population, both among different years as well as among different plots within years (Ouborg, 1993). This suggests that environmental stochasticity was very substantial and important in all populations, both in space and in time. These data were incorporated in the models and the equilibrium finite rate of increase (λ) for the time invariant model (Caswell, 1989) and the mean population growth rate (μ) for the stochastic model (Lande and Orzack, 1988) were determined by means of computer simulations. For both species small populations showed in general slightly lower equilibrium population growth rates than large populations, and six out of eight populations (including large ones) had already a high probability of becoming extinct, even in the absence of inbreeding depression. The effects of inbreeding depression then were included in the models by incorporating a certain (but for each simulation constant) fraction of selfing plants in the population. Inbreeding depression values for the transitions were expressed as the relative performance of selfed progeny compared to progeny derived from random outcrossing. The results for *S. columbaria* are shown in Figs 4 and 5. Fig. 4 shows that an increase in the proportion of selfing individuals in the models (a decrease in outcrossing rate t_m) had a considerable negative effect on λ and μ in some populations like the large population Wrakelberg (WRAK), but did not so much affect these values in the other large population Olst. The effect of inbreeding depression on the extinction probability was determined for different levels of outcrossing by the stochastic model. As is clear from Fig. 5, three out of four populations already are expected to become extinct within the next 100 years, even without inbreeding depression ($t_m = 1$). Incorporating inbreeding depression clearly increases the extinction probability in the populations Kwartierse Dijk (KDYK) and, especially, Wrakelberg (WRAK). In the other two, in which the extinction probabilities in the absence of inbreeding depression were already high,

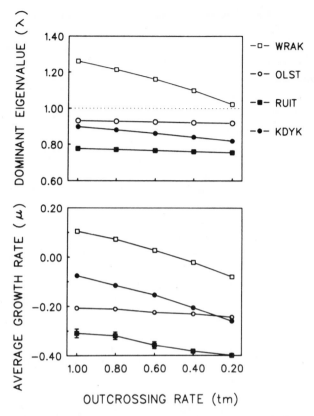

Fig. 4. The effect of inbreeding depression on population growth rate in four *S. columbaria* populations. The top figure shows the dominant eigenvalue (λ) of the time invariant model for different outcrossing rates for each population. The bottom figure shows the mean growth rate (μ) for the stochastic model. Closed symbols indicate small, and open symbols large populations (from Ouborg, 1993).

it has only a minor effect. For *S. pratensis* qualitatively similar results were obtained. The results lead to the conclusion that inbreeding depression does increase the extinction probabilities, but the extent of the increase highly depends on the environmental stochasticity. Moreover, the impact of inbreeding depression strongly depends on the demographic structure of the populations. The impact of inbreeding depression is small when it mainly influences life stages that have only a minor effect on the population growth rate. Conversely, when those life stages that substantially influence the population growth rate are affected by inbreeding (as was the case for WRAK, Figs 4 and 5), the impact of inbreeding depression increases greatly. These results indicate clearly the necessity of integrating the inbreeding depression data with demographic and environmental data.

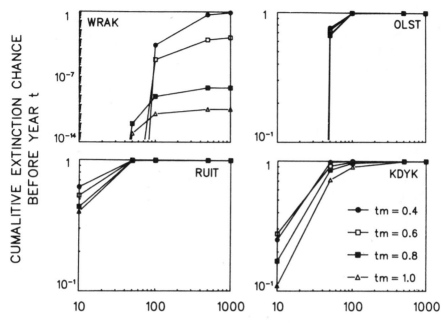

Fig. 5. The cumulative chance of extinction in time for different outcrossing rates (t_m) for four *S. columbaria* populations (see also Fig. 4). Note that the scales of the extinction axes are very different among populations (from Ouborg, 1993).

However, one has to be somewhat careful with these kind of simulations because the validity of many assumptions underlying the models is still unclear, which might have caused both under- and overestimation of the impact of inbreeding depression (Ouborg, 1993). For instance, purging of genetic load, actually decreasing the effect of selfing in the long run, was not accounted for. On the other hand, the inbreeding depression values for *S. columbaria* were determined in the greenhouse, and they may be considerably lower than the values expected under field conditions (Dudash, 1990). Nevertheless, the matrix projection models appear very useful to analyze the combined effects of the different stochastic processes, and seem not only useful to analyze the relative importance of these different stochastic processes, but also may be very helpful in evaluating the possible effects of proposed management measures.

Conclusions

The reduced amount of variation observed for allozymes and for a number of morphological and life history traits suggests that genetic drift is a significant force in the small populations of *S. columbaria* and *S.*

pratensis. This is in agreement with the observations for other plant species (for references see Barrett and Kohn, 1991). The mere fact that loss of genetic variation will reduce the long-term evolutionary potential and adaptability implies that strategies that maintain or increase genetic variation (e.g., genetic exchange between genetically different populations) in these populations are necessary conservation measures.

Both species revealed considerable amounts of genetic load when selfed, indicating that inbreeding, when it occurs, might significantly decrease fitness. Especially the marked negative effects observed for growth-related characteristics may severely limit the competitive ability of the plants, whereas high interspecific competitive ability is required in vegetations that through eutrophication often have become more and more dense (van Treuren, 1993). *S. columbaria*, for instance, prefers a short and not completely closed turf and in dense vegetations with low light conditions seed germination, seedling growth, and seedling survival are significantly decreased (Schenkeveld and Verkaar, 1984).

However, small and large populations showed similar levels of inbreeding depression when selfed and also comparable levels of heterosis when interpopulation crosses were made. This suggests that, although we observed loss of genetic variation in the small populations, the large amount of genetic load has not been or is not (yet) expressed in the small populations. Most probably, these populations became small only recently. As both *S. columbaria* and *S. pratensis* are relatively long-lived perennials (estimated generation time at least 5–7 years), "recently" in terms of the number of generations, therefore, may imply quite a number of years. Although there are indications that most small populations have been small for the last 10 years (Ouborg, 1993), the past history of the populations is unknown. Lack of knowledge of the history of populations in general may seriously hinder a correct interpretation of empirical data, for present findings and subsequent expectations strongly depend on past events. This implies that long-term population surveys and experiments are needed for thorough evaluation of the genetic consequences of genetic erosion.

The matrix projection models indicated that, just because of the high level of environmental variation observed in nature, populations of both species already have a high chance of becoming extinct in the next 100 years. Nevertheless, the models, after integrating the data on inbreeding depression with the demographic data, predicted that inbreeding depression will still significantly increase the extinction probability of some populations. Measures that prevent inbreeding depression to occur in these populations, therefore, would be recommended. The high level of heterosis we observed in the inter-population crosses suggests that artificial gene flow between populations might be an effective management measure. Yet, cautiousness is needed as hybridization between populations might result in outcrossing depression, i.e., loss of fitness because locally adapted gene complexes become disrupted by recombination in the next generations (Templeton, 1986).

Especially in those populations that already showed a high extinction probability in the absence of inbreeding depression, however, inbreeding effects were predicted to be minimal. In these latter cases environmental stochasticity seems to be the prominent cause of extinction. In these cases management measures could firstly focus on reducing the environmental stochasticity of, particularly, those life stages that significantly influence the population growth rate. For *S. columbaria* and *S. pratensis* management, for instance, should aim at improving the conditions for germination, seedling establishment, and juvenile survival, and at adapting mowing and grazing regimes that ensure optimal seed production (Ouborg, 1993; van Treuren, 1993). This does not necessarily mean, however, that inbreeding effects should be ignored in these situations, because they may become manifest when environmental variation, e.g., by effective management measures, is reduced. Furthermore, the contribution of inbreeding depression can strongly depend on the ecological conditions experienced (Dudash, 1990; Koelewijn, 1993). Moreover, it has been suggested that inbreeding may enhance the susceptibility to environmental variation (Menges, 1991b). In this latter case, genetic erosion would increase environmental variance and, consequently, greatly increase the probability of population extinction.

Our results demonstrate that, in the case of *S. columbaria* and *S. pratensis*, genetic erosion, in combination with environmental stochasticity, can significantly increase the extinction probability of populations in the near future. Even though environmental stochasticity in many cases may be a greater and more immediate threat for population persistence, genetic erosion cannot simply be considered as secondary to the other stochastic processes, the more so because genetic and environmental processes might not be independent. Management practices, therefore, preferably should include genetic measures. To fully elucidate the relative impact of genetic erosion, however, much more research is needed. We need to know more about the genetic mechanisms underlying inbreeding depression and the consequences of longterm inbreeding in populations. For evaluation of the consequences of artificial gene flow and population hybridization, transplantation experiments are needed to determine the presence of local adaptations and the likelihood of outcrossing depression occurring. The implication of population substructuring and differences in breeding system have to be studied and longterm demographic population studies are needed for monitoring the responses to population size diminution in nature, and for more precise quantification of the variation in demographic parameters. Essential also is analysis of the supposed interaction between genetic and non-genetic factors to see whether genetic erosion will enhance the negative effect of the other stochastic processes. Research into these problems

270

will not only elevate our understanding of the genetic processes involved, but also will provide knowledge for nature managers to optimize their management strategies for population and species conservation.

Acknowledgements
We thank Els Bijlsma-Meeles for critically reading the manuscript. The investigations were in part supported by the Ministry of Agriculture, Nature Management and Fisheries, and by the Prins Bernhard Foundation.

References

Barrett, S. C. H. and Charlesworth, D. (1991) Effects of a change in the level of inbreeding on the genetic load. *Nature* 352: 522–524.

Barrett, S. C. H. and Kohn, J. R. (1991) Genetic and evolutionary consequences of small population size in plants: implications for conservation. *In:* Falk, D. A. and Holsinger, K. E. (eds), *Genetics and conservation of rare plants.* Oxford University Press, Oxford, pp. 3–30.

Bijlsma, R., Ouborg, N. J. and van Treuren, R. (1991) Genetic and phenotypic variation in relation to population size in two plant species: *Salvia pratensis* and *Scabiosa columbaria. In:* Seitz, A. and Loeschcke, V. (eds), *Species conservation: a population-biological approach.* Birkhäuser Verlag, Basel, pp. 89–101.

Caswell, H. (1989) *Matrix population models.* Sinauer, Sunderland, MA.

Charlesworth, D. and Charlesworth, B. (1987) Inbreeding depression and its evolutionary consequences. *Annu. Rev. Ecol. Syst.* 18: 237–268.

Dudash, M. R. (1990) Relative fitness of selfed and outcrossed progeny in a self-compatible, protandrous species, *Sabatia angularis* L. (*Gentianaceae*): a comparison in three environments. *Evolution* 44: 1129–1139.

Falconer, D. S. (1981) *Introduction to quantitative genetics.* 2nd edition, Longman, London.

Falk, D. A. and Holsinger, K. E. (1991) *Genetics and conservation of rare plants.* Oxford University Press, Oxford.

Frankel, O. H. and Soulé, M. E. (1981) *Conservation and evolution.* Cambridge University Press, Cambridge.

Gilpin, M. E. and Soulé, M. E. (1986) Minimum viable populations: the process of species extinctions. *In:* Soulé, M. E. (ed.), *Conservation biology: the science of scarcity and diversity.* Sinauer, Sunderland, MA, pp. 13–34.

Koelewijn, H. P. (1993) *On the genetics and ecology of sexual reproduction in Plantago coronopus.* Ph.D. Thesis, University of Utrecht.

Lande, R. and Barrowclough, G. F. (1987) Effective population size, genetic variation, and their use in population management. *In:* Soulé, M. E. (ed.), *Viable populations for conservation.* Cambridge University Press, Cambridge, pp. 87–124.

Lande, R. and Orzack, S. H. (1988) Extinction dynamics of age-structured populations in a fluctuating environment. *Proc. Natl. Acad. Sci. USA* 85: 7418–7421.

Menges, E. S. (1991a) The application of minimum viable population theory to plants. *In:* Falk, D. A. and Holsinger, K. E. (eds), *Genetics and conservation of rare plants.* Oxford University Press, New York, pp. 45–61.

Menges, E. S. (1991b) Stochastic modelling of extinction in plant populations. *In:* Fiedler, P. L. and Jain, S. (eds), *Conservation biology: the theory and practice of nature conservation, preservation and management.* Chapman and Hall, London, pp. 253–275.

Mitton, J. B. and Grant, M. C. (1984) Associations among protein heterozygosity, growth rate, and developmental homeostasis. *Annu. Rev. Ecol. Syst.* 15: 479–499.

Nei, M. (1987) *Molecular evolutionary genetics.* Columbia University Press, New York.

Ouborg, N. J. (1993) *On the relative contribution of genetic erosion to the chance of population extinction.* Ph.D. Thesis, University of Utrecht.

Ouborg, N. J., van Treuren, R. and van Damme, J. M. M. (1991) The significance of genetic erosion in the process of extinction. II. Morphological variation and fitness components in populations of varying size of *Salvia pratensis* L. and *Scabiosa columbaria* L. *Oecologia* 86: 359–367.

271

Ralls, K., Ballou, J. D. and Templeton, A. R. (1988) Estimates of lethal equivalents and the cost of inbreeding in mammals. *Cons. Biol.* 2: 185–193.

Schenkeveld, A. J. and Verkaar, H. J. (1984) *On the ecology of short-lived forbs in chalk grasslands*. Ph.D. Thesis, State University of Utrecht.

Schonewald-Cox, C. S., Chambers, S. M., MacBryde, B. and Thomas, L. (1983) *Genetics and conservation: a reference for managing wild animal and plant populations*. Benjamin-Cummings, London.

Shaffer, M. L. (1981) Minimum population sizes for species conservation. *Bioscience* 31: 131–134.

Shaffer, M. L. (1987) Minimum viable populations: coping with uncertainty. *In:* Soulé, M. E. (ed.), *Viable populations for conservation*. Cambridge University Press, Cambridge, pp. 69–86.

Soulé, M. E. (1987) *Viable populations for conservation*. Cambridge University Press, Cambridge.

Templeton, A. R. (1986) Coadaptation and outcrossing depression. *In:* Soulé, M. E. (ed.), *Conservation biology: the science of scarcity and diversity*. Sinauer, Sunderland, MA, pp. 105–116.

van Treuren, R. (1993) *The significance of genetic erosion for the extinction of locally endangered plant populations*. Ph.D. Thesis, University of Groningen.

van Treuren, R., Bijlsma, R., van Delden, W. and Ouborg, N. J. (1991) The significance of genetic erosion in the process of extinction. 1. Genetic differentiation in *Salvia pratensis* and *Scabiosa columbaria* in relation to population size. *Heredity* 66: 181–189.

van Treuren, R., Bijlsma, R., Ouborg, N. J. and van Delden, W. (1993a) The effect of population size and plant density on outcrossing rates in locally endangered *Salvia pratensis*. *Evolution* 74 (4), in press.

van Treuren, R., Bijlsma, R., Ouborg, N. J. and van Delden, W. (1993b) The significance of genetic erosion in the process of extinction. III. Inbreeding depression and heterosis effects due to selfing and outcrossing in *Scabiosa columbaria*. *Evolution*, 74 (5), in press.

van Treuren, R., Bijlsma, R. J., Ouborg, N. J. and Kwak, M. M. (1994) Relationship between plant density, outcrossing rates and seed set in natural and experimental populations of *Scabiosa columbaria*. *J. Evol. Biol.*, 7 (3): 287–302.

Weeda, E. J., Van der Meijden, R. and Bakker, P. A. (1990) Floron-Rode Lijst 1990: rode lijst van de in Nederland verdwenen en bedreigde planten (*Pteridophyta* en *Spermatophyta*) over de periode 1.I.1980–1.I.1990. *Gorteria* 16: 2–26.

Conservation Genetics
ed. by V. Loeschcke, J. Tomiuk & S. K. Jain
© 1994 Birkhäuser Verlag Basel/Switzerland

Effects of releasing hatchery-reared brown trout to wild trout populations

M. M. Hansen and V. Loeschcke

Institute of Biological Sciences, Department of Ecology and Genetics, University of Aarhus, Ny Munkegade, Building 540, DK-8000 Aarhus C, Denmark

Summary. Brown trout is a species showing extreme genetic differentiation among local populations, which to some degree may reflect local adaptation of the stocks. In many countries pollution and destruction of spawning places have led to a significant decline in population sizes. To counteract this decline the traditional solution has been to stock large numbers of hatchery-reared trout. Hatchery-rearing may lead to domestication and loss of genetic variation, and if massive numbers of hatchery fish are stocked these may replace wild populations. Interbreeding of wild and domesticated populations or formerly reproductively isolated wild populations may result in a breakdown of the local genetic population structure, loss of genetic variation among stocks and loss of specific genotype combinations of potential adaptive value. Population genetic monitoring of stocking programs strongly suggests that releases of trout from exogenous hatchery strains cannot be recommended. Conservation and management efforts should primarily be aimed at individual stocks and priority should be given to improving environmental conditions. If stocking is inevitable the preferred option is to release offspring of local wild fish, and only in already heavily stocked populations is the use of exogenous hatchery fish justified.

Introduction

Brown trout (*Salmo trutta* L.) and other salmonids are very vulnerable fish species due to their requirements for unpolluted rivers and suitable spawning localities. In the industrialized and densely populated regions of Europe and North America, such conditions are becoming increasingly rare, and stock enhancement programs based on releases of hatchery-reared fish have therefore been implemented for several decades.

Traditionally, most efforts in conservation genetics have been aimed at describing genetic processes in small populations and developing guidelines for optimal conservation of genetic variation. However, genetic conservation of salmonid fishes confronts us with some additional problems as these species often exhibit a very complex genetic population structure which must be taken into account if rational management guidelines are to be developed. In particular, some of the most severe threats to the natural genetic population structure of salmonid stocks consist in genetic interactions with fish from semi-domesticated strains which are either intentionally released or escapees from commercial aquaculture.

Apart from being interesting and pertinent *per se*, problems concerning gene flow from captive to wild populations are expected to become increasingly important in the future. The accelerating development in the field of genetic engineering will probably soon confront us with problems of escaped genetically modified fish and other organisms interbreeding with natural conspecifics. The study of interactions between hatchery-reared and wild salmonid populations may thus be useful in providing us with knowledge about if, and how, such gene flow will take place and perhaps point out measures which could minimize the risks of spreading transgenic fish to natural populations.

Brown trout population genetics

Salmonid fishes are famous for their well-developed homing instinct, which is probably a decisive factor leading to the strong population subdivision often observed in these species (Ryman, 1983). The most pronounced population subdivision is found in brown trout which has its natural distribution in Europe, North Africa, and northwestern Asia. Two life-history types, resident and migratory, respectively, are commonly found and the significance of such variability in relation to conservation is discussed by Hindar (this volume).

Several allozyme studies (Ferguson, 1989; Karakousis and Triantaphyllidis, 1990) and sequencing of a part of the mitochondrial DNA D-loop (Bernatchez et al., 1992) have clearly shown the presence of genetically divergent brown trout races in the Mediterranean and northern parts of Europe. Perhaps even more remarkable is the genetic differentiation which has been found at microgeographic levels. Allozyme studies from all over Europe have revealed both large amounts of polymorphism and extensive genetic differentiation into local populations, and even *within* one river or lake system genetically different stocks may be found in the tributaries (e.g., Ryman et al., 1979; Ryman, 1983; Crozier and Ferguson, 1986; Ferguson, 1989). Furthermore, several studies (e.g., Ryman, 1983; Crozier and Ferguson, 1986) show that within regions where gene flow is likely to take place no obvious correlation exists between genetic and geographical distance among populations. The G_{ST} values from such studies are generally very high. For instance, Crozier and Ferguson (1986) found a G_{ST} value of 31% in a study of tributary populations of Lough Neagh, Northern Ireland. Even higher values, 55% and 61%, have been reported from France and nothern Spain, respectively (Krieg and Guyomard, 1985; Garcia-Marin et al., 1991), but it should be noted that the French study comprises a large geographical area and possibly more than one brown trout race may have been included.

There are at least two reasons for the observed high levels of genetic differentiation. Firstly, the distribution of alleles at the eye-specific lactate dehydrogenase locus *LDH-C1** indicates that at least two colonization events by genetically different brown trout have taken place in northern Europe since the last glaciation (Hamilton et al., 1989). Some remarkable phenomena including a so-called fine-spotted brown trout in Norway (Skaala and Jørstad, 1987) and three coexisting but genetically and morphologically very different trout morphs from the Irish Lough Melvin (e.g., Ferguson and Taggart, 1991) may thus be due to colonization by allopatrically diverged genotypes. Secondly, probably the majority of brown trout stocks are a result of genetic differentiation which has taken place *after* the initial colonization event. In this chapter we will primarily focus on the importance of this latter sort of population subdivision in relation to conservation.

Does the fact that brown trout shows such pronounced genetic differentiation mean that individual stocks are adapted to their specific environments and/or may contain unique genetic variation? According to Wright (1931), gene flow exceeding one migrant *per* population *per* generation should prevent local genetic differentiation in terms of fixation of different alleles in different populations. However, even in the case of significant migration a specific allele may be maintained at a high frequency in a population if the selection favoring the allele exceeds the rate at which other alleles are being brought into the population by migration (Haldane, 1930). Furthermore, as discussed by Allendorf (1983), significant genetic differentiation in terms of differences in allelic frequencies may still be found even if gene flow exceeds one migrant. Possible local adaptations are most likely to consist in polygenic, quantitative traits and the emphasis should consequently be on genotypes and allelic combinations among different loci rather than single alleles. Local adaptations may therefore be found if substantial genetic differentiation among populations has taken place even though gene flow exceeds one migrant *per* generation. There are no well-defined limits for which levels of population subdivision should be considered "substantial," but Wright (1978, p. 85) states that differentiation is "...by no means negligible if F ($\approx G_{ST}$) is as small as 0.05 or even less..." G_{ST} estimates from studies of brown trout within areas of small to moderate size would in some cases suggest an exchange of *less* than 1 migrant (e.g., Crozier and Ferguson, 1986), and in other cases (e.g., Allendorf, 1983; Ferguson and Taggart, 1991; Hansen et al., 1993) *more* than 1 migrant per population per generation, but in all of the latter studies G_{ST} values are of a "not negligible" order of magnitude.

Effective population numbers of salmonid stocks appear to be mainly in the range of a few hundreds to a few thousands (L'Abée-Lund, 1989;

Altukhov, 1990). In terms of inbreeding and loss of genetic variation such numbers should not be critical (e.g., Frankel and Soulé, 1981). Short-term temporal stability of allelic frequencies within populations also seems to be the rule (e.g., Ryman, 1983). Thus, information about the genetic population structure of brown trout does not rule out the possibility of unique genetic variation and local adaptation in individual stocks. This is, however, not the same as to say that all identified stocks are genetically unique; brown trout may spawn in very small rivers and it is obvious that a lot of extinctions and recolonizations must take place in such localities.

Another approach is to look at direct evidence for adaptation in local stocks. It has been asserted that traits connected with fitness and survival are expected to show little or no variation, but as pointed out by, for example, Barker and Thomas (1987) this is based on oversimplified assumptions. More importantly, as is well-known, it is often problematic to attribute variation in a specific trait to adaptation. Nevertheless, there does seem to be a certain body of "evidence" for local adaptations in stocks of salmonid fishes, even on a microgeographical scale, most of which has been reviewed by Taylor (1991). Specifically for brown trout we would like to mention the transplanation experiments by Svärdson and Fagerström (1982) which have demonstrated remarkable genetic differences in migration and homing behavior among stocks. All in all, we find it likely that brown trout stocks *may* show some local adaptation and contain unique genetic variation, and conservation efforts should consequently be aimed at conserving individual stocks.

Stocking activity

All over Europe, pollution and other environmental changes in rivers have resulted in poor spawning conditions and eradication of many natural trout stocks. As for many other organisms, the response to the decline of the natural populations has been to substitute natural reproduction with the release of artificially bred offspring. Brown trout is of little importance to commercial fisheries, but of immense value to recreation fisheries and thereby also to the tourist industry. Unfortunately, as we shall see, this has been the focus of most stocking programs as many attempts have been made to increase the number of fish irrespective of their origin, while few efforts have aimed at conserving native stocks.

Brown trout stocking is a very old practice which, in Denmark, dates back to at least the 19th century. However, as described by Larsen (1972), stocking activity for the first many years was rather planless. Fish were sometimes stocked regardless of need and no attention was made to ensure that the chosen localities were suitable habitats. Therefore, probably many of these stockings have not had much effect in terms of survival and reproduction of the stocked fish. Only more recently has stocking activity become well-planned and systematized, and the present situation in Denmark may be representative of other countries: Specific stocking programs are made for each river system, i.e., the density of trout in individual tributaries is evaluated by electrofishing, and if found too low compared to an estimated carrying capacity, the population is stocked with a corresponding number of hatchery-reared trout (Larsen, 1972). In order to further increase the number of stocked fish without exceeding the carrying capacity in individual rivers, for several years it has been common practice to stock large numbers of smolts (i.e., young fish ready to migrate to the sea) in the river mouths. A more recent development is to stock smolts directly in the sea to benefit coastal angling which is very popular in Denmark. The latter two stocking methods do not serve any environmental purposes and must be considered as pure "put-and-take". Approximately 2.1 million trout are stocked each year in Denmark, of which 210 000 are stocked in river mouths, and 75 000 directly on the coast.

The fish used for stocking may be of two different types. In many countries a number of commercial hatchery strains supply a significant proportion of the stocking material. Such strains are often of more or less unknown origin, and in Denmark, there are examples of hatchery populations which have been kept in captivity for up to 100 years (Simonsen and Rasmussen, 1989). Several independent allozyme studies have shown close genetic relationships among many of the hatchery strains (Krieg and Guyomard, 1985; Simonsen and Rasmussen, 1989; Garcia-Marin et al., 1991). The explanation is probably that new hatchery strains are founded by fish from other hatcheries rather than by taking in wild fish. Thus, many strains share a common ancestry and, consequently, represent a very narrow genetic variability.

An alternative to the commercial hatchery strains consists of using offspring of local wild fish as stocking material. One way to do this is simply to found a gene bank from local fish and use the offspring both for stocking material and for perpetuating the strain. In Denmark, a slightly different strategy is employed: Wild parent fish are caught *each year* by net or electrofished in the spawning season and the offspring are reared in a hatchery and stocked at the appropriate age. In most cases no attempts are made to discriminate among different stocks *within* a river system as the parent fish are caught at one or a few locations, usually in the main river, while the offspring are stocked in all the tributaries of the system.

Genetic problems in connection with stocking

In recent years there has been a growing concern about the possible consequences of the immense "genetic manipulation" many salmonid species are experiencing, mainly as a result of stocking activity and interbreeding between escaped, cultured fish and wild conspecifics. This subject has been discussed in great detail by, for example, Nelson and Soulé (1987) , Hindar et al. (1991), and Waples (1991). For simplicity, we will focus in the following on three main subjects which are the most relevant in connection with brown trout.

Intensive stocking with fish from one source may lead to displacement of native populations by fish from the source population

This is the rather trivial outcome of oneway gene flow from one population to another. The problem is particularly pertinent if fish from hatchery strains are used for stocking material and, especially, if smolts are stocked directly into the sea. In the latter case the fish are not imprinted to home to any river and, if they spawn, they may in principle spawn anywhere and thus even interbreed with natural trout stocks which are not otherwise affected by stocking.

Stocking activity may cause a breakdown in genetic population structure and, perhaps, outbreeding depression

If fish from one source, e.g., a hatchery strain, are used for stocking different natural populations it is obvious that the genetic differentiation in terms of G_{ST} among these stocks will decline. If hatchery trout exhibit less efficient homing than wild trout it is plausible that even a few stocked fish may mediate gene flow among otherwise reproductively isolated populations (Ferguson, 1989). Even if offspring of "local" wild fish are used for stocking the end result may be a decline of genetic differentiation. As mentioned previously, no discrimination is usually made among stocks *within* river systems and, accordingly, the parent fish may belong to different stocks. In terms of allelic variation interbreeding among different stocks may lead to an increase in intrapopulational variability which, from a superficial viewpoint, would even be beneficial. However, if the stocks are distantly related a loss of genotypic variation of potential adaptive value, perhaps leading to outbreeding depression, may also be the outcome (Nelson and Soulé, 1987; Hauser et al., this volume). It should be noted that the available empirical data strongly suggest that potential beneficial effects are vastly outnumbered by drawbacks (e.g., Hindar et al., 1991).

Hatchery-rearing may result in loss of genetic variation due to low effective population number and adaptation to the hatchery environment

A number of cases of loss of allozyme variation and even inbreeding depression in salmonid hatchery strains caused by use of too few parent fish have been reported (reviewed by Allendorf and Ryman, 1987). Another problem in connection with hatchery-rearing is the immense differences between living conditions in hatchery and natural environments. It is probably an inevitable outcome of many generations of captive breeding that some sort of selection for adaptation to the hatchery environment takes place, thus gradually leading to some degree of domestication.

At least in Denmark, it is appropriate also to be concerned about the procedures of raising stocking material from wild parent fish. In some rivers no more than 10–20 parent fish are caught and stripped each season, and the sex ratio is often very skewed because of lack of larger males. Under natural conditions small precocious males probably make a significant contribution to the spawning (L'Abèe-Lund, 1989), but unfortunately this is not taken into consideration in artificial propagation. In addition, some examples of intentional selection for larger fish are known to take place. A potential loss of variation may to some extent be counterbalanced by natural spawning, but if a large proportion of the population originates from very few parent fish the contribution from natural spawning may be almost negligible (Ryman, 1991).

Genetic monitoring of brown trout stocking programs

It is of the utmost importance to decide to what degree stocked and native fish will interbreed and if the effects of introgression will persist over longer periods of time. For brown trout this has been the focus of several studies employing allozyme electrophoresis, but the results are not unambiguous. For instance, Taggart and Ferguson (1986) monitored the influence of stocked trout in the tributaries to the Northern Irish Lower Lough Erne by using allelic variation at the lactate dehydrogenase locus *LDH-C1** as a genetic tag. Their results were basically that massive interbreeding between the hatchery and natural stocks took place and in some tributaries the genetic contribution of stocked fish was as high as 90%. In contrast to this study, Moran et al. (1991) used the same genetic tag to monitor a stocking program in rivers in Asturias, Spain. Their observations showed no genetic contribution of hatchery fish in rivers where native stocks were already present, and the only locality which could be confirmed as being inhabited by descendants of stocked fish was one river which had previously been devoid of trout. The latter observation could perhaps indicate some sort of competition between stocked and native fish playing a role. A similar unpredictability of the effects of stocking and transplantation activity has been reported for several other salmonids. For instance, Gyllensten et al. (1985) reported massive introgression between a native and an introduced cutthroat (*Oncorhynchus clarki*) subspecies while Altukhov and Salmenkhova (1987) have described cases of transplantation of chum salmon (*O. keta*) stocks where the genetic contribution of the introduced stocks was small and gradually disappeared.

In Denmark, some studies have been specifically aimed to investigate the effects of rehabilitating stocks by means of stocking offspring of local trout relative to stocking trout from hatchery strains. The first study concerned brown trout spawning in tributaries to the small Lake Hald (for more details, see Hansen et al., 1993). One of the tributaries had been devoid of trout because of lack of suitable spawning conditions, but following habitat restoration 1200 trout were transplanted to this river from one of the other tributaries. Electrofishing results showed that the transplanted fish were indeed able to survive in their new environment and 6 years later an allozyme study confirmed the close genetic relationships between this established population and the original donor population.

In contrast to this successful transplantation, in the late 1970s, 3000 trout, 2 years old, from a commercial hatchery strain were stocked directly into the lake. A comparison of allelic frequencies of this strain and of the Lake Hald tributary populations showed that *if* some introgression had taken place the genetic input to the native populations must be very modest, below 5–10%, and overall much less than would be expected. Thus, trout from a tributary population of the lake were able to recolonize a vacant river while hatchery trout were seemingly unable to survive and reproduce. This could be interpreted as an example of local adaptation of native trout and lack of adaptation of exogenous trout, but it should also be taken into account that the hatchery trout were stocked directly into the lake. This means that the fish have not been imprinted to any of the tributaries and would consequently be expected to show less efficient spawning behavior than the native trout. To circumvent this problem a river system with spawning and nursery areas directly affected by stocking was chosen for the next study.

Karup River

Karup River (Fig. 1) used to be famous among anglers for its very large-sized sea trout which weighted up to approximately 15 kg. However, over the years pollution and other sorts of habitat destruction led

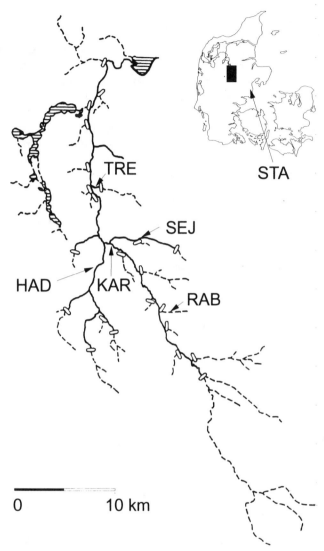

Fig. 1. Approximate location of the Karup River system and Stampemølle Bæk in Denmark, and detailed map of the Karup River system with the sampled localities indicated. Impassable dams are indicated by bars and hatched lines indicate stretches of the river system where upstream migration of trout is restricted. Abbreviations: TRE = Trevad Møllebæk, SEJ = Sejbæk, KAR = Karup main river, RAB = Rabis Bæk, HAD = Haderis River, and STA = Stampemølle Bæk. Not shown on map: HAT = Hatchery strain.

to the eradication of most indigenous populations within the river system. Only a stretch of the main river and very few of the tributaries were suitable for spawning. Not surprisingly, natural reproduction in the river system declined drastically and an intensive stocking program was therefore undertaken. From 1980 to 1989, approximately 240 000 trout originating from one specific hatchery strain were stocked in the river. Concurrently, in 1984 the local anglers started to electrofish wild parent fish and produce their own stocking material and since 1990 this practice has almost completely substituted the use of hatchery fish. Meanwhile, the environmental conditions have been vastly improved and today a large number of very large sea trout once again ascend the river.

Our main interest was to investigate whether the reappearing sea trout are a result of the large number of stocked hatchery trout. In 1992, samples of trout were therefore collected by electrofishing from the following localities (indicated in Fig. 1):

Karup main River: A stretch of the main river with good spawning opportunities which was never severely affected by pollution. This part of the river has mainly been stocked with larger fish (of 2 years of age or more). The sample consisted of sea trout of lengths ranging from approximately 45 cm to 90 cm (1–10 kg) caught in the spawning season.

Rabis Bæk: A non-stocked shallow tributary to the Karup River system. Gene flow to the river is restricted by a dam built in connection with a rainbow trout (*O. mykiss*) hatchery.

Sejbæk, Haderis River, and Trevad Møllebæk: Three tributaries to the Karup River system which have been heavily stocked over the years.

Stampemølle Bæk: A tributary to the completely non-stocked Odder River system in Eastern Jutland (location indicated in Fig. 1) which was included as a reference population.

Finally, a sample was collected from the *Hårkær Hatchery* strain which was used for stocking the Karup River system until recently. Several hundred parent fish are used each year for breeding and an allozyme study did not reveal significant differences in allelic frequencies among age classes (Simonsen and Rasmussen, 1989). We therefore assume that the genetic composition of this hatchery strain has remained almost unchanged over the years.

The applied method was mitochondrial DNA RFLP, which has been recommended for such studies by, for example Hynes et al. (1989). Seven different restriction endonucleases (*Xba*I, *Ava*II, *Hin*fI, *Hae*III, *Mbo*I, *Eco*RV, *and Rsa*I) were routinely used, resulting in the identification of eight mtDNA lineages. The frequencies of these in the sampled populations are shown in Tab. 1. No variation was found in the confined Rabis Bæk population, which is perhaps an indication of a low effective population number. The differences in mtDNA lineage frequencies among all sampled populations were significant at the 0.1%

Table 1. Observed mitochondrial DNA lineage frequencies, unbiased nucleon diversity, and sample sizes (N) of the sampled populations. For a list of sample locality abbreviations, see legend for Fig. 1

Restriction morph	HAT	STA	RAB	KAR	SEJ	TRE	HAD
Type 1	0	0.10	0	0.18	0.04	0.29	0.39
Type 2	0.06	0.15	1	0.58	0.56	0.39	0.39
Type 3	0.58	0.10	0	0.22	0.32	0.32	0.14
Type 4	0	0.30	0	0.02	0	0	0
Type 5	0.36	0.10	0	0	0.04	0	0.07
Type 6	0	0.20	0	0	0	0	0
Type 7	0	0.05	0	0	0	0	0
Type 8	0	0	0	0	0.04	0	0
Nucleon diversity	0.54	0.86	0	0.59	0.60	0.69	0.69
N	36	20	23	50	25	28	28

level ($\chi^2 = 146.8$, 18 d.f.), while in contrast the differences among samples from the stocked localities (Karup River, Trevad Møllebæk, Sejbæk, Haderis River) within the Karup River system were non-significant ($\chi^2 = 11.8$, 6 d.f.). Pairwise χ^2 tests for homogeneity of mtDNA lineage frequencies between populations (Tab. 2) showed that the hatchery strain differed significantly from all other populations and both this pattern and the close relationships among the stocked populations was further supported when a maximum likelihood tree was constructed (Fig. 2).

These results strongly suggest that the large anadromous trout from Karup River are not merely descendants of the hatchery strain and, similarly, the trout populations in the stocked tributaries appear mainly to be descendants of Karup River parent fish. The 240 000 stocked hatchery trout have clearly had a surprisingly small effect on the present gene pools in the river system, although we cannot preclude a modest

Table 2. Pairwise χ^2-tests for homogeneity of mitochondrial DNA lineage frequencies between samples. Numbers in parentheses designate degrees of freedom. For a list of sample locality abbreviations, see legend for Fig. 1

	STA	RAB	KAR	SEJ	TRE	HAD
HAT	$32.8_{(2)}$***	$51.3_{(2)}$***	$48.7_{(3)}$***	$25.9_{(2)}$***	$31.5_{(3)}$***	$36.4_{(3)}$***
STA		$32.3_{(1)}$***	$26.6_{(3)}$***	$21.6_{(3)}$***	$25.0_{(3)}$***	$13.7_{(2)}$***
RAB			$13.6_{(2)}$**	$13.1_{(1)}$**	$20.9_{(2)}$***	$20.9_{(2)}$***
KAR				$3.9_{(2)}$	$2.6_{(2)}$	$4.4_{(2)}$
SEJ					$5.7_{(2)}$	$9.6_{(2)}$*
TRE						$1.1_{(2)}$

***$p < 0.001$, **$p < 0.01$, *$p < 0.05$

284

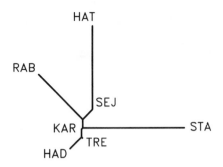

Fig. 2. Restricted maximum likelihood tree based on mitochondrial DNA lineage frequencies, constructed with the program CONTML from Felsenstein's (1989) PHYLIP package, to summarize the genetic relationships among the sampled populations and the hatchery strain used for stocking Karup river. For a list of sample abbreviations, see legend for Fig. 1.

genetic input from the hatchery. The type 5 mtDNA lineage may be a useful marker in this context as it was found at a frequency of 0.36 in the hatchery, but was absent in the Karup River sample. In Haderis River, two trout with the hatchery specific type 5 mtDNA lineage were actually recorded. However, both fish deviated from the rest by being large (≈ 40 cm) "hatchery-looking" resident trout. Some "put-and-take" stocking with large trout from the hatchery strain still takes place in this river, so this is most likely the explanation for the presence of these two fish.

There may be several factors, genetic as well as non-genetic, responsible for the modest influence of the hatchery fish. It should be considered that the switch to using offspring of Karup River fish as stocking material may have purged the genetic input from the hatchery trout. However, this does not explain why the main river population itself has been so little affected by hatchery fish. Perhaps, the use of larger trout (2 years old) for stocking this locality has resulted in a higher initial mortality due to angling pressure than would otherwise have been expected. It should also be pointed out that stocking with larger trout (in contrast to younger fish) may result in a lowered potential for adapting to the specific environment. Nevertheless, even though such circumstances may have been important, they are unlikely to suffice as sole explanations for so little effect by so many stocked fish. It is therefore plausible that genetic factors such as lack of adaptations to local conditions and semi-domestication of the hatchery strain have played a significant role.

One important conclusion may be drawn: Both in terms of conserving genetic diversity and of maximizing yield, offspring of native trout rather than hatchery trout should be preferred for stocking material. Apart from this point, the very diverse results of monitoring of stocking programs do not really allow for any generalizations to be made. We would especially like to stress that apparent lack of influence of stocked hatchery trout should not be regarded as indicative of lack of harmful effects. If stocked and native fish interbreed, and if strong selection is acting against stocked fish and "hybrids", the result may be a lowered average fitness of the affected populations. In any case, the release of hatchery trout is expected to intensify competition for food and space which may have a negative influence on native fish.

We would like to point out the need for more controlled long-term stocking experiments, and we find it especially important to clarify why some stocking programs result in massive introgression between stocked and native trout while other programs have virtually no effect in terms of survival and reproduction of stocked fish. Is the apparent failure of some stocking programs due to gradual selection against stocked fish, or are the fish not at all able to survive and reproduce? Is there a relationship between the number of generations a hatchery strain has been artificially propagated (i.e., its potential degree of domestication) and its ability to perform well in nature? The answers to such questions could provide valuable information for future management programs.

Management recommendations

It is evident that all sorts of stocking activity may have profound effects on the affected populations and should therefore be considered a far from optimal management strategy. On the other hand, it would be unrealistic to ignore the demand for fish to sustain commercial and recreational fisheries. In densely populated areas some degree of stocking is probably inevitable and will remain so for the foreseeable future. The aim must consequently be to develop an integrated strategy which ensures the presence of utilizable fish resources, but also conserves as much genetic diversity within the species as possible. As stated earlier, for a species showing very pronounced genetic differentiation the emphasis should be put on conserving individual stocks.

It is important to recognize that one common strategy applicable to all populations cannot be developed (e.g., Waples, 1991). Different scenarios require different measures and it would therefore be appropriate to classify populations according to their previous stocking history and need for management. As much care as possible should be taken to protect populations which have only been slightly or not at all affected by stocking with exogenous fish; if such populations are already doing well, they should simply be left to themselves. If the number of fish has declined due to negative changes in environmental conditions, management efforts should be aimed at reversing this process, for instance, by restoring and facilitating access to the spawning places. If such measures do not suffice, temporary rearing and release of local fish could be considered. Another situation arises if a non-stocked population has been subject to inbreeding and loss of genetic variation which may justify introduction of new genetic variation. Then, stocking should be regarded as a method for creating artificial gene flow and, accordingly, the number of stocked fish should be equivalent to the gene flow which would occur under normal circumstances (Ryman, 1991). That is, only *very few* individuals should be stocked each generation instead of the usual thousands and, preferably, the stocked fish should originate from other natural populations.

Even if a population has been moderately affected by stocking, this does not mean it is not worth conserving. The fact that it is still able to exist and may contain valuable genetic variation should justify a certain degree of consideration. Again, facilitating natural reproduction and otherwise not interfering would be the optimal choice, but if stocking is inevitable, local fish should be used to provide the stocking material. The potentially negative effects connected with stocking exogenous fish from hatchery strains may eventually lead to eradication of local gene pools, which is obviously not compatible with the aims of conservation genetics.

Finally, many populations have been heavily affected by stocking and may even have been founded by fish introduced to previously void rivers. If the ultimate goal is to naturalize such populations, it will be inexpedient to stock them with exogenous fish, but otherwise the use of hatchery fish may be justified in such localities.

A very important issue is to minimize unintentional influence of stocked hatchery fish on natural, non-stocked populations. In this context, stocking smolts directly into the sea is completely inappropriate and should be abandoned. It would, however, perhaps be possible to draw advantage of the well-developed homing instinct exhibited by salmonid fishes. As an example, a large number of smolts could be stocked in mouths of rivers which are not otherwise suitable for spawning or are inhabited by already heavily stocking influenced populations, *but we want to stress the necessity for confirming that the stocked fish do indeed exhibit homing and do not spread at random to other rivers.*

As long as stocking with hatchery fish takes place, influence on unstocked populations by natural gene flow is probably unavoidable in the long run. As previously mentioned, this is a particularly pertinent problem for brown trout because many hatchery strains are seemingly very closely related. In order to maintain genetic diversity it would be highly recommendable to found new hatchery strains from verified genetically different populations in order to avoid this massive one-way gene flow from one source population.

A final issue is the rearing and perpetuation of captive populations. We have previously argued in favor of supporting natural populations by releasing offspring of indigenous fish in contrast to hatchery fish. There are basically two options for doing that: *to found a gene bank once and for all and stock the offspring*, or *to sample wild fish each spawning season and rear and release the offspring*. We will certainly recommend the latter. This will ensure that altered selection regimes as a result of the hatchery environment may still be counterbalanced by natural selection and the reared fish will be still integrated in other natural processes like gene flow. For practical purposes it may be inconvenient to catch a sufficient number of wild fish each year, but we still recommend that the captive population be supplemented by as many wild fish as possible. The subject of minimal effective population number has been covered in great detail elsewhere (e.g., Frankel and Soulé, 1981), but we would like to point out the fundamental problem of the time-scale of conservation efforts. The usual recommendations of an effective population number of 50–60 (Soulé, 1980; Ryman and Ståhl, 1980) are specifically aimed at minimizing inbreeding and are only applicable for short periods of time, but a realistic time-scale for conservation of salmonids will probably range from centuries to indefinitely. In order to avoid a depletion of evolutionary potential by loss of alleles, the recommendation by Allendorf and Ryman (1987) to use at least 100 parent fish of each sex, and preferably more, is therefore more appropriate.

Conservation of salmonid fishes and, not least, brown trout, is admittedly a complicated issue, but we feel this is indeed an area where vast improvements could be obtained through rather modest changes of existing management procedures. From the perspective of the discussion about the significance of conservation genetics, this may also provide a good example of the necessity of integrating population genetics in conservation programs.

Acknowledgements
We are grateful to Kjetil Hindar, Subodh Jain, Jürgen Tomiuk, Carlo Largiader, Gorm Rasmussen, and Peter Geertz-Hansen for critical comments on the manuscript and/or stimulating discussions, and to Christian Dieperink for assistance in preparing the figures. MMH would like to direct special thanks to Rosaleen Hynes for much help and good advice on mitochondrial DNA techniques, and to Andrew Ferguson, John Taggart, and Paulo Prodöhl for providing a very pleasant stay in Belfast in spring 1992 in order to learn DNA techniques. Finally, we would like to thank the Danish Natural Science Research Council for providing financial support (11-9343-1 PD and 11-9634-1).

References

Allendorf, F. W. (1983) Isolation, gene flow, and genetic differentiation among populations. *In:* Schonewald-Cox, C. M., Chambers, S. M., MacBryde, B. and Thomas, W. L. (eds), *Genetics and conservation.* The Benjamin/Cummings Publishing Co., Menlo Park, CA, pp. 51–65.

Allendorf, F. W. and Ryman, N. (1987) Genetic management of hatchery stocks. *In:* Ryman, N. and Utter, F. M. (eds), *Population genetics and fishery management.* University of Washington Press, Seattle, pp. 141–159.

Altukhov, Yu. P. (1990) *Population genetics: Diversity and stability*. Harwood Academic Publishers, London.

Altukhov, Yu. P. and Salmenkhova, E. A. (1987) Stock transfer relative to natural organization, management, and conservation of fish populations. *In:* Ryman, N. and Utter, F. M. (eds), *Population genetics and fishery management*. University of Washington Press, Seattle, pp. 333–343.

Barker, J. S. F. and Thomas, R. H. (1987) A quantitative genetic perspective on adaptive evolution. *In:* Loeschcke, V. (ed), *Genetic constraints on adaptive evolution*. Springer Verlag, Berlin, pp. 3–23.

Bernatchez, L., Guyomard, R. and Bonhomme, F. (1992) DNA sequence variation of the mitochondrial control region among geographically and morphologically remote European brown trout *Salmo trutta* populations. *Mol. Ecol.* 1: 161–173.

Crozier, W. W. and Ferguson, A. (1986) Electrophoretic examination of the population structure of brown trout (*Salmo trutta*) from the Lough Neagh catchment, Northern Ireland. *J. Fish Biol.* 28: 459–477.

Felsenstein, J. (1989) PHYLIP – Phylogeny Inference Package (Version 3.2). *Cladistics* 5: 164–166.

Ferguson, A. (1989) Genetic differences among brown trout (*Salmo trutta*) stocks and their importance for the conservation and management of the species. *Freshw. Biol.* 21: 35–46.

Ferguson, A. and Taggart, J. B. (1991) Genetic differentiation among the sympatric brown trout (*Salmo trutta*) populations of Lough Melvin, Ireland. *Biol. J. Linn. Soc.* 43: 221–237.

Frankel, O. H. and Soulé, M. E. (1981) *Conservation and Evolution*. Cambridge University Press, Cambridge.

Garcia-Marin, J. L., Jorde, P. E., Ryman, N., Utter, F. and Pla, C. (1991) Management implications of genetic differentiation between native and hatchery populations of brown trout (*Salmo trutta*) in Spain. *Aquaculture* 95: 235–249.

Gyllensten, U., Leary, R. F., Allendorf, F. W. and Wilson, A. C. (1985) Introgression between two cutthroat trout subspecies with substantial karyotypic, nuclear, and mitochondrial genomic divergence. *Genetics* 111: 905–915.

Haldane, J. B. S. (1930) A mathematical theory of natural and artificial selection. Part IV. Isolation. *Proc. Cambridge Philos. Soc.* 26: 220–230.

Hamilton, K. E., Ferguson, A., Taggart, J. B., Tomasson, T., Walker, A. and Fahy, E. (1989) Post-glacial colonization of brown trout, *Salmo trutta* L.: Ldh-5 as a phylogeographic marker locus. *J. Fish Biol.* 35: 651–664.

Hansen, M. M., Loeschcke, V., Rasmussen, G. and Simonsen, V. (1993) Genetic differentiation among Danish brown trout (*Salmo trutta*) populations. *Hereditas* 118: 177–185.

Hindar, K., Ryman, N. and Utter, F. M. (1991) Genetic effects of cultured fish on natural fish populations. *Can. J. Fish. Aquat. Sci.* 48: 945–957.

Hynes, R. A., Duke, E. J. and Joyce, P. (1989) Mitochondrial DNA as a genetic marker for brown trout, *Salmo trutta* L., populations. *J. Fish Biol.* 35: 687–701.

Karakousis, Y. and Triantaphyllidis, C. D. (1990) Genetic structure and differentiation among Greek brown trout (*Salmo trutta* L.) populations. *Heredity*, 64: 297–304.

Krieg, F. and Guyomard, R. (1985) Population genetics of French brown trout (*Salmo trutta* L.): Large geographical differentiation of wild populations and high similarity of domesticated stocks. *Génét. Sél. Evol.* 17: 225–242.

L'Abée-Lund, J. H. (1989) Significance of mature male parr in a small population of Atlantic salmon (*Salmo salar*). *Can. J. Fish. Aquat. Sci.* 46: 928–931.

Larsen, K. (1972) New trends in planting trout in lowland streams. *Aquaculture* 1: 137–171.

Moran, P., Pendas, A. M., Garcia-Vazquez, E. and Izguierdo, J. (1991) Failure of stocking policy of hatchery reared brown trout (*Salmo trutta*) in Asturias, Spain, detected using *LDH-5** as a genetic marker. *J. Fish Biol.* 39 (Suppl. A): 117–122.

Nelson, K. and Soulé, M. E. (1987) Genetical conservation of exploited fishes. *In:* Ryman, N. and Utter, F. M. (eds), *Population genetics and fishery management*. University of Washington Press, Seattle, pp. 345–368.

Ryman, N. (1983) Patterns of distribution of biochemical genetic variation in salmonids: Differences between species. *Aquaculture* 33: 1–21.

Ryman, N. (1991) Conservation genetics considerations in fishery management. *J. Fish Biol.* 39 (Suppl. A): 211–224.

Ryman, N. and Ståhl, G. (1980) Genetic changes in hatchery stocks of brown trout (*Salmo trutta*). *Can. J. Fish. Aquat. Sci.* 37: 82–87.

289

Ryman, N., Allendorf, F. W. and Ståhl, G. (1979) Reproductive isolation with little genetic divergence in sympatric populations of brown trout (*Salmo trutta*). *Genetics* 92: 247–262.

Simonsen, V. and Rasmussen, G. (1989) Undersøgelse af genetisk variation hos ørred (*Salmo trutta*) som funktion af tid og dambrug. (Genetic diversity of brown trout (*Salmo trutta*) in Danish fish farms). (In Danish). *DFH-Rapport* No. 367.

Skaala, Ø. and Jørstad, K. E. (1987) Fine-spotted brown trout (*Salmo trutta*): Its phenotypic description and biochemical genetic variation. *Can. J. Fish. Aquat. Sci.* 4: 1775–1779.

Soulé, M. E. (1980) Thresholds for survival: Maintaining fitness and evolutionary potential. *In:* Soulé, M. E. and Wilcox, B. A. (eds), *Conservation biology: An evolutionary ecological perspective*. Sinnauer Associates, Sunderland, MA, pp. 151–170.

Svärdson, G. and Fagerström, Å. (1982) Adaptive differences in the long-distance migration of some trout (*Salmo trutta* L.) stocks. *Rep. Inst. Freshw. Res., Drottningholm* 60: 51–80.

Taggart, J. B. and Ferguson, A. (1986) Electrophoretic evaluation of a supplemental stocking programme for brown trout (*Salmo trutta* L.). *Aquacult. Fish. Manage.* 17: 155–162.

Taylor, E. B. (1991) A review of local adaptation in Salmonidae, with particular reference to Pacific and Atlantic salmon. *Aquaculture* 98: 185–207.

Waples, R. S. (1991) Genetic interactions between hatchery and wild salmonids: Lessons from the Pacific Northwest. *Can. J. Fish. Aquat. Sci.* 48 (Suppl. 1): 124–133.

Wright, S. (1931) Evolution in Mendelian populations. *Genetics* 16: 97–159.

Wright, S. (1978) *Evolution and the genetics of populations. Volume IV: Variability within and among natural populations*. The University of Chicago Press, Chicago and London.

Conservation Genetics
ed. by V. Loeschcke, J. Tomiuk & S. K. Jain
© 1994 Birkhäuser Verlag Basel/Switzerland

Genetics and demography of rare plants and patchily distributed colonizing species

S. K. Jain

Department of Agronomy and Range Sciences, University of California, Davis, CA 95616, USA

Summary. Several population genetic and ecological studies are presented to illustrate how learning about the fate of endemics and small populations is important in conservation. Numerous California endemics are listed as rare and highly threatened plants as their habitats are being destroyed and fragmented. Two herbaceous plant genera of vernal pools, *Limnanthes* and *Orcuttia*, have been studied for genetic structure, breeding systems, and life histories under population size changes over several seasons. The *Limnanthes* taxa show genetic variation patterns to be quite predictable from the geographical range of various "metapopulation units" and breeding systems; the wide-ranging species (*L. alba, L. douglasii*) are more variable and heterozygous than the inbreeding and narrowly distributed species (*L. floccosa, L. bakeri*). Life history comparisons showed a critical role of seed bank and early flowering in the lower amplitude of population numbers, correlated with greater persistence of even small populations in *L. floccosa* than its closest outbreeding relative *L. alba*. Heritability estimates of several life history traits showed concordance with the allozyme variation patterns, and reciprocal transplant experiments suggested that *L. alba* populations might evolve toward more selfing and safer bet-hedging strategy under some environments. Populations of several *Orcuttia* species provided evidence for the critical role of early seeding establishment in their persistence. Conservation programs have focussed on the habitat protection by setting up several nature reserves and on monitoring the vernal pool communities.

Populations of rose clover (*Trifolium hirtum*), an introduced forage legume, have been studied during their colonizing phase on pastures and along roadsides. Their patch dynamics allowed some detailed observations on the relative success of colonies. Pollinators (bumblebee, honeybee) account for gene flow among such colonies and genetic variation is significantly ubiquitous even under higher rates of selfing. An experiment with founding of new colonies showed that input high levels of variability helped in colony success. More recently, gynodioecy seems to have evolved and increased in relative frequency in the areas of new range expansion. These findings provide some clues to the genetic and demographic features of rare species stands struggling to persist and to evolve under human-impacted environments.

The serpentine sunflower (*Helianthus exilis*) has become a sort of cause célèbre due to its biosystematic status changes. Early floras listed it as a bonafide species, endemic on serpentine habitats. In 1949, C. B. Heiser reported it as a variety of recently spreading wild-weedy taxon *H. bolanderi*. This has lately allowed land use around its localities threatening its extinction. Detailed morphological, cytogenetic, and molecular studies, along with some work on crossability and special agronomic attributes, are now available to recommend its original species status. Several populations have been lost and the remaining few are small and need protection.

These conservation-oriented projects involve long-term population studies, but since habitat protection is the first step, community approach to biodiversity monitoring is equally important. Population genetic information is useful in sampling materials for gene banks (*ex situ* conservation) as well as for planning restoration work.

Introduction

Many small, poorly drained depressions in the Great Valley and foothills of cismontane California are called vernal pools which are of great interest for their distinctive and highly endangered floristics. Periodic inundation due to winter season precipitation provides an extremely interesting zonation pattern of species distributions. Several symposia and conservation-oriented agency reports have described vernal pool habitat and biology in detail (e.g., Jain and Moyle, 1984). Several plant genera including *Allocarya*, *Lasthenia* and the endangered list member *Pogogyne* have been studied for biosystematic and adaptational features. We shall briefly review studies on two vernal pool genera, *Limnanthes* and *Orcuttia*, with many taxa now requiring serious conservation efforts.

Species distribution of *Limnanthes* taxa and several other common genera (e.g., *Allocarya*, *Lasthenia*) showed (a) an island biogeographic pattern with a rather low z-value in the species-area curve, (b) marked zonation along the moisture gradient within pools, (c) marked regional patterns of species replacement such that very few congeners live sympatrically, and (d) the fact that commonness or rarity of most taxa is highly dependent on the time-space scales. In fact, rarity of vernal pool taxa needs some discussion. Some rare species (e.g., *Pogogyne abramsii*, *Tuctoria mucronata*) are restricted to a very small geographical area or a single population; some endemics are more widely distributed (e.g., *Gratiola heterosepala*, *Legenere limosa*). As noted by Stone (1990), the concepts of rarity and endemism are related but not interchangeable. Many but not all endemic vernal pool taxa are habitat specialists or stress-tolerators, and genetic uniformity is not to be taken for granted. In fact, local populations might often have population size (N) larger than several thousands in most *Limnanthes* or *Allocarya* taxa. Drury (1980), in a classic article on rare plants, discussed relative rarity in terms of frequency (% plots occupied), consistency (multiyear presence), and "density" (N). To this, Rabinowitz (1981) added the criteria of geographical range and habitat specificity. For the species mentioned in this chapter, we have tabulated rarity classes following Rabinowitz's criteria (Tab. 1). We should add here that with the continued losses of vernal pools and intermittent streams, almost all of the listed taxa are quickly sliding toward increased rarity and extinction risks.

Ecological genetics and conservation of Limnanthes

The prospects of developing a new industrial crop by domesticating meadowfoam (*Limnanthes* spp.), an annual spring flowering herb, led to an extensive research project for collection, evaluation, and conservation of its genetic resources. Fifteen out of 16 known taxa are endemic to California and Oregon, largely restricted to temporarily wet habitats (vernal pools, shallow ditches and intermittent streams). Nearly 125 sites have been sampled over the past 18 years of which nearly 30–40%

Table 1. Limnanthaceae, Orcuttieae and *Helianthus* spp. taxa following the Rabinowitz (1981) typology of rare species

Geographic range Habitat specificity	Large		Small	
	Wide	Narrow	Wide	Narrow
Local Population Size: Large (N ≫ 10³); Dominant	*L. nivea* *L. alba* *H. bolanderi*	*L.d. douglasii* *L. montana* *L. striata* *L.d. rosea*	*L.g. gracilis*	*L. vinculans* *L.f. pumila* *L.f. grandiflora* *L.g. parishii* *L.d. sulphurea*
Small (N often <10³); Non-dominant and very small patches (<200 m²)	*L.f. floccosa*	*O. californica* *O. tenuis* *O. pilosa* *T. greenei*‡		*L. macounii* *L.f. californica* *L. bakeri* *H. exilis* *O. viscida* *O. inaequalis* *T. mucronata*‡ *T. fragilis*‡

‡Several *Orcuttia* species have recently been renamed under *Tuctoria*.

are lost or threatened by human impact. At least 10 taxa should be listed as rare and endangered (e.g., *Limnanthes bakeri*, *L. vinculans*, *L. floccosa* var. *californica* with only less than five populations extant and all within a small geographical range). Even the more widespread taxa (e.g., *L. alba*, *L. douglasii* var. *rosea*) appear to have declined in most areas by the urbanization as well as water flow pattern changes. Therefore, several ecological genetic studies were initiated for long-term observations as summarized below.

Population structure and breeding system

Nearly 30 to 50 plants sampled from five or more populations (fewer in rarer taxa) were scored for phenotypic variation in plant, seed, and flower traits. Progenies were grown in a greenhouse for scoring between- and within-family components of variation, along with several morphological polymorphisms (calyx and petal hairiness, anther color, nutlet texture). Brown and Jain (1979) found in a comparative study of outbreeder *L. alba* and inbreeder *L. floccosa* that overall, *L. floccosa* has less variability, as expected, than *L. alba*, but there were significant exceptions; for several traits including flowering time, *L. floccosa* has larger interpopulation as well as interfamily variation. Moreover, phenotypic plasticity for several traits was higher in *L. floccosa* than *L. alba*. Surveying all of the other taxa also (Tabs 2 and 3), two generalizations appeared: (1) rarer and narrowly distributed taxa have less genetic

Table 2. Genetic variation in relation to the breeding systems

Inflexae members	Breeding systems‡	No. of alleles/locus Popn.	No. of alleles/locus Taxon	Nei's D_{popn}	Geog. range based on mean-inter popn. dist. (kms)
L.g. parishii	2	1.33	1.77	.026	<16
L.g. gracilis	2	1.46	1.92	.059	32
L. montana	2–3	1.19	2.37	.077	240
L.f. floccosa	1	1.02	1.13	.109	290
L.f. californica	1–2	1.08	1.10	.079‡	70
alba	3	1.60	2.53	.062	500

Reflexae members	Breeding system	No. of alleles/locus Popn.	No. of alleles/locus Taxon	Nei's D_{popn}	Geog. range based on mean-inter popn. dist. (kms)
L. bakeri	2	1.3	1.3	.013	12
L. vinculans	2	1.5	1.8	.013	16
L.d. sulphurea	3	1.4	1.6	.016	10
L.d. douglasii (California)	3	1.9	3.1	.038	150
L.d. rosea	3	1.8	3.1	.063	400
L.d. nivea	3	2.1	3.4	.095	240
L. striata	3	1.6	2.6	.159	320

‡Estimate from Kesseli (1992).

Table 3. Summary descriptors of variation and population dynamics in *Limnanthes* taxa

Taxon & approx. population sizes	Approx. geographic range (km)	No. of "metapopulations"	Variation in quantitative traits h² Rank‡	CV Rank‡	Yearly variation in population size; CV Rank‡	Habitat and life history features
L. alba N ≫ 1000	500	5+	1 (1–3)	3 (2–4)	245	Late flowering; variable seed production
L. floccosa var. *california* N < 500	135	2	4 (2–5)	2	1182	Highly disturbed & failed recruitment in most sites
L. floccosa var. *floccosa* N = 65–500	290	7+	6	4 (2–5)	77	Good seed bank; early flowering
L. douglasii N ≫ 1000	150	4+	2	1	207	High seedling mortality; low seed bank
L. bakeri N ≫ 1000	12	Only 2 sites	3	5 (4–6)	?	Pollination limiting; seed quality highly varied
*L. macounii**	40	1	5	6	150+	Germination varies; disturbed

(*Data from A. Ceska, personal communication).
‡h², estimates of heritability, based on between/within family variances, from 3–5 populations per taxon. h² values greater than 0.45 rank 1, h² lower than 0.20 was ranked on this relative scale as 6, etc. Similarly, phenotypic coefficients of variation (CV), higher than .80 ranked as 1, and lower than .25 as 6, etc.

and phenotypic variation than the widespread ones; (2) interpopulation differentiation might be comparable for both the outbreeding and inbreeding members.

Allozyme variation and morphological variation were generally concordant (McNeill and Jain, 1983; Kesseli and Jain, 1984), although a more detailed study of *L. alba* sampled from the same geographical region with life history traits added to this comparison showed this concordance to be less consistent (Ritland and Jain, 1984a,b). The next logical step was to estimate specific levels of selfing or outcrossing for each of the populations using a multilocus estimation procedure. Large amounts of variation in outcrossing rates within and between populations, as noted in *L. alba* (McNeill and Jain, unpublished data), gives a caveat against broad generalizations about the genetic structure. A few attempts have also been made to estimate gene flow by pollen flow, e.g., marking pollinators (Thorp, Kesseli and Jain, unpublished data) and through seed movement using denuded plots or markers released in a source upstream patch. Typical leptokurtic curves gave average values of mean dispersal distance as measured by the neighborhood size parameter σ ranging between 0.5 to 1.2 m (Kesseli, 1992). However, these gene flow measures markedly underestimate the gene flow rate among populations within a pasture or along a stream path due to water-dispersed seed. In fact, genetic distances among neighboring populations were found to be an order smaller than the interregional estimates. *Limnanthes* taxa seem to have a metapopulation structure in terms of series of interconnected stands within each of the water flow channels (coarsely mapped using creeks and roadways for each region), but no data are available to verify local extinction-colonization processes. Especially due to seed carryover in the soil it is particularly difficult to locally observe an extinction event without detailed and highly intrusive experiments. Past anecdotal records suggest, however, that in a typical pasture or creek watershed with dozens of pools within a range of 10 to 100 meters from each other, local recolonization regularly occurred after a pool was heavily disturbed. Now, with many pools extirpated and even birds or animals as seed dispersers reduced in numbers, metapopulation dynamics might be losing to an island or stepping stone models with reduced gene flow. Monitoring of allozyme variation during the recent drought years (1986–91) suggested that total genic diversity is significantly reduced in *L. douglasii* populations (low seed carryover), whereas no significant allelic frequency changes occurred in *L. alba* populations (seed carryover more than 10% of a year's seed crop). The overall situation nevertheless is precarious since 5 or 10 years' monitoring does not tell the whole story, as judged by the statewide losses of habitat and increased human impact.

Alternative life histories in *Limnanthes*

Besides variation in seed carryover rates among taxa, as noted earlier, there are at least three more variables in life history: (a) flowering time, (b) seedling mortality, often density dependent as most *Limnanthes* patches have high plant densities (up to 1500 plants/m² in *L. alba*), and (c) seed output per plant (and per flower). Heritability estimates for several seed yield components and life history features were derived from estimates of between- and within-family variance components (Ritland and Jain, 1984a; Krebs and Jain, 1985). Both *L. alba* and *L. floccosa* showed significant levels of genetic variation in most of these traits. Ritland and Jain (1984b) found in a comparative study of *L. alba* and *L. floccosa* that early flowering time, rapid seed maturation combined with a low number of high quality seed (a safe strategy with assured reproduction in wet and drier years) accounted for relatively higher persistence (judged by lower CV in population sizes over years and localities) in *L. floccosa*, than *L. alba* with the opposing set of life history traits (a risky strategy with high seed output only in wet years). This does seem anomalous given that *L. floccosa* is often less abundant, grows in isolated patches, selfs largely, and has lower genetic variation. Thus, population census and genetic data should be looked at in terms of life history variation over several years. It should be noted emphatically that extinctions of most vernal pool stands are often due to wholesale destruction of the habitat with the biological terms of persistence or evolutionary responses having little role.

A long-term colonization experiment in *Limnanthes*

An experiment involving artificial transplanting of two species was started in 1979 with a three-fold objective: (a) to test the potential colonization by outbreeder *L. alba* (needs insect pollination) and selfer *L. floccosa* outside their ranges of distribution, (b) to test if input genetic variation (based on known allozyme diversity level) would be correlated with colonizing success, and (c) to study the demographic and breeding behavior of such small colonies. Using eight locations, a total of 128 *L. alba* and 128 *L. floccosa* colonies were planted with 200 seeds per colony; for *L. alba* two different levels of genetic diversity were used by choosing two populations from the genetic resource collection (gene bank) and for *L. floccosa*, we sought to compare a mixture of partial selfers *L.f. grandiflora* and *L.f. californica*, with the complete selfer *L. floccosa*. Results are summarized in Tab. 4. Three observations are relevant here: (1) Very few colonies got started (unfavorable habitat, or a numbers game of low probability event?) of which even fewer (only *L. alba*) have survived to date; (2) population with higher

Table 4. Fate of founded *Limnanthes* colonies (data shown only for two out of eight localities with relatively higher success)

Founder populations	Location*		No. of colonies planted	No. of colonies after 1 year	No. of colonies after 10 years	Comments
L. alba						
(a) Highly Polymorphic (Popn. 312)	(A)	N. of Burney	8	6	3	Polymorphic seed stock gave greater success; colonies have not expanded at N. of Burney site but several have expanded at Bear Creek
	(B)	Bear Creek	8	4	3	
(b) Less Polymorphic (Popn. 308)	(A)	N. of Burney	8	6	1	
	(B)	Bear Creek	8	3	1	
L. floccosa						
(a) var. *Californica & grandiflora* mixed	(A)	N. of Burney	8	3	0	After survival for the past 2–3 years, all except two colonies failed to recruit after 6 years
	(B)	Bear Creek	8	5	1	
(b) var. *floccosa* (monomorphic)	(A)	No. of Burney	8	2	1	
	(B)	Bear Creek	8	2	0	

*North of Burney location is outside the range of both *L. alba* and *L. floccosa*; cooler and very few pollinators; Cayton Creek is at the northern margin of their ranges.

genetic variation accounted for nine out of the 11 surviving colonies (in *L. floccosa* this could not be assessed); (3) survival of *L. alba* colonies did seem to require some shift toward autogamy and earlier flowering time (toward *L. floccosa* strategy?). These data are preliminary until we have enough material to collect and analyze (we did not wish to lose colonies by extensive sampling!). Demographic data have not yet been analyzed using population viability projection matrices which might help us make some prognosis for the surviving few colonies. Many authors have recommended such transplant experiments (e.g., Holsinger and Gottlieb, 1991; Barrett and Kohn, 1991), but it must be realized that the protocol has many limitations, e.g., it requires the use of a large number of colonies (founding attempts), a long waiting time and yet possible failure due to disturbance, or drought. Moreover rarity does not allow extensive experimental work, and, often, permission to manipulate natural areas is largely ruled out.

Potential role of inbreeding depression

Experimental work has been conducted on *L. alba* and *L. douglasii* var. *rosea*, the latter with more pronounced protandry and virtually no delayed selfing mechanism under conditions of limited pollinators' visits. Naturally out-pollinated progenies and selfed progenies provided evidence for greater inbreeding depression in *L. d. rosea*, as expected (unpublished data of McNeill, Kesseli and Jain). Admittedly, these estimates are limited by several experimental design restrictions. However, during our efforts to domesticate these two taxa, several years of efforts to select for higher self-fertility suggested that selfing most members of section Reflexae (to which *L. d. rosea* belongs) is far less feasible than developing highly inbred lines of Inflexae taxa, including, of course, the autogamous *L. floccosa*. We mapped out five small populations each of *L. d. rosea*, *L. alba*, and *L. f. floccosa* within the Butte and Colusa region, which has extensive disturbance due to agricultural and urban developments. Most vernal pools have been modified by cultivation and water deficits. These small dwindling populations have suffered in both fecundity and seedling reestablishment. All five *L. d. rosea* populations have continued to decline with very poor seed production and inbreeding depression levels higher than those in *L. alba*. For example, viability and fecundity-based estimates of inbreeding depression (δ) ranged between 0.66 and 0.79 in *L. d. rosea*, and between 0.75 and 0.82 in *L. alba*. Three other populations of *L. d. rosea* in the same vicinity with gynodioecy and higher heterozygosis appear to have normal seed output and persistence.

Population studies in *Orcuttia*, another endangered genus

The grass genus *Orcuttia* is unique in the gross family biosystematically and ecologically, and its habitat specialization is adapted to the summer annual cycle in the larger pools. Griggs and Jain (1983) reported on population census data and life history variation among several taxa. Genetic and demographic information gathered over the years is summarized in Tab. 5. Similar to the *Limnanthes* pattern, although much less genetic information is at hand, different taxa have characteristic strategies for survival with a minimum threshold of occupancy (cf. Lande, 1987), and with natural gene exchange patterns. *Orcuttia mucronata*, known from a single stand, has been monitored only cursorily, and no seed sampling was allowed, thereby ruling out any genetic monitoring. Genetic variation for certain life history features (e.g., tiller production, seed germination requirements) should be studied. Here, census data over an 8-year period strongly suggested that populations smaller than 100 to 150 had higher proneness to extinction; over 20% of our sampled populations have been extirpated.

Conservation of vernal pool plants, *in situ* and *ex situ*

Several reserves (five officially protected and others with variable protection status) have been established during the past 15 years. Most of these sites do not cover more than a few of the endangered taxa; for example, *Limnanthes* is found only in one-third of these areas. Not all *Limnanthes* or *Orcuttia* taxa are included in these reserves, nor is there variation among populations within taxa, which are in several cases

Table 5. Summary of *Orcuttia* data on genetic variation and demography

Taxon	GR	RI	No. of popns.	CV_N (yrs)	Genetic variation Taxon	Genetic variation Popn.	Demography SB	Demography SS	Demography RR
Tenuis	200	5	5	L–H	L	VL	H	M	L–H
Greenei	270	3–4	14	L	H	L	L	L	M
Pilosa	350+	3–4	10	L–H	H	H	—	H	L–H
Inaequalis	60+	3–4	13	?	—	—	—	—	—
Viscida	5	5	1	H	H	L	M	H	H
Californica	220	4–5	3	L	H	H	M	H	M–H
Mucronata	1	5	1	L	L	L	L	H	L

GR = Geographic range (kms), RI = rarity index high (5), Moderate (3–4), SB = Percent seed carryover, SS = seedling surv., RR = mean no. seed/plant. VL (very low), L (low), M (medium) and H (high) refer to the relative ranks for within *Orcuttia* comparisons. Note that rarity index (RI) and GR are correlated; RI and CV_N are correlated; and different life cycle stages are critical in population persistence. (CV_N were based on 8–10 years of approximate census data).

well-defined at the varietal or subspecific levels. Local efforts for *L. floccosa* var. *californica* (by the Chico city government) and for *L. vinculans* (by the Sonoma State University researchers) have generated interest in protection and research work. On the other hand, despite many years of careful monitoring and genetic work, *L. macounii* in Victoria Island (Canada) has not made the endangered list (A. Ceska, personal communication). Our data on genetics and life histories have been of interest to the conservationist, but the crop prospect of *Limnanthes* has played the crucial role in generating much of the conservation interest. Collections of *Limnanthes* represent an *ex situ* effort to conserve variation, but with very little funding, seed stocks have not been replenished by collecting or proper seed multiplication.

Colonization dynamics of rose clover

A colonizing species has to pass through the phases of dispersal, initial establishment or failure of numerous small colonies, and eventual success through a pattern of threshold occupancy of habitat (cf. Lande, 1987). Genetic variation could play a role in dispersibility and some microevolutionary changes during establishment. Such a patch dynamic process of colonization is illustrated here for an introduced forage legume *Trifolium hirtum* (rose clover), introduced into California nearly 50 years ago. It is a successful colonizer annual herb, now living in distributed sites along the roadsides and grasslands (Jain and Martins, 1979). During the 1970s, we selected three pastures in Northern California in which grazing animals were primary dispersers of the seed into new patches. A total of 175 patches were observed for 5 years in terms of colony failure or success, polymorphism at four morphological marker loci, and the role of pollinator bees in interpatch pollen movement. Data are summarized in Tab. 6. Three points should be noted: (1) Patches with initially some minimal plant numbers ($N \sim 20$) and located close to each other (less than 5 m) had a higher rate of success; (2) patch size and polymorphism index values were significantly correlated, taking potential pollen flow distances into account. It appears highly likely that these patches had the advantage of attracting pollinators and of interpatch gene flow; (3) thus, colonization, or conversely local extinction, seem to require a minimum threshold number of sufficiently large, adjacent patches (thus, metapopulation dynamics) and this includes the role of genetic variation. However, we still do not have sufficient genetic information for a long enough time (due to large seed carryover, generation time is effectively longer than 2 years) to test how genetic variation shifted in these patches. In another experiment using artificially founded colonies with three levels of initial variability in seed stocks, more variable founders had a higher success in establishing

Table 6. Fate of rose clover patches selected for two interpatch distances and two initial size classes

Location	Status*	Near (<5 m)		Distant (>15 m)		Comments‡
		$N_0 \leq 5$	$N_0 \geq 20$	$N_0 \leq 5$	$N_0 \geq 20$	
	No. of patches	25	20	25	20	
Auburn	A	8	12	5	7	Larger patches in the "near" category had significantly higher success ($P < .05$)
	B	4	5	0	2	
	Initial No. of patches	25	20	25	15	
Placerville	A	11	15	8	6	Same as above except that at this site seed output was lower and patches did not expand
	B	4	1	2	0	

*Status recorded and after 5 years in terms of A = number of patches surviving without size increase, and B = number of expanded patches. The remainder patches disappeared.
‡Polymorphic index (same as H_S) calculated for four marker loci was correlated ($P < .05$) with patch size: r = .46 for the patches B in the "Near" category (both sites pooled).

colonies (Martins and Jain, 1979). More recently, we find that some of the most successful new colonies have gynodioecy which is associated with higher outcrossing rats and higher heterozygosity (Molina-Freaner and Jain, 1992a,b). Gynodioecy is under nucleocytoplasmic control and accounts for increased levels of outcrossing in this predominantly autogamous species. It would be interesting to follow the models and field studies reported for thyme (*Thymus vulgaris* L.) by Belhassen et al. (1987) and Couvet et al. (1990). Both genetic and demographic data on a large number of small populations, along with more details on the migration patterns, would be clearly needed in many more rare plant conservation and reintroduction projects.

Biosystematic status of an endangered *Helianthus* species

A wild sunflower known as serpentine sunflower (*Helianthus exilis* Gray) has drawn controversial accounts in several monographs and conservation agency reports. We use this example to illustrate the important role of genetic and new biosystematical approaches in conservation biology (see chapters by Arctander and Fjeldså, and by Crozier and Kusmierski, this volume).

Helianthus exilis Gray, the serpentine sunflower in California, was reclassified by C. B. Heiser as a serpentine race of weedy *H. bolanderi* on the basis of an introgressive hybridization model. Various treatments of California flora and endemism underscore this change and the resulting neglect of further research on these taxa. Recent morphological, karyotypic, agroecological, and molecular biosystematic studies are reviewed to show the unique and distinctive features of *H. exilis* as a rare endangered genetic resource. Our data show that weedy *H. bolanderi* has two races (*riparian* and *valley*), both separable from the serpentine "native" taxon, *H. exilis*.

Principal components analysis was used on 212 individuals representing four putative groups sampled from a total of 14 populations. Characters scored were plant height, leaf number, leaf shape index, disk diameter, size and shape of phyllary (involucral bract), ray number leaf pubescence, seed color and size. Original untransformed values of the variance-covariance matrix were used in the stepwise discriminant and principal components analyses (PCA) in which the first two axes accounted for 77% variation. Most riparian *bolanderi* populations were found near the serpentine *exilis* populations (e.g., near Knoxville and Middletown). Thus, as expected, the cluster of riparian *bolanderi* populations occurs between the *exilis* and valley *bolanderi* clusters. Overall, the four groups defined in our earlier work are reconfirmed here. A discriminant function based on *a priori* grouping of various populations in the three taxa (*H. annus, H. bolanderi, H. exilis*) showed >80% correct classification of individuals. In an allozyme variation study of 21 populations, most loci showed polymorphism, with large within-taxon variation and rather low variation between taxa. Unique alleles (11 out of 64) differentiated the riparian and valley races of *bolanderi*. Overall, there was little evidence that *exilis* and *bolanderi* are more related than *exilis* and *annuus*. We expected riparian *bolanderi*, collected from the localities often close to the serpentine *exilis* in many cases, to show greater genetic affinity, but these data did not fit this expectation.

Rieseberg's (1991) work, using allozyme, cpDNA and rDNA analyses, clearly ruled out the origin of weedy *bolanderi* from *exilis-annuus* introgression; in fact, it supported the distinct status of *H. exilis*, as we had independently reported.

A primary motivation for this work comes from the need to conserve *exilis* populations *in situ*. Several sources of rare and endangered species lists in California do not adequately emphasize this serpentine endemic and the serpentine habitats in general. Several populations have been regularly visited for monitoring population numbers, seed output, and habitat status. In general, *exilis* populations are small in habitat size as well as effective neighborhood size as estimated from the small gene flow distances, but densities of 30 to 110 plants/m^2 suggest that often several hundred plants occur on these sites. Seed output *per* plant varies between 7 and 23; progeny sizes fit a platykurtic curve, with means of different populations varying between 10.6 and 16.3 seeds *per* plant.

Plants are short and often have one head, only rarely exceeding four heads. Survivorship in certain years was as low as 35%, but small effective population size do not seem to threaten their viability. In contrast, total loss of a population must have occurred by habitat alteration or destruction at many sites. No data are available, however, on the modes or effective dispersal distances, or on the rates of recolonization, which must be low.

Conservation genetics and agriculture in partnership

Three of the four genera discussed in this chapter are of agricultural interest. Domestication of *Limnanthes* for industrial oil has been making steady progress with higher seed yield and greater agronomic efficiency. Our work has identified populations or their hybrids as sources of some important traits needed by plant improvement projects (Tab. 7). This also emphasizes the need to conserve not just a taxon or one metapopulation somewhere, but also several natural populations both *in situ* and *ex situ*. Studies on rose clover as a colonizing species have paralleled the current extensive uses in revegetation work with this species; more likely, its different genetic resources for marginal adaptability and evolutionary responses will soon be recognized as well. In *Helianthus*, at least two breeding projects have utilized *H. exilis* for its

Table 7. Examples of traits useful and their sources in *Limnanthes* collections

Trait	Many populations	Few populations	Inter-population hybrids	Inter-specific hybrids	NA[a]
1. Non-shattering[b]	—	X	—	X	
2. Early flowering[b]	—	X	X	X	
3. Loss of dormancy	—	X	?	?	
4. Upright, taller habit	—	—	—	—	X
5. Fewer synchronous flowers	—	—	—	X	X
6. Natural selfing:					
a. Cleistogamy	—	X	?	?	
b. Lacking protandry	—	—	X	X	
c. High autofertility	X	—	X	X	
7. High relative growth rate	—	X	X	X	
8. Determinate habit[c]	X	—	X	X	
9. Large seed size with high percent oil[b]					
10. Male sterility[c]	—	X	X	X	

[a]NA = not available in any accession.
[b]High h^2 values have been found in some populations and crosses. Selection response in preliminary efforts appears encouraging.
[c]Major genes were identified for only these traits. In addition, we have found major genes controlling flower color, anther color, gynodioecy, leaf and stem pubescence, various plant part pigmentation patterns, earliness, and possibly nutlet morphology.

unique germination requirements, oil quality, and short stature characteristics. Thus, genetic variation research can mutually reinforce the interests of the plant breeders and the conservation biologists.

Some final comments

Numerous genetic studies on rare plants have appeared in the recent literature. From their thorough work in *Eucalyptus*, Moran and Hopper (1987) and Sampson et al. (1988) developed specific recommendations for conservation action. Ouborg et al. (1991), van Treuren (1993), and Bjilsma et al. (this volume) have initiated elegant work on the genetics and ecology of *Salvia* and *Scabiosa* taxa. This work is uniquely designed for a comparative study of small and large populations for their response to inbreeding and demographic stochasticity. For rare plants Falk and Holsinger (1991) have many nice examples which illustrate critical needs for ecological genetic research.

In this chapter several long-term studies on plant populations were presented with the objective of learning about the genetics and ecology of small populations. Generally speaking, these results are not precise or based on elegant experimental designs, but a few of my collaborators with genuine field biology or agricultural interest participated in them (I have been rather fortunate on that score). In a discussion of research needs, Soulé (1987), referring to MVP, stated that "a mature viability analysis is one that integrates all of the factors in a biologically realistic way." To ask whether genetics is less or more important than demography is futile, as he recommends long-term monitoring of demographic (N, r) and genetic parameters (gene frequencies, inbreeding), with metapopulation aspects well in mind. Our studies have included many of these ideas. To quote Holsinger and Gottlieb (1991) from a synthesis chapter: "surprisingly little is known about the overall biology of any individual plant species growing in the wild." Most longterm studies on variation and population sizes would be labeled nowadays as descriptive, nonrigorous or horizontal (=unfocussed) research and one without a Popperian hypothesis-testing paradigm! Conservation in practice is likely to change this attitude even in academia. Finally, what advice did I give for the conservation of vernal pools and their plants? More than 20 sites have been selected and regularly monitored by several research groups; genetics has not yet made any direct impact except to recognize local varieties, or specific genetic resources (in the case of *Limnanthes*). Nevertheless, all research and educational activities have helped raise the conservation issue widely and forcefully.

Acknowledgements
I am grateful to J. Dole, R. Holland, R. Kesseli, C. McNeill, F. Molina-Freaner, and K. Ritland for technical and intellectual help in my research projects, and to C. Qualset and K. Rice for stimulating discussions. I sincerely thank Marilee Schmidt for her cordial help in manuscript preparation.

References

Barrett, S. C. H. and Kohn, J. R. (1991) Genetic and evolutionary consequences of small population size in plants: implications for conservation. *In*: Falk, D. A. and Holsinger, K. E. (eds), *Genetics and conservation of rare plants*. Oxford Univ. Press, New York, pp. 3–30.

Belhassen, E., Dockes, A. C., Gliddon, C. and Gouyon, P. H. (1987) Gene dispersal and neighborhood in a gynodioecious species, the case of *Thymus vulgaris* L. *Genet. Sel. Evol.* 19: 307–320.

Brown, C. R. and Jain, S. K. (1979) Reproductive system and pattern of genetic variation in two *Limnanthes* species. *Theor. Appl. Genet.* 54: 181–190.

Couvet, D., Atlan, A., Belhassen, E., Gliddon, C., Gouyon, P. H. and Kjellberg, F. (1990) Coevolution between two symbionts: The case of cytoplasmic male-sterility in higher plants. *Oxford Surv. Evol. Biol.* 7: 225–249.

Drury, W. H. (1980) Rare species of plants. *Rhodora* 82: 3–48.

Falk, D. A. and Holsinger, K. E. (1991) *Genetics and conservation of rare plants*. Oxford Univ. Press, New York.

Griggs, F. T. and Jain, S. K. (1983) Conservation of vernal pool plants in California. II. Population biology of a rare and unique grass genus *Orcuttia*. *Biol. Cons.* 27: 171–193.

Heiser, C. B. (1949) Study in the evolution of the sunflower species *Helianthus annuus* and *H. bolanderi*. *Univ. Calif. Publ. Bot.* 23: 157–196.

Holsinger, K. E. and Gottlieb, L. D. (1991) Conservation of rare and endangered plants: principles and prospects. *In*: Falk, D. A. and Holsinger, K. E. (eds), *Genetics and conservation of rare plants*. Oxford Univ. Press, New York.

Jain, S. K. and Martins, P. S. (1979) Ecological genetics of the colonizing ability of rose clover (*Trifolium hirtum* All.). *Amer. J. Bot.* 66: 361–366.

Jain, S. K. and Moyle, P. (1984) *Vernal pools and intermittent streams*. Institute of Ecology, Davis, CA.

Kesseli, R. V. (1992) Population biology and conservation of rare plants. *In*: Jain, S. K. and Botsford, L. W. (eds), *Applied population biology*. Kluwer, Dordrecht, pp. 69–90.

Kesseli, R. V. and Jain, S. K. (1984) New variation and biosystematic patterns detected by allozyme and morphological comparisons in *Limnanthes* sect. Reflexae (*Limnanthaceae*). *Plant Syst. Evol.* 147: 133–136.

Krebs, S. and Jain, S. K. (1985) Variation in morphological and physiological traits associated with yield in *Limnanthes* spp. *New Phytol.* 10: 717–729.

Lande, R. (1987) Extinction thresholds in demographic models of territorial populations. *Am. Nat.* 130: 624–635.

Martins, P. S. and Jain, S. K. (1979) Role of genetic variation in the colonizing ability of rose clover (*Trifolium hirtum* All.). *Am. Nat.* 114: 591–595.

McNeill, C. I. and Jain, S. K. (1983) Genetic differentiation and phylogenetic analysis in the plant genus *Limnanthes* (section Inflexae). *Theor. Appl. Genet.* 66: 257–269.

Molina-Freaner, F. and Jain, S. K. (1992a) Isozyme variation in California and Turkish populations of the colonizing species *Trifolium hirtum*. *J. Hered.* 83: 423–430.

Molina-Freaner, F. and Jain, S. K. (1992b) Breeding systems of hermaphroditic and gynodioecious populations of the colonizing species *Trifolium hirtum* All. in California. *Theor. Appl. Genet.* 84: 155–160.

Moran, G. F. and Hopper, S. D. (1987) Conservation of the genetic resources of rare and widespread Eucalyptus in remnant vegetation. *In*: Saunders, D. A., Arnold, G. W., Burbidge, A. A. and Hopkins, A. J. M. (eds), *Nature conservation: The role of remnants of native vegetation*. Surrey Beatty, CSIRO, pp. 151–162.

Ouborg, N. J., van Treuren, R. and van Danume, J. M. M. (1991) The significance of genetic erosion in the process of extinction. II. Morphological variation and fitness components in populations of varying size of *Salvia pratensis* L. and *Scabiosa columbaria* L. *Oecologia* 86: 359–367.

Rabinowitz, D. (1981) Seven forms of rarity. *In*: Synge, H. (ed.), *The biological aspects of rare plant conservation*. Wiley, New York, pp. 205–218.

Rieseberg, L. H. (1991) Hybridization in rare plants: insights from case studies in *Cercocarpus* and *Helianthus*. *In*: Falk, D. A. and Holsinger, K. E. (eds), *Genetics and conservation of rare plants*. Oxford Univ. Press, New York, pp. 171–181.

Ritland, K. R. and Jain, S. K. (1984a) A comparative study of floral and electrophoretic variation with life history variation in *Limnanthes alba* (*Limnanthaceae*). *Oecologia* 63: 243–251.

Ritland, K. R. and Jain, S. K. (1984b) The comparative life histories of two annual *Limnanthes* species in a temporally varying environment. *Am. Nat.* 124: 656–679.

Sampson, J. F., Hopper, S. D. and James, S. H. (1988) Genetic diversity and the conservation of *Eucalyptus crucis* Maiden. *Aust. J. Bot.* 36: 447–460.

Soulé, M. E. (1987) *Viable populations for conservation.* Cambridge Univ. Press, Cambridge.

Stone, R. D. (1990) California's endemic vernal pool plants: some factors influencing their rarity and endangerment. *In*: Ikeda, D. H. and Schlising, R. (eds), *Vernal pool plants – their habitat and biology.* Calif. State Univ., Chico, pp. 88–107.

van Treuren, R. (1993) *The significance of genetic erosion for the extinction of locally endangered plant populations.* Ph.D. Thesis, Rijksuniversiteit Groningen.

Conservation Genetics
ed. by V. Loeschcke, J. Tomiuk & S. K. Jain
© 1994 Birkhäuser Verlag Basel/Switzerland

Response to environmental change: Genetic variation and fitness in *Drosophila buzzatii* following temperature stress

R. A. Krebs and V. Loeschcke

Department of Ecology and Genetics, University of Aarhus, Ny Munkegade, Bldg. 540, DK-8000 Aarhus C, Denmark

Introduction

How do organisms adapt to environmental change? This question is not new, and was the focal point of studies of natural animal populations by Andrewartha and Birch (1954), Levins (1968), Parsons (1983), and Hoffmann and Parsons (1991). For the conservation of rare species, how populations respond to environmental change is an important consideration when estimating the level of care and monitoring that should be expended to guarantee their survival. Few environments are truly constant, either in time or space, and therefore environmental change, at some scale, will affect the majority of animal and plant species (Grime, 1989). Many recent changes to the environment have been induced by humans over a much shorter time frame than normal without human disturbance (Dobson et al., 1989; Holt, 1990; Kareiva et al., 1983). Such changes may cause physiological stress that is expressed by a reduction in growth or performance. Where genetic variation is present for resistance to stress factors, populations may adapt according to the stress experienced. Without this variation, no genetic response is possible, although stress effects will continue to act on the physiology, influence population size, and may cause extinction. Therefore, knowledge of the effects of stress and of genetic variation for stress resistance is necessary for decision-making with respect to choices of reserves designed to protect particular species, choosing the individuals to introduce to a reserve, and estimating long-term performance of populations in environments that may change unpredictably.

Organisms may encounter many types of stress, although the study of thermal stresses has been one of the most prominent (see Hoffmann and Parsons, 1991), as possible effects of local and global temperature changes are frequently debated (e.g., Dobson et al., 1989). Experiments maintaining populations in constant thermal environments have shown that organisms rapidly respond to stable conditions (Cavicchi et al., 1989; Huey et al., 1991). With environmental fluctuation, whether stochastic or predictable, adaptation to all possible temperatures is not possible, but organisms need to be able to respond to a range of conditions. Environmental uncertainty may select for less specialized phenotypes, or may lead to a physiological switch, with the development changing with the environment (Levins, 1968, pp. 21–22). Such a switch occurs in almost all organisms for resistance to extremes of heat. A specific group of stress proteins, which confer protection against cell damage caused by short-term exposure to temperatures 10–15 °C above that for normal growth, are transcribed at temperatures well below those causing damage. Importantly, exposure to low levels of heat stress induces resistances to otherwise damaging or lethal temperature stress levels, and this induction is transient, being lost over time (Lindquist, 1986). Studies of genetic variation in heat resistance, both within and among populations, are important to provide a foundation about the biology of thermal tolerance, to show how the environment of population origin relates to variation in thermal resistance of those populations, and eventually to provide information on how habitats and populations should be matched when stocking a reserve.

The biology of *Drosophila buzzatii*

In nature, many *Drosophila* species that utilize cactus (see Barker and Starmer, 1982) typically may encounter very high temperatures, at least for part of the year. For one of these species, *D. buzzatii*, larvae are confined to necrotic cladodes (rots) of *Opuntia* cactus and pupate within them. Although adult flies may depart from and return to rots, these rots are the primary locations for feeding and mating, and the only suitable sites for oviposition.

Drosophila buzzatii inhabits cactus over a wide geographic distribution, and those localities we compare encompass much of its distribution, climatically and geographically. All of the Australian populations originated from what was probably a single introduction 60–65 years ago from Argentina: Oxford Downs and Dixalea in Queensland, Metz Gorge from New South Wales, and Bulla from Victoria. In Australia, the more northern localities tend to be warmer, although the Metz Gorge site is at an elevation of almost 1000 m, and is therefore the coolest of the four. The only Argentinian population available was from El Chañar, in Túcuman province, and although the species originated from Argentina and Brazil, the amount of variation within this region is unknown. Additionally, we obtained one population from the Canary Islands (southwestern Tenerife), which was the probable source for secondary colonization throughout the Mediterranean region, 300–400 years ago, after cactus was brought from the New World.

Rot Temperatures

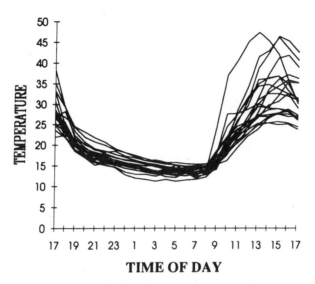

Fig. 1. Temperature recorded on the hour for necrotic cactus cladodes on Tenerife, in the Canary Islands, November, 1992.

We provide a case study of research on resistance of *Drosophila buzzatii* to short-term exposure to extremes of temperature. First, we measured rot temperatures in cactus cladodes on Tenerife over five climatically similar days in November, 1992, to determine what temperatures may be naturally encountered. Means for 10 min, over a 24-h period, were determined by placing electrodes, which were connected to a data logger, into the rots (Fig. 1). Recorded temperatures ranged from 12 °C to 47 °C. Consequently, within the same day, this species may encounter both cold and warm stress temperatures, and temperature change may occur rapidly within cactus rots. Growth and activity of *D. buzzatii* is greatly reduced at temperatures below 18 °C (Watt, 1987), while constant temperatures above 30 °C are stressful to larvae (Krebs and Barker, 1993), as were higher fluctuating temperatures in preliminary experiments. By comparing lines, seven each from six climatically different localities, we have examined how genetic variation in heat stress resistance is maintained. To link survival to fitness, offspring production of males and females that were exposed to different temperatures was compared to determine whether differences are present between survivors of stressed and unstressed flies, and between flies encountering different stress levels.

Genetic variation for stress resistance

For all lines, resistance to thermal stress was measured on 6–7-day-old virgin adults that were held at 25 °C in fresh food vials with added live yeast before being exposed to high temperature stress. All flies were pretreated by exposure to 38.0 °C (incubator temperature, ca. 36.5 °C within vials, which activates the heat shock genes), for 75 min to acclimate the flies to heat, 24 h before being exposed to the higher stress temperature (ca. 40.4 °C, vial temperature, for 100 min, under saturated humidity to avoid confounding effects of desiccation).

Significant variation for resistance to heat stress was present among lines, and among localities of *D. buzzatii* (Tab. 1), with higher significance levels for the locality effect than for effects of line variation. However, Loeschcke et al. (1994) also found much variation among isofemale lines from three other Australian localities. The order of survival of males from these localities, from highest to lowest, was Bulla, Canary Islands, Dixalea, Oxford Downs, Metz, and Argentina (Tab. 2). The order for females was Canary Islands, Metz, Dixalea, Bulla, Oxford Downs, and Argentina. The survival of males was significantly higher than that of females (Tab. 1), with a significant locality x sex interaction ($P < 0.001$). Theory predicts that populations derived from more stressful localities, i.e., those with higher temperatures, will be more resistent to stress from high temperatures (Lynch and Gabriel, 1987), although stochasticity of environmental temperatures may influence selection for resistance, with rare, extreme events dominating selection intensities (Kingsolver and Watt, 1983).

The best test of environment-of-origin effects may be made on the Australian populations. For these, the ranked order of resistance, averaging across males and females, was Dixalea = Bulla ≥ Metz Gorge = Oxford Downs, a result not in accord with the expectation from comparisons of climate (Tab. 2), as measured from mean temper-

Table 1. Analysis of variance for the comparison of survival to heat stress among localities (Loc), sex, and isofemale lines (considered random) nested within localities (Line). Denominator mean squares for this mixed-model ANOVA are indicated in parentheses. Data were percent survival, arcsin-square-root transformed, with type III mean squares presented

Source	Denominator mean square	df	Mean square	F value
1. Loc	(2)	5	0.613	5.02**
2. Line	(6)	36	0.122	1.47*
3. Sex	(5)	1	3.833	45.6***
4. Loc × Sex	(5)	5	0.084	0.79
5. Line × Sex	(6)	36	0.106	1.27
6. Error		407	0.083	

$*p < 0.05$, $**p < 0.01$, $***p < 0.001$.

Table 2. Mean percent survival ($\pm$ se) to heat stress of males and females from six localities of *D. buzzatii*, measured for seven isofemale lines per locality

Locality	Lat.	Climate	Males	Females
Canary Islands	28°10N	warm open	81.6 $\pm$ 3.3	70.4 $\pm$ 3.0
Dixalea	23°56S	warm shaded	79.2 $\pm$ 4.1	64.5 $\pm$ 4.1
Bulla	38°00S	cool open	81.9 $\pm$ 3.7	62.8 $\pm$ 5.4
Oxford Downs	21°50S	warm shaded	78.8 $\pm$ 4.0	58.5 $\pm$ 4.2
Metz	30°21S	cool shaded	74.2 $\pm$ 4.2	65.3 $\pm$ 4.0
Argentina	26°48S	warm ?	58.9 $\pm$ 4.4	50.0 $\pm$ 4.4

atures at the nearest meteorological site. Because a test for correlation between environments and stress resistance across a number of localities is, in actuality, a test of the effects of many environmental differences, inferring natural selection to a particular climatic factor, such as temperature may be difficult using spatial comparisons. Air temperatures at the Bulla locality are cooler than at either the Dixalea or Oxford Downs localities, but the cactus are more exposed to direct sunlight. Flies from the Metz locality, which is cool and shaded, had the lowest resistance for males, and in another study using mass populations from each of these localities, resistance to high temperature in both sexes was below that of all other Australian populations (R. Krebs, unpublished). However, these results on lines showed survival of Metz females to be higher for all populations except for that from the Canary Islands. The influence of exposure to sunlight may have contributed to the high resistance of Canary Island flies, where the habitat is both warm and open. No habitat description was available for the collection site near El Chañar, where the Argentina flies were collected. However, mean temperature is, on average, 2 °C cooler at El Chañar than at Dixalea, which is only a small difference considering the great difference between these populations in thermal resistance. For the three Australian localities examined by Loeschcke et al. (1994), higher resistance also was observed for *D. buzzatii* from the locality with the highest summer maximum temperature (Westwood, Queensland, near Dixalea) than for two others, an earlier collection from Metz Gorge, and a coastal population, which was warm, but with less extremes (Hemmant, near Brisbane). Therefore, we expect that selection has influenced divergence in thermal resistance among populations of *D. buzzatii*, although effects are unclear, and the contribution of selection versus other alternatives, such as genetic drift among those localities that vary in their degree of isolation, cannot be distinguished clearly.

Resistance and fitness

Mortality is the usual trait measured in studies of stress resistance (Hoffmann and Parsons, 1991), or where it is impractical to measure mortality, such as for lizards, a trait is measured that is presumed relevant to survival or offspring production (Huey and Bennett, 1987). Offspring production may be affected by lower stresses than those affecting survival, or a stress may reduce fitness of those organisms that survive. We compared offspring production for flies from a mass population made by combining offspring from many isofemale lines from the Canary Islands locality, as these flies were the most stress resistant. Water baths were used to better provide control over the precise stress temperatures: 41.0°, 41.1°, 41.3°, 41.6°, 41.7°, 42.0°, and 42.2 °C, with between day variation leading to seven different temperatures, although only four water baths were used at any one time. Vial temperatures will increase much more rapidly in water baths than in the incubators used in the first analysis, with a fast rate of change reducing the ability of the flies to acclimate to change. One digital thermometer was used to measure all water tanks. Therefore, temperature differences were exact. After the last set of flies was exposed to heat stress, at 41.3 °C or 41.7 °C, 50 males and 50 females from each treatment, along with 50 of each sex that had been pretreated 24 h before, were placed individually in vials with a young, mature virgin mate which had not been stressed. Fifty unstressed pairs were used as controls. All pairs were transferred to new food vials after 3, 6, 9, 12, and 15 days.

Survival of *D. buzzatii*, judged as being able to walk, varied from almost 100% to almost zero, across the narrow temperature range of 41.0° to 42.2 °C, for flies exposed for only 30 min (Fig. 2). For the three intermediate temperatures used, 41.3°, 41.6°, and 41.7 °C, male survival was greater than that of females. At higher or lower temperatures, most individuals either lived (≤ 41.1 °C) or died (≥ 42.0 °C). Apparently, there is almost a threshold, below which mortality may be very low, but above which all individuals in a population will die. For survivors, the total number of progeny produced was affected greatly by exposure to stress (Tab. 3), with significantly fewer total progeny produced by flies that survived exposure to 41.3° or 41.7 °C than were produced by controls or those pretreated only. Pretreatment, to acclimate flies, did not reduce male or female progeny production relative to controls, although activating proteins in large concentrations to defend against stress is expected to impose an energetic cost on the organism (Huey and Bennett, 1990). If such a cost is small, larger sample sizes would be

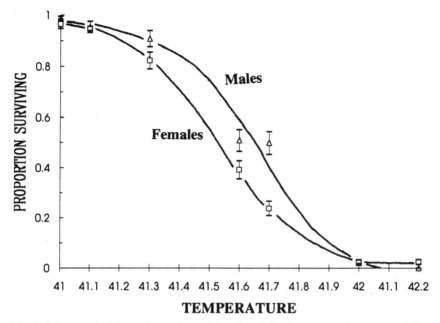

Fig. 2. Mean survival ($\pm$ se) for males and females of the Canary Island mass population exposed to different temperatures for 30 min in water baths.

needed to show a difference, especially within a high nutrition environment, as provided here. Comparing survivors between the two high stresses, females paired with males that survived the lesser stress (41.3 °C) produced more progeny than those paired with males exposed to 41.7 °C (significant by separate t-tests for days 1–3, 4–6, and 7–9, and for total progeny produced). Females stressed at 41.7 °C produced more progeny than those at 41.3 °C, but the difference was not significant.

Male fertility and female fecundity that are reduced due to stress damage can recover with time. Male fertility increased significantly between days 7–9 and days 10–12, after which the number of progeny produced was no longer significantly different from controls. For females, significant improvement in fecundity occurred between days 1–3 and days 4–6, but no further improvement was observed after this time. Therefore, stressed females always produced fewer progeny than controls. These effects on progeny produced were due, in part, to both offspring produced per female and to the number of pairs producing offspring on each set of days. Fewer pairs produced offspring on any given 3-day period than across all time periods, and significant changes in this number generally were in parallel with those for total progeny production. However, mean offspring number *per* fly increased over time at a very large rate (Tab. 3), suggesting physiological repair. Death

Table 3. Progeny produced by D. buzzatii males and females either held only at 25 °C, preconditioned only (37.0 °C), or preconditioned and stressed (41.3° or 41.7 °C), with all temperature-treated flies provided with an untreated mate

Treatment	N	Progeny per day						Total progeny
		days 1–3	days 4–6	days 7–9	days 10–12	days 13–15	days 16–22	
Unstressed controls	47	8.8 ± 1.2 A	9.6 ± 1.9 A	10.6 ± 1.5 A	10.3 ± 1.5 A	7.4 ± 1.3 A	5.5 ± 0.8 A	173 ± 20 A
Males								
Preconditioned only (37 °C)	44	6.6 ± 1.2 A	9.5 ± 1.4 A	12.1 ± 1.5 A	13.6 ± 1.7 A	6.3 ± 1.5 A	7.5 ± 1.1 A	190 ± 22 A
Heat stressed (41.3 °C)	36	1.3 ± 0.4 B	1.5 ± 0.7 B	2.8 ± 1.2 B	7.5 ± 1.5 B	5.8 ± 1.4 A	6.2 ± 1.2 A	94 ± 17 B
Heat stressed (41.7 °C)	26	0.0 ± 0.0 B	0.3 ± 0.2 B	0.3 ± 0.1 B	5.5 ± 1.4 B	4.9 ± 1.5 A	4.2 ± 1.1 A	58 ± 12 B
Females								
Preconditioned only (37 °C)	47	8.4 ± 0.9 A	13.7 ± 1.5 A	14.0 ± 1.5 A	14.6 ± 1.7 A	9.8 ± 1.6 A	6.8 ± 1.1 A	222 ± 23 A
Heat stressed (41.3 °C)	26	0.6 ± 0.2 B	2.6 ± 0.5 B	1.5 ± 0.6 B	2.2 ± 0.9 B	0.7 ± 0.4 B	0.3 ± 0.1 B	24 ± 6 B
Heat stressed (41.7 °C)	21	0.5 ± 0.3 B	1.9 ± 0.5 B	2.5 ± 0.9 B	2.9 ± 1.1 B	1.2 ± 0.4 B	1.4 ± 0.6 B	35 ± 9 B

Difference in letters denote statistical significance among groups based on Scheffe's multiple range test. Data were log transformed for statistical analysis, but untransformed means are presented above.

rates also were higher for flies in the stressed groups, with significantly more stressed flies dying during the experiment (after having survived the stress) than controls.

In nature, adult flies possibly can leave exposed rots and cladodes as temperature increases, and thereby avoid heat stress. Such behavioral avoidance requires that cooler habitat lies close to the cactus, and leaf litter might provide this cooler zone. However, higher resistance would still increase the length of time a fly could remain active (Levins, 1969). For small insects, the lag between air and body temperatures is very small (Huey et al., 1992). Movement during the day would require the organism to enter the sun-exposed areas around the host plant, risking rapid thermal increase. Low success rates for attempts to collect cactophilic *Drosophila* on banana bait in the late morning/early afternoon suggest that flies move little during this time period (J. S. F. Barker, personal communication).

Resistance to stress and population persistence

Given the assumption that, for most organisms, genetic variation is present for proteins involved in resistance to stress, and that fitness of organisms is affected by even short-term exposure to this stress, predictions may be made for the effect of stress on long-term persistence of populations. The response of populations will relate both to the evolution of regulatory proteins, and to the structure of the stress proteins which, once translated, are important for protecting organisms against damage from environmental change. Generally, the environment is considered thermally heterogeneous in space. If it were uniform, all individuals would be exposed equally to stress, and in a hard selection model, only those surviving the stress would have any opportunity to contribute offspring to the following generation. Selection for tolerance to the specific conditions would rapidly fix adaptations. Such a trend is found for thermal tolerance of laboratory populations of *Drosophila*, where resistance to heat stress and morphology varies among populations maintained at different temperatures (Cavicchi et al., 1989; Huey et al., 1991).

Regulatory functions affect the timing and quantity of the proteins produced. The quantity of a stress protein expressed is expected to relate directly to energy expenditure, as a cost for protecting the organism. The more protein produced, the greater the cost. Therefore, costs of inefficiency may be incurred by individuals that have either too sensitive a regulatory system, i.e., which may express the heat shock response at times when it is not required, or by individuals that overproduce these proteins beyond the quantity necessary to survive undamaged. Of course, organisms that either fail to produce these proteins when they are needed, or produce insufficient quantities will likely suffer damage, thus reducing reproductive output or causing death.

The variation in the structure of the proteins is unlikely to have similar constraints. Any change of amino acids that improves the ability of these proteins to protect the organism will be beneficial, and unlikely to affect the cost of their production. Therefore, such changes in protein composition would be selectively advantageous in all environments where the organisms may undergo exposure to high stress levels, and be approximately neutral in the absence of stress, or in uniform environments (for example, at temperatures normal for physiological growth). The only possible trade-off which may relate to the structure of these proteins is if there are differences associated with their function at normal physiological temperatures and their function in protection against stress. As heat shock proteins primarily function for transport of other proteins or for maintaining protein stability and preventing degradation (Gething and Sambrook, 1992), most changes that improve the performance of these proteins under high temperatures may be expected to improve their performance under normal conditions. Therefore, the structure of the proteins may be under continuous directional selection in ability to stabilize other proteins, the factor possibly responsible for their great structural similarity across taxonomically very different groups of organisms (Lindquist, 1986).

For small populations, the genetic variation for the stress proteins may be affected stochastically, as would variation for any trait, by effects of genetic drift shifting allele frequencies independently of, and possibly in the opposite direction to changes from selection. Chance will effect which individuals are exposed, when exposure occurs, and what temperature extremes individuals encounter. We suggest that only certain exposure levels are important from the standpoint of a selection response. Above some upper temperature extreme, determined relative to the length of time that the organism is exposed, the genotype of the organism is unimportant, because all individuals will succumb to that stress (Fig. 2). Stress effects of this magnitude affect population size, but do not have direct effects on gene frequencies relating to resistance. Below some lower level, all organisms will survive unaffected, except perhaps by any costs associated with activating the heat shock response. These costs are small enough to be ignored for consideration of population size effects, although regulation of the stress response could change as a result of unnecessary activation such as for the temperature at which stress protein production begins.

Within the temperature range where some mortality or a reduction in offspring production occurs, selection may act on variation for stress resistance, and as the proportion of individuals exposed to stress temperatures may vary, variation among populations for stress resistance may evolve. Such variation, either among populations or species, will be affected by the breadth of the temperature range where fitness differences occur. The more narrow this range is, the smaller will be the proportion of the organisms under selection. Similarly, if individuals or populations are rarely exposed to stress, adaptation to that stress will evolve slowly. An important difference is present, however, in that a stress that occurs rarely has a small effect on population size, while a common stress that also may kill many exposed individuals can cause large and frequent fluctuations in population size and directly increase the possibility of extinction. Additionally, inbreeding, which can increase when population sizes are small, reduces stress tolerance of individuals (J. Dahlgaard, unpublished), an effect that could influence the probability of extinction. If reproductive output declines following exposure to thermal stress, as seen for *D. buzzatii*, then effects of stress on population size could be greater than that estimated from measurements of mortality alone, and the potential for evolved resistance greatly reduced, as almost all of the next generation will be produced by those individuals that, by chance, were not exposed to stress. Because the response to selection is slowed, fluctuations in the population size due to stress may continue for a long time, and the population may not evolve tolerance before the possibility of extinction is realized.

Relating this model to the results presented here for *D. buzzatii* could explain the discrepancy between the expected and observed order of thermal resistance among populations. Directional selection is expected to increase stress resistance in most populations, with the climate affecting the rate. A correlation between mean temperature at collection localities and resistance would hold only if mean temperature was strongly correlated with the probability of individuals encountering stress levels that select for resistance. Thus, populations derived from warmer or more variable habitats should have greater resistance than those from cooler or less variable habitats. At higher temperatures, the severity of exposure increases, and stress levels above the upper threshold temperature may occur, thus reducing the response to selection in populations most in need of rapid adaptation.

Summary and Conclusions

Drosophila buzzatii typically may encounter high temperatures in nature, and this species is genetically variable for resistance to stress, both within and among populations. Fitness of survivors to stress, however, was reduced, and observed as a reduction in male fertility and female fecundity. With time following exposure to severe stress, reproductive capacity improved, but lifetime offspring production still was reduced significantly. This effect would greatly reduce a population's recovery

320

from small size, which could occur following exposure to some man-made or environmental extreme. Although the results presented here were obtained for effects of heat stress, such consequences likely apply to a wide range of natural and man-made environmental stresses, including heavy metal toxicity or other pollutants. Low levels of these pollutants may not cause an observable effect on populations, even if some individuals are killed or offspring production is decreased. If genetic variation for resistance is present, higher tolerance may evolve. However, if concentrations are permitted to rise too far, some stress threshold may be reached, as observed for thermal stress, causing mass die-off or sterility and, possibly, local extinction.

Understanding the effects of stress is important when preparing programs for the conservation of species. Organisms generally do not become extinct when resources are abundant and the climate benign, but unfortunately, no guarantee can be made that environmental conditions in any locality will remain stable over a long time. Consequently, a high possibility of exposure to an extreme stress in an area would greatly reduce its usefulness as a reserve. Likewise, when choosing organisms for reintroduction, stress resistance of the chosen individuals and high levels of genetic variation within a population would be valuable. The organisms placed there must be able to change. Analysis of stress resistance (at non-lethal levels) among either family groups or for different populations would be very useful when deciding which individuals to reintroduce to an area or place in a reserve. Additionally, analysis may suggest that stress resistance is very low in a population or species. If identified, particular care may be taken to monitor the occurrence of that stress in the environment, and to take action to protect the population from that stress. Preservation of species should be designed for the future, with the goal to preserve, not simply to postpone extinction.

Acknowledgements
We thank Doth Andersen, Stuart Barker, Jesper Dahlgaard, Subodh Jain, Jennifer Krebs, and Jürgen Tomiuk for their various assistance in data collection, discussion, and comments on the manuscript. This research was supported by the Danish Research Council for Natural Sciences (grant no. 11-9719-1 and 11-9639-1).

References

Andrewartha, H. L. and Birch, L. C., (1954) *The distribution and abundance of animals.* University of Chicago Press, Chicago.

Barker, J. S. F. and Starmer, W. T. (1982) *Ecological genetics and evolution. The cactus-yeast-Drosophila model system.* Academic Press Australia, Sydney.

Cavicchi, S., Guerra, D., Natali, V., Pezzoli, C. and Georgii, G. (1989) Temperature-related divergence in experimental populations of *Drosophila melanogaster*. II. Correlation between fitness and body dimensions. *J. Evol. Biol.* 2: 235–251.

Dobson, A., Jolly, A. and Rubenstein, D. (1989) The greenhouse effect and biological diversity. *Trends Ecol. Evol.* 4: 64–68.

Gething, M.-J. and Sambrook, J. S. (1992) Protein folding in the cell. *Nature* 355: 33–45.

Grime, J. P. (1989) The stress debate: symptom of impending synthesis. *Biol. J. Linn. Soc.* (London) 37: 3–17.

Hoffmann, A. A. and Parsons, P. A. (1991) *Evolutionary genetics and environmental stress*. Oxford Science Publications, Oxford.

Holt, R. D. (1990) The microevolutionary consequences of climate change. *Trends Ecol. Evol.* 5: 311–315.

Huey, R. B. and Bennett, A. F. (1987) Phylogenetic studies of coadaptation: Preferred temperatures versus optimal performance temperatures of lizards. *Evolution* 41: 1098–1115.

Huey, R. B. and Bennett, A. F. (1990) Physiological adjustments to fluctuating thermal environments: An ecological and evolutionary perspective. *In: Stress proteins in biology and medicine*, Morimoto, R. I., Tissiéres, A. and Georgopoulos, C. (eds), Cold Spring Harbor Laboratory Press, N.Y., pp. 37–59.

Huey, R. B., Crill, W. D., Kingsolver, J. G. and Weber, K. E. (1992) A method for rapid measurement of heat or cold resistance of small insects. *Funct. Ecol.* 6: 489–494.

Huey, R. B., Partridge, L. and Fowler, K. (1991) Thermal sensitivity of *Drosophila melanogaster* responds rapidly to laboratory natural selection. *Evolution* 45: 751–756.

Kareiva, P. M., Kingsolver, J. G. and Huey, R. B. (1993) *Biotic interactions and global change*. Sinauer, Sunderland, MA.

Kingsolver, J. G. and Watt, W. B. (1983) Thermoregulatory strategies in *Colias* butterflies: Thermal stress and the limits to adaptation in thermally varying environments. *Am. Nat.* 121: 32–55.

Krebs, R. A. and Barker, J. S. F. (1993) Coexistance of ecologically similar colonising species. II. Population differentiation in *Drosophila aldrichi* and *D. buzzatii* for competitive effects and responses at different temperatures, and allozyme variation in *D. aldrichi*. *J. Evol. Biol.* 6: 281–298.

Levins, R. (1968) *Evolution in changing environments*. Princeton Univ. Press, Princeton, N.J.

Levins, R. (1969) Thermal acclimation and heat resistance in *Drosophila* species. *Am. Nat.* 103: 483–499.

Lindquist, S. (1986) The heat-shock response. *Annu. Rev. Biochem.* 55: 1151–1191.

Loeschcke, V., Krebs, R. A. and Barker, J. S. F. (1994) Genetic variation for resistance and acclimation to high temperature stress in *Drosophila buzzatii*. *Biol. J. Linn. Soc.* (London), in press.

Lynch, M. and Gabriel, W. (1987) Environmental tolerance. *Am. Nat.* 129: 283–303.

Parsons, P. A. (1983) *The evolutionary biology of colonizing species*. Cambridge Univ. Press, Cambridge.

Watt, A. W. (1987) *Temperature tolerance in cactophilic Drosophila*. Master of Science Thesis, University of Sydney.

Conservation Genetics
ed. by V. Loeschcke, J. Tomiuk & S. K. Jain
© 1994 Birkhäuser Verlag Basel/Switzerland

Alternative life histories and genetic conservation

K. Hindar

Norwegian Institute for Nature Research (NINA), Tungasletta 2, N-7005 Trondheim, Norway

Summary. Ecologically and morphologically polymorphic species pose several difficulties for conservation genetics. First, the genetic relationships among the different phenotypes must be established. Second, the ecological factors maintaining the polymorphism must be understood. Third, intervention to protect threatened populations must integrate population genetic and ecological considerations. Here, I have discussed these issues based on information about the biology and management of salmonid fishes in particular. A number of salmonid species contain coexisting phenotypes which differ in life history, habitat use, and morphology. Population genetic studies of these species typically show that there is less genetic differentiation between coexisting life-history types than between the same life-history type sampled from geographically separate localities. The life-history polymorphisms appear to be one way by which salmonid populations adapt to heterogeneous environments. How can this genetic and ecological knowledge be integrated in the conservation of polymorphic populations? First, anthropogenic activities that lead to the loss of genetic variation within and between populations should be discouraged. Second, anthropogenic activities should be evaluated with respect to their effect on the life history polymorphism itself, even though this aspect is less well understood. Knowledge about the evolutionary response of natural populations to novel selection regimes may lead to a more detailed view about how populations should be managed from the perspective of conservation genetics.

Introduction

This chapter deals with the genetic conservation of populations and species where strikingly different phenotypes coexist in time and space. Examples of such intraspecific polymorphisms are the life history and body morphology of insects, fishes, birds, and mammals, and the habitat use and trophic morphology of fishes and birds. These polymorphisms appear to reflect adaptations to local environmental conditions; some of them are among the best documented cases of natural selection in action. Conservation of species with polymorphic life histories requires considerable knowledge of their ecology and genetics. First, the genetic basis of the phenotypic variation must be established. Second, the ecological factors maintaining the polymorphism must be understood. Third, it follows naturally that intervention to protect populations must integrate population genetic and ecological considerations. In this chapter, I discuss each of these issues based on information about salmonid fishes, which amply illustrate both the phenomenon of alternative life histories and the consequences of various management strategies on polymorphic populations.

Alternative life histories

Throughout this study, the term "alternative life histories" is used to refer to the occurrence of a discontinuous phenotype in some life-history, ecological, behavioral or morphological trait within a population. I will cover several cases where the discontinuity is not absolute, that is, where the frequency distribution of a phenotypic trait is bimodal and yet continuous. Additional examples will cover cases where the different morphs appear to be reproductively isolated and not part of a single population. Accordingly, not much attention is given here to the situations where polymorphism is the result of a sexual dimorphism, unless the sexual dimorphism manifests itself as one sex being polymorphic for a trait which is monomorphic in the other, as is often the case in male and female reproductive tactics (e.g., Gross, 1991).

A taxonomic overview

Polymorphic life histories have been found in animal species from a variety of taxa (Tab. 1). It was first brought to wide attention by Sage and Selander (1975), who showed that markedly different trophic phenotypes of a Neotropical cichlid fish were morphs of the same species, and not a group of separate species. Later studies have documented cases of polymorphisms in several other vertebrate classes, as well as in invertebrates (Dingle, 1985; Roff, 1992). Sometimes the difference in phenotype between the conspecific morphs is larger than the difference between related taxa at the species or genus level, as for example regarding the variation in bill size of African finches (Smith, 1990). The nature of the polymorphism can be trophic morphology, reproductive morphology and behavior, habitat use, migratory habits, development time, or age and size at maturity. Often, these polymorphisms are interconnected, as will be seen below when comparing the life histories of coexisting phenotypes of salmonid fishes.

Salmonid life histories

A number of salmonid species contain coexisting life-history types. One common polymorphism is the coexistence of freshwater resident and anadromous (sea-run migratory) individuals in the same coastal river system, as found in several species within the genera *Salmo*, *Oncorhynchus* and *Salvelinus*. In Atlantic salmon, this polymorphism occurs with a few exceptions among males only; in other species such as brown trout, sockeye salmon, and Arctic charr, it occurs commonly among both sexes. Typical for these morphs is that accompanying the resident

325

Table 1. Polymorphism in morphology, ecology, behavior and life history of select, vertebrate species

Species	Nature of the polymorphism	References*
FISHES		
Atlantic salmon, *Salmo salar*	Resident and migratory males, occasionally both sexes	1
Brown trout, *S. trutta*	Resident and migratory individuals of both sexes	2
Arctic charr, *Salvelinus alpinus*	Dwarf- and normal-sized morphs of both sexes	3
Coho salmon, *Oncorhynchus kisutch*	Precocious and adult males	4
Sockeye salmon, *O. nerka*	Resident and migratory individuals of both sexes	5
Lake whitefish, *Coregonus clupeaformis*	Dwarf- and normal-sized morphs; trophic apparatus	6
Bluegill sunfish, *Lepomis macrochirus*	Parental and cuckolder males; benthic and limnetic morphs	7–8
Scale-eating cichlids, *Perissodus* spp.	Rightward and leftward mouth opening	9
A Neotropical cichlid, *Cichlasoma cyanoguttatum*	Trophic morphology	10
Stickleback, *Gasterosteus aculeatus*	Benthic and limnetic morphs	11
AMPHIBIANS		
Spadefoot toad, *Scaphiopus hammondi*	Dimorphism in larval mouthparts	see ref. 13
Tiger salamander, *Ambystoma tigrinum*	Cannibalistic and non-cannibalistic individuals	12
BIRDS		
Ruff, *Philomachus pugnax*	Lekking and satellite males	see ref. 8
Black-bellied seedcracker, *Pyrenestes ostrinus*	Small- and large-sized bills in both sexes	13
Blackcap, *Sylvia atricapilla*	Resident and migratory individuals	14
MAMMALS		
Red deer, *Cervus elaphus*	Antlered and antlerless males	see ref. 8

*The numbered references are: 1. Verspoor and Cole (1989); 2. Jonsson (1989); 3. Hindar and Jonsson (1993); 4. Gross (1985); 5. Foote et al. (1989); 6. Bodaly et al. (1992); 7. Wilson (1989); 8. Gross (1991); 9. Hori (1993); 10. Sage and Selander (1975); 11. Schluter and McPhail (1993); 12. Pfennig and Collins (1993); 13. Smith (1990); 14. Berhold (1991)

and migratory life histories is a striking difference in body size, the migratory morph being up to a thousand-fold larger by weight than the resident morph. Not surprisingly, these morphs differ in mating behavior and breeding coloration: small fish sneak opportunities for fertilization whereas bigh fish fight for them; small fish are cryptically colored, often resembling juveniles, whereas big fish are brightly colored and show other secondary sexual characteristics such as the hooked jaws of male salmon. The size and behavior of the migratory phenotype may also be bimodally distributed, as in male coho salmon where small, early-maturing males coexist with large, late-maturing ones (Gross,

1985). It must be emphasized that the resident-migrant polymorphism among salmonid populations is not always a true dichotomy; in iteroparous species such as Atlantic salmon, males commonly spawn as small, freshwater resident parr and thereafter migrate to the sea to return as large-sized spawners. In the semelparous sockeye salmon, males spawn either as freshwater residents or as migrants. In both species, some localities have residents of both sexes coexisting with migrants of both sexes. This is quite common in sockeye salmon, where the nonanadromous form is known as "kokanee" (Foote et al., 1989).

Another common polymorphism is dwarf- and normal-sized individuals in the same lake, as in Arctic charr and lake whitefish. Apart from the body-size difference, these morphs differ in habitat use, food habits, trophic morphology, and (for Arctic charr) also in spawning coloration and mating behavior. The dwarf-normal polymorphism may not be very different from the resident-migrant one: dwarf Arctic charr dwell in benthic habitats throughout life, whereas the normal morphs show a habitat shift from being benthic as juveniles to being pelagic as sub-adults (Hindar and Jonsson, 1993). This habitat shift is accompanied by changes in body shape and coloration which parallel the transformation (smoltification) from the stream-dwelling to the migratory phase of anadromous salmonids. It is known that all of this variation can occur within a single salmonid population, both for the resident-migrant and dwarf-normal polymorphism. However, when coexisting morphotypes are compared over a broader geographic and environmental range, a more complex picture emerges regarding their genetic relationships.

Patterns of gene flow between coexisting morphotypes

Molecular genetic studies suggest that among the salmonids, there are cases where coexisting morphotypes appear to be reproductively isolated, and other cases where they are parts of the same population. As a measure of genetic differentiation, Wright's fixation index (F_{ST}) is a useful statistic when the goal is to infer patterns of gene flow between the morphotypes (Slatkin, 1991). Wright (1951) showed that at equilibrium in an island model of population structure

$$F_{ST} = 1/(4N_e m + 1),$$

where N_e is the effective population size and m is the migration rate. F_{ST} (or the multiallelic equivalent, which is designated G_{ST}) can be readily estimated from enzyme electrophoretic or other molecular genetic data (e.g., Barton and Clark, 1990). Using such information, the genetic differentiation between coexisting morphotypes of salmonid fishes turns out to be quite variable among localities (Fig. 1). Coexisting dwarf and normal Arctic charr morphs show G_{ST} values ranging from 0.001 to 0.13, which translate into estimates of gene flow between the morphs

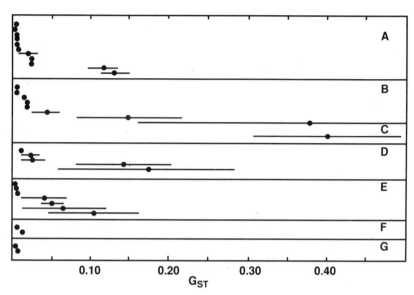

Fig. 1. Distribution of estimates of genetic differentiation (G_{ST}) between coexisting morpho-types of various fish species, based on multiple-locus enzyme electrophoretic studies. (A) Arctic charr (references 1–4), (B) Brown trout (5–9), (C) Atlantic salmon (10), (D) Sockeye salmon (11), (E) Lake whitefish (12–13), (F) *Cichlasoma* sp. (14), (G) *Ilyodon* sp. (15). The references are: (1) Hindar et al. (1986) and unpublished; (2) Sandlund et al. (1992); (3) Partington and Mills (1988); (4) Hartley et al. (1992); (5) Ryman et al. (1979); (6) Krueger and May (1987); (7) Skaala and Nævdal (1989); (8) Ferguson and Taggart (1991); (9) Hindar et al. (1991); (10) Verspoor and Cole (1989); (11) Foote et al. (1989); (12) Kirkpatrick and Selander (1979); (13) Bodaly et al. (1992); (14) Sage and Selander (1975); (15) Turner and Grosse (1980).

from 250 to 1.7 effective migrants ($N_e m$) per generation, respectively. In the former case, the G_{ST} value is indistinguishable from zero and most probably reflects that the coexisting morphotypes are different pheno-types of the same population. In the latter case, the coexisting morpho-types are as genetically divergent as are local populations of salmonids in geographically separate localities (cf. Ryman, 1983; Møller Hansen and Loeschcke, this volume).

The range of G_{ST} values shown by dwarf- and normal-sized Arctic charr covers most of the range observed for coexisting life-history types from a variety of teleost taxa, including resident and anadromous brown trout and sockeye salmon, dwarf and normal lake whitefish, and different trophic morphs of non-salmonid species (Fig. 1). Larger ge-netic differentiation has been documented between different morpho-types of lake-resident brown trout (Ryman et al., 1979), and between lake resident and anadromous Atlantic salmon (Verspoor and Cole, 1989). In each case, $N_e m$ estimates are less than 0.5 effective migrants per generation; in the brown trout study, positive evidence for current reproductive isolation (i.e., fixed genetic differences) between the morphotypes was found.

It is not the purpose of this study to explain why the different morphotypes appear to belong to one population in some localities and be reproductively isolated in others. Rather, it documents that this variation exists and discusses how it should be incorporated in conservation programs. Two other generalizations deserve special mention. First, population genetic analyses of salmonid species typically show that there is less genetic differentiation between coexisting life-history types than between the same life-history type sampled from geographically separate localities (Hindar et al., 1986, 1991; Foote et al., 1989; Bodaly et al., 1992). The simplest explanation for this observation is that most coexisting life-history types were derived from the same evolutionary lineage. Challenging this explanation is a study of mitochondrial DNA variation in lake whitefish, where Bernatchez and Dodson (1990) showed that a previously reported case of sympatric divergence of dwarf- and normal-sized morphs was the result of immigration to the same lakes of two allopatrically formed morphs. Although their study does not disprove that sympatric divergence may occur, it illustrates that historical factors must be included when analyzing the genetic relationships between the coexisting morphs.

Second, when combining genetic and ecological information from a number of studies, it appears that life-history polymorphism and genetic population differentiation may occur independently in salmonids (Hindar et al., 1986). That is, life-history polymorphism can occur without detectable genetic differentiation, as well as being associated with genetic differences between the coexisting phenotypes (cf. Fig. 1). This observation is consistent with theoretical studies suggesting that natural selection can act against intermediate phenotypes in at least three different ways, (1) by evolution of closer linkage so that the morphs are controlled by a single "supergene", (2) by evolution of dominance and other epistasis modifiers so that heterozygotes resemble homozygotes, and (3) by evolution of assortative mating which in the extreme may become a speciation event (Wilson, 1989). On this background, we should not be surprised to see that life-history polymorphisms occur over a wide range of levels of genetic divergence between the coexisting morphs.

Causes of life-history polymorphisms

Proximate causes

It is often suggested that phenotypic polymorphisms have a simple genetic basis. However, apart from some color polymorphisms in snails and butterflies, the evidence for this is sparse. On the one hand, Smith (1993) showed that the polymorphism in bill size of the black-bellied seedcracker is being produced by a single di-allelic locus with complete dominance for large bill. Likewise, Zimmerer and Kallman (1989) showed a major gene effect on body size polymorphism in the pygmy swordtail (*Xiphophorus nigrensis*). On the other hand, several studies of birds, amphibians and fishes suggest that intrapopulation polymorphisms in migratory behavior, trophic morphology and body size are threshold traits with polygenic inheritance, where environmentally induced variation contributes to the expression of the trait (Berthold, 1991; Hindar and Jonsson, 1993; Pfennig and Collins, 1993). In birds, genetic factors appear to be more important than environmental factors, whereas the opposite holds true for fishes and amphibians. This may be related to the fact that fishes and amphibians have indeterminate growth, which favors a fine-tuning of several aspects of their life histories (e.g., size at maturity) to environmentally determined variation in growth rate. For example, rearing studies of Arctic charr demonstrate that offspring of the various morphotypes have very similar growth and maturation patterns under similar growing conditions (Nordeng, 1983; Hindar and Jonsson, 1993). In this and other salmonid fishes, each maturation phenotype probably reflects the best option for individuals experiencing a particular growth rate. The conditional nature of salmonid morphotypes is moreover illustrated by the observation that individuals having spawned as one morphotype may enter a period of growth and later mature as another morphotype (Nordeng, 1983). Equally striking effects of the environment are shown by the trophic apparatus of cichlid fishes, which develops according to the prey items eaten (Meyer, 1987), and by the occurrence of the cannibalistic morph in tiger salamanders, which is highly responsive to density and kinship of conspecifics (Pfennig and Collins, 1993).

Genetic factors underlying trait variation may be more apparent when populations from different locations are compared under similar environmental conditions. For example, salmonid populations from different geographic locations vary with respect to what proportions of the population follow the various life-history trajectories (Nordeng, 1983; Jonsson, 1989). Thus, there may be genetic variation between populations for the nature and range of phenotypes that are realized under some set of environmental conditions (Stearns, 1992).

Selective forces

In order to predict the long-term response to changes in the environment, we need to know the ultimate causes for the polymorphisms. This requires an understanding of the selective forces acting on them. A number of abiotic and biotic environmental factors are believed to affect intraspecific polymorphisms: case studies indicate that physical factors, conspecifics, food quality, prey behavior, predators and competitors may all be important. Some of these studies are among a few where the causative agents of natural selection are well documented (e.g., Hori, 1993; Smith, 1993); in other cases our knowledge is of a much more tentative nature.

Productivity in fresh water and adjacent oceanic habitats is a major determinant of the resident-migrant polymorphism in salmonid fishes (Gross et al., 1988). Predation in the two habitats further molds the proportion of resident and migratory individuals in the population. Physical factors such as waterflow and river size affect the body size of the migrant morph: rivers with high waterflow having larger-sized Atlantic salmon than rivers with low waterflow (Jonsson et al., 1991).

Conspecifics constitute another selective agent, primarily by favoring alternative mating behaviors through male-male competition. If fighting between males for breeding opportunities is common in the population, as is the case in many salmonids, then an alternative breeding strategy based on sneaking behavior may be favored as long as it is rare. A stable situation may develop where frequency- or density-dependent selection acts to maintain an equilibrium point where the breeding success of the two mating types is equal (Gross, 1985, 1991). Some of the selective factors listed above, for example, productivity and predation in the two habitats, also influence the dwarf-normal polymorphism which however is not as well understood as the resident-anadromous one. Comparisons between several lakes with or without the dwarf-normal polymorphism suggest that the niche structure of the lake is important: two, and in extreme cases, up to four morphs of Arctic charr develop where the lake is deep and has a depauperate fish community (Sandlund et al., 1992).

Food quality is the most probable selective agent causing disruptive selection on bill size in the polymorphic African finch *Pyrenestes ostrinus* (Smith, 1993). The hardness of seeds that these birds are capable of handling is primarily determined by bill width. During the dry season when competition for food is most intense, birds with large or small bills survive better than birds having intermediate bill sizes. They do so by feeding more efficiently on either a hard-seeded sedge (the large-billed morph) or a soft-seeded sedge (the small-billed morph). Prey behavior (i.e., alertness towards predator type) appears to be the selective agent causing the scale-eating cichlid fish to reach an equilibrium frequency of right-handed and left-handed mouth types in the population (Hori, 1993). Again, the frequency-dependent equilibrium is reached because as long as one morph is rare, its success as a scale-eater is higher than that of the more common morph.

Resilience

The various modes of inheritance and selective forces acting on the polymorphisms have different implications for the resilience of the polymorphism to human disturbance. Lande and Barrowclough's (1987) findings regarding the maintenance of genetic variation are particularly relevant for evaluating this resilience, which can be expressed in terms of the effective population size needed to maintain significant quantities of genetic variation. First, Lande and Barrowclough (1987) noted that single-locus variation must be considered separately from polygenic variation; the former is more easily lost in small populations than is the latter. In polygenic traits, mutations will maintain a significant amount of genetic variation even when the long-term effective population size does not exceed 500. Single-locus polymorphisms, on the other hand, may easily be affected by a population bottleneck. Evidence for this has been provided for the weevil *Sitona humeralis* (Roff, 1992), which in its native European range features both winged and wingless forms, controlled by genetic variation at a single locus. Only winged forms have been found in a population introduced to New Zealand, supposedly because the founding population consisted only of homozygotes for the winged form.

Second, neutral variation must be considered separately from selected variation; in the latter case the form of selection must also be taken into consideration. This suggests that the necessary population size to maintain genetic variation will vary between traits. Polygenic traits under stabilizing or fluctuating directional selection will retain genetic variation at the same population sizes as expected from neutral models for quantitative variation (Lande and Barrowclough, 1987). Even though these forms of selection are considered to be the commonest in the wild, we should perhaps be more concerned with the effects of strong directional selection which may be effected through human disturbance (e.g., global warming). Under such a scenario, it is conceivable that genetic variation and adaptive polymorphisms will be lost even in populations with relatively large effective population sizes, because the optimum phenotype changes faster than can be tracked by the genetic variation in the population (Lynch and Lande, 1993).

Some of the polymorphisms reported above, for example, in mating behavior or in mouth handedness, seem at first sight quite robust to anthropogenic influence, because the major selective agents are conspecifics or coexisting species. But most of the factors which cause large-scale reductions in biological diversity, such as habitat degradation, overharvesting, and species introductions, may well affect life-history polymorphisms by manipulating reproduction and survival of the various types. The potential for anthropogenic influence on these polymorphisms is thus manifold and is discussed below.

Conservation of polymorphic populations

The goal for genetic conservation may be stated as "striving towards maintaining as much of the genetic variation within and between populations as possible" (Ryman, 1991). Implementation of this goal has proven difficult even for short-term maintenance of variation in managed populations, and even for easily monitored loci which are selectively neutral or nearly so. How then can the genetic and ecological knowledge about polymorphic populations be applied to implement the goal for genetic conservation? In the following, three aspects of the conservation of polymorphic populations are discussed.

Genetic relationships

One of the major contributions of genetic analyses to conservation biology is to delineate the inter- and intraspecific relationships among the taxa in question (Ryman and Utter, 1987; Avise, 1989). Well-intended conservation action can have disastrous consequences when the genetic relationships are ignored. Often, this has led to large-scale admixture of genetically distinct populations. Experience from reintroductions in general, and from releases of salmonid fishes in particular, suggests that this will not lead to anything but poorer performance of the recipient population (Templeton, 1986; Ryman et al., in press). In the case of polymorphic life histories, the evidence presented in Fig. 1 suggests a highly variable amount of genetic differentiation between the coexisting morphs. It has recently been suggested that the genetically effective number of migrants between populations ($N_e m$), for example, as estimated above from G_{ST} values, be used as a quantitative guideline for the number of individuals which can be transferred between populations (Ryman, 1991; Ryman et al., in press). It is clear from Fig. 1 that genetic studies are required before any specific recommendation can be made regarding forced migration between populations. Moreover, even when the genetic differentiation between the coexisting morphotypes is modest, for example, as estimated by a G_{ST} value of 0.04, this would result in the recommendation that migration rates should not exceed six individuals per generation, or about one individual per year in many salmonid populations. It must be pointed out that even this apparently conservative guideline should be used with caution, as it is based on the equilibrium between genetic drift and migration, whereas considerations of the maintenance of adaptive genetic variation must take the balance between selection and migration rate into account (Barton and Clark, 1990; Møller Hansen and Loeschcke, this volume). This does not, however, negate the usefulness of a guideline based on the dynamics of neutral variation. First, the amount and distribution of neutral variation are easily monitored by enzyme electrophoresis and other molecular genetic techniques. And second, in the very small populations where conservation action is of utmost importance, genetic drift will overpower even strong selection so as to render most of the genetic variation effectively neutral. Of course, it is always preferable to know whether the various phenotypes have evolved adaptive differences in spite of little genetic divergence as estimated by molecular methods.

Some authors (e.g., Bodaly et al., 1992; Schluter and McPhail, 1993) contend that most of the morphs presented in Fig. 1 are reproductively isolated and should be treated as separate species whether or not this is supported by published genetic analyses. Such a view might seem to be even more conservative than the one taken above, and could be justified with reference to the possible existence of genetic differences that we are unaware of. But this view could also counteract genetically based recommendations to maintain high population sizes. For example, Norwegian legislation has allowed netting for resident Arctic charr in lakes where anadromous charr are present (and belong to the same population; cf. Nordeng, 1983). This fishing depresses the genetical effective size of the entire population rather than favoring one "population" over another. Even more important are cases where realization that the morphotypes are parts of the same population would facilitate the collection of broodstock for captive propagation.

Altered selective regimes

Anthropogenic activities should also be evaluated with respect to their long-term selective effect on the life-history polymorphism, even though this aspect is less well understood. For example, selective harvesting of migratory and large-sized individuals alters the selective forces acting on these populations. Ricker (1981) showed a significant decreasing trend in the average body size of several species of Pacific salmon, most probably due to fishing pressure acting more severely on large-sized individuals. The strength with which selective fishing can alter salmonid life histories was questioned by Riddell (1986). But increasing evidence seems to indicate that strong fishing pressure on specific age- or size-classes has a high potential for altering life histories of migratory fish (Law and Grey, 1989; Stearns, 1992). Watercourse regulation is another anthropogenic activity which may have significant effects on polymorphic fish populations. First, this practice often involves the building of dams which may block upstream migration by anadromous fish and effectively "landlock" populations on the upstream side of the dam. Second, in the downstream part of the river, the relative reproductive success of large-sized individuals may decrease following reduced waterflow, and exert a selective pressure towards smaller body size. At any rate, even though our knowledge is incomplete regarding the selective factors molding salmonid life histories, the available evidence suggests that when the environment is dramatically altered, local populations will change genetically in spite of captive breeding and other conservation efforts.

A number of other anthropogenic activities are known to alter the selective advantage of various fish life histories. We have ourselves evidence that acidification affects one normal-sized morph of Arctic charr more severely than the dwarf morph, because the most acidic water of that lake flows near the surface where spawning grounds of the normals but not the dwarfs are located (T. Hesthagen et al., NINA, unpublished results). In future, it is also conceivable that climate change through alteration of waterflow and water temperture regimes will alter salmonid life histories, for example, the resident-migrant polymorphism.

Captive propagation

Artificial reproduction and reintroduction to the wild are increasingly being used to protect endangered populations and species. In salmonid fishes, this same strategy has been used in stock enhancement for over a century and can therefore be evaluated on its merits (Ryman et al., in press). Apart from the potential, above-mentioned genetic problems related to limited population size and/or interbreeding of genetically distinct populations, captive propagation also involves possible selection to the captive environment, as well as lack of continuing adaptation to natural environments. Several breeding strategies have been suggested to slow down the rate of adaptation to the captive environment (Ryman and Utter, 1987). One such strategy, that of reducing the variance in the number of offspring (e.g., equalizing family size), can be shown to slow down the rate of adaptation (Allendorf, 1993). For anadromous salmonids in particular, where the captive breeding program may involve releases to the sea at the smolt stage, the point at which family size should be equalized is the time of release. This would slow down the rate of adaptation to the captive phase of the life cycle while upholding selective forces acting on the free-living, oceanic phase (Allendorf, 1993). This finding has implications for conservation biology in general, because it is increasingly recognized that the captive phase must be as short as possible or as similar to the wild as possible. Recognition that separate strategies are needed for the artificial and natural environment may make implementation of conservation genetic strategies easier in the future.

Acknowledgments
This study was undertaken while I was on sabbatical to the University of California, Berkeley, and was supported by grants from the Research Council of Norway and the Norwegian

Institute for Nature Research. I am grateful to M. Slatkin and D. Wake for hospitality during the writing of the manuscript.

References

Allendorf, F. W. (1993) Delay of adaptation to captive breeding by equalizing family size. *Cons. Biol.* 7: 416–419.

Avise, J. C. (1989) A role for molecular genetics in the recognition and conservation of endangered species. *Trends Ecol. Evol.* 4: 279–281.

Barton, N. and Clark, A. (1990) Population structure and processes in evolution. *In:* Wöhrmann, K. and Jain, S. K. (eds), *Population biology. Ecological and evolutionary viewpoints.* Springer-Verlag, Berlin, pp. 115–173.

Bernatchez, L. and Dodson, J. (1990) Allopatric origin of sympatric populations of lake whitefish (*Coregonus clupeaformis*) as revealed by mitochondrial-DNA restriction analysis. *Evolution* 44: 1263–1271.

Berthold, P. (1991) Genetic control of migratory behaviour in birds. *Trends Ecol. Evol* 6: 254–257.

Bodaly, R. A., Clayton, J. W., Lindsey, C. G. and Vuorinen, J. (1992) Evolution of lake whitefish (*Coregonus clupeaformis*) in North America during the Pleistocene: genetic differentiation between sympatric populations. *Can. J. Fish. Aquat. Sci.* 49: 769–779.

Dingle, H. (1985) Migration and life histories. *Contrib. Mar. Sci.* (Suppl.) 27: 27–42.

Ferguson, A. and Taggart, J. B. (1991) Genetic differentiation among the sympatric brown trout (*Salmo trutta*) populations of Lough Melvin, Ireland. *Biol. J. Linn. Soc.* 43: 221–237.

Foote, C. J., Wood, C. C. and Withler, R. E. (1989) Biochemical genetic comparison of sockeye and kokanee, the anadromous and nonanadromous forms of *Oncorhynchus nerka*. *Can. J. Fish. Aquat. Sci.* 46: 149–158.

Gross, M. R. (1985) Disruptive selection for alternative life histories in salmon. *Nature* 313: 47–48.

Gross, M. R. (1991) Evolution of alternative reproductive strategies: frequency-dependent sexual selection in male bluegill sunfish. *Phil. Trans. R. Soc. Lond.* B 332: 59–66.

Gross, M. R., Coleman, R. C. and McDowall, R. (1988) Aquatic productivity and the evolution of diadromous fish migration. *Science* 239: 1291–1293.

Hartley, S. E., McGowan, C., Greer, R. B. and Walker, A. F. (1992) The genetics of sympatric Arctic charr [*Salvelinus alpinus* (L.)] populations from Loch Rannoch, Scotland. *J. Fish Biol.* 41: 1021–1031.

Hindar, K. and Jonsson, B. (1993) Ecological polymorphism in Arctic charr. *Biol. J. Linn. Soc.* 48: 63–74.

Hindar, K., Jonsson, B., Ryman, N. and Ståhl, G. (1991) Genetic relationships among landlocked, resident, and anadromous Brown Trout, *Salmo trutta* L. *Heredity* 66: 83–91.

Hindar, K., Ryman, N. and Ståhl, G. (1986) Genetic differentiation among local populations and morphotypes of Arctic charr, *Salvelinus alpinus*. *Biol. J. Linn. Soc.* 27: 269–285.

Hori, M. (1993) Frequency-dependent natural selection in the handedness of scale-eating cichlid fish. *Science* 260: 216–219.

Jonsson, B. (1989) Life history and habitat use of Norwegian brown trout (*Salmo trutta*). *Freshw. Biol.* 21: 71–86.

Jonsson, N., Hansen, L. P. and Jonsson, B. (1991) Variation in age, size and repeat spawning of adult Atlantic salmon in relation to river discharge. *J. Anim. Ecol.* 60: 937–947.

Kirkpatrick, M. and Selander, R. K. (1975) Genetics of speciation in lake whitefishes in the Allegash basin. *Evolution* 33: 478–485.

Krueger, C. C. and May, B. (1987) Stock identification of naturalized brown trout in Lake Superior tributaries: differentiation based on allozyme data. *Trans. Am. Fish. Soc.* 116: 785–794.

Lande, R. and Barrowclough, G. F. (1987) Effective population size, genetic variation, and their use in population management. *In:* Soulé, M. E. (ed.), *Viable populations for conservation.* Cambridge University Press, Cambridge, pp. 87–123.

Law, R. and Grey, D. R. (1989) Evolution of yields from populations with age-specific cropping. *Evol. Ecol.* 3: 343–359.

336

Lynch, M. and Lande, R. (1993) Evolution and extinction in response to environmental change. *In:* Kareiva, P. M., Kingsolver, J. K. and Huey, R. B. (eds), *Biotic interactions and global change.* Sinauer, Sunderland, MA, pp. 234–250.

Meyer, A. (1987) Phenotypic plasticity and heterochrony in *Cichlasoma managuense* (Pisces, Cichlidae) and their implications for speciation in cichlid fishes. *Evolution* 41: 1357–1369.

Nordeng, H. (1983) Solution to the "char problem" based on Arctic char (*Salvelinus alpinus*) in Norway. *Can. J. Fish. Aquat. Sci.* 40: 1372–1387.

Partington, J. D. and Mills, C. A. (1988) An electrophoretic and biometric study of Arctic charr, *Salvelinus alpinus* (L.), from ten British lakes. *J. Fish Biol.* 33: 791–814.

Pfennig, D. W. and Collins, J. P. (1993) Kinship affects morphogenesis in cannibalistic salamanders. *Nature* 362: 836–838.

Ricker, W. E. (1981) Changes in the average size and average age of Pacific salmon. *Can. J. Fish. Aquat. Sci.* 38: 1636–1656.

Riddell, B. E. (1986) Assessment of selective fishing on the age at maturity in Atlantic salmon (*Salmo salar*): a genetic perspective. *Can. Spec. Publ. Fish. Aquat. Sci.* 89: 102–109.

Roff, D. A. (1992) *The evolution of life histories. Theory and analysis.* Chapman & Hall, New York.

Ryman, N. (1983) Patterns of distribution of biochemical genetic variation in salmonids: differences between species. *Aquaculture* 33: 1–21.

Ryman, N. (1991) Conservation genetics considerations in fishery management. *J. Fish Biol.* 39 (Suppl. A): 211–224.

Ryman, N. and Utter, F. (1987) *Population genetics and fishery management.* University of Washington Press, Seattle.

Ryman, N., Allendorf, F. W. and Ståhl, G. (1979) Reproductive isolation with little genetic divergence in sympatric populations of brown trout (*Salmo trutta*). *Genetics* 92: 247–262.

Ryman, U., Utter, F. and Hindar, K. (in press) Introgression, supportive breeding, and genetic conservation. *In:* Ballou, J. D., Foose, T. and Gilpin, M. (eds), *Population management for survival and recovery.* Columbia University Press, New York.

Sage, R. D. and Selander, R. K. (1975) Trophic radiation through polymorphism in cichlid fishes. *Proc. Nat. Acad. Sci. USA* 72: 4669–4673.

Sandlund, O. T., Gunnarsson, K., Jónasson, P. M., Jonsson B., Lindem, T., Magnússon, K. P., Malmquist, H. J., Sigurjónsdóttir, H., Skúlason, S. and Snorrason, S. S. (1992) The arctic charr *Salvelinus alpinus* in Thingvallavatn. *Oikos* 64: 305–351.

Schluter, D. and McPhail, J. D. (1993) Character displacement and replicate adaptive radiation. *Trends Ecol. Evol.* 8: 197–200.

Skaala, Ø. and Nævdal, G. (1989) Genetic differentiation between freshwater resident and anadromous brown trout, *Salmo trutta*, within watercourses. *J. Fish Biol.* 34: 597–605.

Slatkin, M. (1991) Inbreeding coefficients and coalescence times. *Genet. Res.* 58: 167–175.

Smith, T. B. (1990) Patterns of morphological and geographic variation in trophic bill morphs of the African finch *Pyrenestes*. *Biol. J. Linn. Soc.* 41: 381–414.

Smith, T. B. (1993) Disruptive selection and the genetic basis of bill size polymorphism in the African finch *Pyrenestes*. *Nature* 363: 618–620.

Stearns, S. C. (1992) *The evolution of life histories.* Oxford University Press, Oxford.

Templeton, A. R. (1986) Coadaptation and outbreeding depression. *In:* Soulé, M. E. (ed.), *Conservation biology: the science of scarcity and diversity.* Sinauer, Sunderland, MA, pp. 105–116.

Turner, B. J. and Grosse, D. J. (1980) Trophic differentiation in *Ilyodon*, a genus of stream-dwelling goodeid fish: speciation versus ecological polymorphism. *Evolution* 34: 259–270.

Verspoor, E. and Cole, L. J. (1989) Genetically distinct populations of resident and anadromous Atlantic salmon, *Salmo salar*. *Can. J. Zool.* 67: 1453–1461.

Wilson, D. S. (1989) The diversification of single gene pools by density- and frequency-dependent selection. *In:* Otte, D. and Endler, J. A. (eds), *Speciation and its consequences.* Sinauer, Sunderland, MA, pp. 366–385.

Wright, S. (1951) The genetical structure of populations. *Ann. Eugen.* 15: 323–354.

Zimmerer, E. J. and Kallman, K. D. (1989) Genetic basis for alternative reproductive tactics in the pygmy swordtail, *Xiphophorus nigrensis*. *Evolution* 43: 1298–1307.

Conservation Genetics
ed. by V. Loeschcke, J. Tomiuk & S. K. Jain

The principles of population monitoring for conservation genetics

Yu. P. Altukhov

N.I. Vavilov Institute of General Genetics, Russian Academy of Sciences, Gubkin Str. 3, 117809, Moscow W B-333, Russia

Introduction

In the past century, human activity has caused the extinction of approximately 25 000 higher plant species and more than 1000 vertebrate species. Hundreds of unique breeds of domestic animals are on the verge of extinction (Wilson, 1988). The rate of extinction exceeds anything that is known from the paleontological records.

Conservation of biodiversity is one of the important contemporary global problems. However, in order to solve this problem it is necessary to understand that, besides directly unfavourable effects, such as environmental pollution and the destruction of habitats, at least one more factor is responsible for the reduction of diversity at the level of species and populations. This factor is the irrational commercial utilization of living resources which ignores the characteristic genetic structure of individual species. This conclusion is based on long-term monitoring of genetic processes at the population level and forecasting of corresponding negative consequences (Altukhov, 1990). The main task of this chapter is to demonstrate such effects for some natural and agricultural populations. But before doing this it is necessary to emphasize at least two fundamental principles of population monitoring.

Firstly, many native undisturbed populations are, as a rule, not homogeneous but form metapopulations, i.e., historically formed systems of subpopulations (Altukhov and Rychkov, 1970). Sampling procedures must therefore be organized in a way to allow for complete characterization of the subpopulation structure by a complex of genetic, demographic and biological traits.

Secondly, this subpopulation structure of the metapopulation is ordered in space and time (Altukhov, 1974) and the ratio of intra- and interpopulation components of genetic diversity is remarkably constant for representatives of numerous taxonomic groups (see, e.g., Nei, 1975). Knowledge of this ratio provides us with a unique possibility to organize the genetic monitoring of biological systems and this information

338

makes it possible to understand and estimate the state of the genetic process at the population level. Following the major goal of this chapter, let us consider the particular results of the application of these principles.

Genetic processes in intact metapopulations

The population of the non-exploited mollusk, *Littorina squalida*, from the Bousse lagoon in South Sakhalin, can be used as an example of the intact metapopulation structure. This species has a pronounced subpopulation structure and a well-expressed shell pattern polymorphism which is treated as the manifestation of a diallelic system with incomplete dominance (Altukhov and Kalabushkin, 1974). As well as for the thoroughly studied land snail, *Cepea nemoralis*, differences in genotypes (phenotypes) are preserved in fossil forms. It is therefore possible to estimate gene frequencies in samples from presently living and ancient populations separated by a time interval of about 4500–5000 years corresponding to approximately 2000–2500 successive generations. Such an evaluation on the basis of the character of fossil deposits and accompanying thermophilic fauna and with respect to the age structure of littorines, seems to be highly reliable (Kalabushkin, 1976). In 1969–1974 the distributions of genotypes in three modern and five fossil samples were studied. The estimates of gene frequencies presented in Fig. 1 show that a comparison of individual modern and fossil samples may bring us to rather contradictory conclusions revealing both genetic similarity and difference in time and space. However, despite the variability in parts, the system as a whole preserves the genetic composition inherited from the ancestral population.

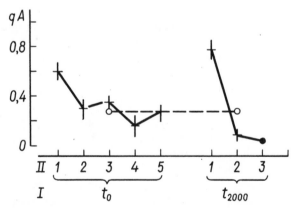

Fig. 1. Gene frequency variation in a conventional zero generation (t_0; n = 479; q_0 = 0.293 ± 015) and in the approximately 2000th generation (t_{2000}; n = 1252; q = 0.280 ± 01) of the population system of *Littorina squalida* mollusk as a whole (I) and in comparison with the variability for individual localities (II).

Similar effects were shown in experimental populations of *Drosophila melanogaster* (Altukhov, 1990) and for computer simulations of genetic processes in population systems corresponding to the elementary circular stepping-stone model (Altukhov and Blank, 1991, 1992; Blank and Altukhov, 1992): With equal population sizes the subdivided population was more stable than the panmictic in terms of preserving genetic diversity. For instance, in the latter experiment 8 of 10 models of panmictic populations, each consisting of 500 individuals, became completely homozygous by the 1000th generation while none of the subdivided populations of the same size ($N = 20$; 25 subpopulations; migration rate $= 0.03$) lost their genetic diversity. We calculated the "lifetime" of panmictic and subdivided populations, each with total population sizes of 500, as the number of generations corresponding to the loss of 99% of the initial level of heterozygosity. This "lifetime" value was 2301 generations for panmictic populations ($N = 500$) and 5341 generations for subdivided populations ($N = 20$; 25 subpopulations; migration rate $= 0.005$). In other words, the presence of just the simplest subpopulation structure of the circular type with a limited rate of gene flow (about 0.5–1%) slows down the loss of genetic diversity to a level that would amount to a doubling of the effective number of the panmictic population.

This conclusion has important consequences for conservation biology: In one case, the size of a "minimal viable population" (Soulé, 1987) may be only hundreds of individuals due to inner fragmentation, while in another case it may be thousands or more.

Genetic processes in natural metapopulations evolving under anthropogenic pressure

Since 1968, we have carried out long-term monitoring studies for species of Pacific salmon (genus: *Oncorhynchus*) by taking their specific genetic population structure into account (for details see: Altukhov, 1974, 1990; Altukhov and Salmenkova, 1991; Salmenkova, 1989). Since then, similar approaches have been implemented for other species of salmonid fishes (for details see: Hindar et al., 1991; Gall et al., 1992). One of the main findings in these monitoring programs is the loss of genic diversity in artificially maintained populations compared to natural populations. The discovery of negative effects connected with an increase of intrapopulation genic diversity in the process of selective fishing and gene pool transplantation is, however, not less important.

Fishing

Investigation of natural fish populations ("shoals") reveals their pronounced heterogeneity and differentiation into smaller genetically distinct subpopulations ("stocks") which are connected by migration links. For instance, this pattern is very typical for sockeye (*O. nerka* Walbaum) reproducing in the Kamchatka lakes. An important biological feature of sockeye salmon populations is a unique pattern of body length variation of spawners: Females are characterized by a unimodal distribution whereas males display a clear-cut bimodality (Altukhov, 1990).

Within this species three groups of spawning fish are easily distinguished: Small males, large males, and females of intermediate body length. This relates to a rather conservative system of so-called selective matings: Females prefer old, larger males, but in years with low water levels in the rivers and in shallow spawning sites where large males are not able to penetrate, fast growing young males are reproductively successful. Large males exhibit low levels of heterozygosity, small males are very heterozygous, while females are characterized by an intermediate level of heterozygosity. Heterotic animals are usually characterized by the largest body length, but this is not the case in sockeye as the fish reproduces only once in life and dies after spawning. Instead, the fast growth of heterotic males leads to sexual maturity at a young age.

It is a common observation that the proportion of small sexually mature males in the studied populations increases in time. Usually, these 3-year-old males ("grilse," "jacks," or "kayurki" in Kamchatka) are found in low frequency in native intact stocks. Contrary to this, in populations exposed to systematic fishing pressure the proportion of small early maturing males increases drastically. This process is at present more or less typical for all sockeye populations from both sides of the North Pacific which have been intensively exploited by commercial fishing from the start of this century.

An illustrative example of this situation is the sockeye shoal from Lake Dal'neye (Kamchatka). Its biology has been studied in detail by F. V. Krogius: In the 1930s, the spawning population of this shoal included approximately 100 000 spawners and the proportion of grilse among sexually mature males did not exceed 0.2%. In the 1960–70s the number of spawners decreased to approximately 2000 and the proportion of grilse increased to 38% (Krogius, 1979; Tab. 1).

What is the cause of such dramatic changes? We have found that selective marine fishing, by disproportionately catching large (and more homozygous) old males with gill nets, is the principal responsible fact in this process. Other sockeyes which differ genetically from large males are exploited either evenly (females: average size, average heterozygosity level) or underexploited (small males: the maximum level of heterozygosity) (Altukhov and Varnavskaya, 1983).

As a consequence, commercial fishing disturbs the system of matings and more heterozygous small males pass their genes to the next generations to an increasing extent leading to an overall more heterozygous

Table 1. The average values of some biological parameters in the Lake Dal'neye sockeye salmon population in successive periods (from Krogius, 1979)

Parameters	Intervals			
	1935–46	1947–56	1957–65	1966–76
Spawners entering lake ($\times 10^3$)	62.6	10.0	5.7	1.6
Females in anadromous population (%)	52	54	59	68
Jacks among anadromous males (%)	0.2	0.6	4.3	37.5
Dwarfs spawning in lake ($\times 10^3$)	14.9	7.6	5.8	5.3
Dwarf among males (%)	26.1	49.4	74.2	88.8

population. This leads to a decrease of the proportion of large fish in the stocks, the optimal sex ratio is distorted, the age at maturation decreases, average life-time is reduced, and the average generation time decreases. Simultaneously, the population size of the shoals decreases since small females have a lower fecundity. Fishing of a constant intensity, which should otherwise be compatible with the initial production abilities of the population, may therefore lead to a reduction of the reproductive ability of the shoal only because of disproportional catches of fish with certain genotypes. These processes are characteristic not only for sockeye salmon populations, but also for other fish species which are objects of fishing (Altukhov, 1992). However, as we shall see, intrapopulational genetic diversity may also be reduced in the process of artificial propagation when selection is directed in favor of homozygotes.

Artificial reproduction

Pink salmon (*O. gorbuscha* Walbaum) is characterized by a fixed life cycle and therefore the populations spawning in odd and even years in the same locality may be genetically quite different. In the context of the following, it should also be noted that there is a high frequency of males in the first part of the spawning run while the later arriving fish are dominated by the presence of females. For the last 20 years three closely located pink salmon populations in South Sakhalin have been monitored. Two of them are natural populations (Firsovka and Bakhura Rivers) while the third (Naiba River) is artificially propagated. The latter is maintained by two hatcheries by means of catching spawners and rearing and releasing the offspring. These hatcheries have previously specialized in reproduction of the Naiba chum (*O. keta* Walbaum) shoal, but since 1972, after a sharp decline of its population size (Altukhov, 1981), they have been engaged in the reproduction of pink salmon. Judging from catches, the size of the local population has increased by several times due to the fish-breeding activity, but in recent years the biological structure of the hatchery stock has changed. The fish have become considerably larger, the frequency of males has increased, and the total number has decreased. In order to investigate the mechanisms of this process, we chose the following approach: (1) the dynamics of sex ratio and body length in successive generations of the artificially reproduced Naiba population were traced; (2) the genetic characteristics of male and female spawners used and rejected in the fish-breeding process were compared by a set of allozyme loci; (3) the genetic and biological parameters of the hatchery population and of the two native shoals reproducing in the neighboring rivers were compared. Additionally, to determine the relationships between heterozygosity and biologically important characters in males the frequency of anomalies of gill rakers was used as an indicator of developmental stability.

The temporal dynamics of body length and sex ratio in the Naiba pink salmon population were estimated from the records of the Sokolovsky hatchery. Analogous data for the self-reproducing pink salmon population from the Firsovka River have been collected in the course of our own work.

The comparison of males used in the fish-breeding process with the rejected males ("control" group) indicates that fish breeders prefer large fish (Tab. 2). For females such selectivity has not been revealed. Of particular importance is the fact that the differences between the control group of males and males used for breeding are also observed both at the level of allozyme heterozygosity and in the frequency of fish with gill raker anomalies (Tab. 2). Large males taken for hatchery purposes turned out to be more homozygous as compared to the control group;

Table 2. Morphological and genetic differences between two groups of pink salmon males

Parameter	Males for breeding at the hatchery	Males not used for hatchery breeding	Significance of differences, t_d-test
Body length (cm)	52.6 ± 0.2	47.8 ± 0.1	$p < 0.001$
Standard deviation	3.87	2.64	$p < 0.001$
No. of fish	300	293	
Proportion homozygous for all loci*	0.433 ± 0.029	0.362 ± 0.028	$p < 0.05$
No. of fish	300	293	
Proportion with severe anomalies of gill rakers	0.17 ± 0.03	0.09 ± 0.02	$p < 0.05$
No. of fish	203	210	

*Loci analyzed: Mdh-3, -4; Pgd; Me-2; and Pgi-2.

Table 3. Genotypic diversity (μ) in samples of spawning pink salmon from natural (Firsovka and Bakhura) and artificial (Naiba) reproduction

| Year | River | Number of | | μ average per locus | Significance of difference u-test |
		Fish	Loci		
1977	Naiba	234	4	1.920 ± 0.039	$p < 0.001$
	Firsovka	250		2.129 ± 0.045	
1979	Naiba	214	4	1.992 ± 0.040	n.s.
	Firsovka	243		2.073 ± 0.043	
1981	Naiba	474	4	2.148 ± 0.047	$p < 0.01$
	Firsovka	449		2.300 ± 0.050	
1985	Naiba	318	5	2.157 ± 0.040	$p < 0.001$
	Firsovka	250		2.394 ± 0.046	
1985	Naiba	318	5	2.157 ± 0.040	$p < 0.05$
	Bakhura	400		2.295 ± 0.037	
1985	Firsovka	250	5	2.394 ± 0.046	n.s.
	Bakhura	400		2.295 ± 0.037	

the frequency of individuals with gill raker anomalies among them was also higher. If we assume this selection was more or less systematic during 17 years of artificial reproduction of the Naiba pink salmon, the latest arriving spawners of the shoal characterized by an excess of females was not involved in reproduction. It is therefore possible to expect quite definite shifts in the biological structure of the population. Firstly, a decrease in heterozygosity and genotypic diversity; secondly, an increase in body length; thirdly, the distortion of the optimal sex ratio due to an increase in the portion of males (since the collection of eggs is carred out mainly for early migrating fish groups with a high frequency of males).

This suggestion is supported by other results: We have observed that the number of females falls and the body length increases in successive generations of the Naiba population in more "productive" odd years when the artificial selection among males is more pronounced (Altukhov and Salmenkova, 1991). In contrast to this observation, the estimates of sex ratio and average body length for five odd-year generations of naturally reproducing pink salmon from the Firsovka River demonstrate the steady state of this shoal. We have also found that, compared to the Firsovka and Bakhura shoals, the mean body length of males and females in the Naiba River is significantly higher while the dispersion of the traits is lower (Altukhov and Salmenkova, 1991). The same is also characteristic for the index of genotypic diversity determined by the formula of Zhivotovsky (1980; see Tab. 3).

Transplantation

Transfers of artificially fertilized salmon eggs between rivers were a widely used practice in the 1950–70s in the Russian Far East, when in the fishing industry, as well as in other branches of economy of the USSR, a plan was pursued only for the sake of a plan. In order to fulfill the plans concerning production of fry, some hatcheries propagating chum salmon imported huge amounts of eggs from exogenous populations (for details see Altukhov and Salmenkova, 1991).

We found a set of allozyme loci which could be used for identifying the populations in mixed accumulations and thereby it was possible to evaluate the results of the transplantations. They appeared to be practically ineffective. Return of spawners, if any, was observed only in the first generation. The return rates, as integral indices of population fitness, were approximately 10 times lower than in the native river and, later, no traces of the transplanted population were found. Moreover, the return of the populations to their native rivers decreased as a considerable portion of the gene pool had been withdrawn for acclimatization in a different part of the area (Altukhov and Salmenkova, 1991).

From the point of view of population genetics acclimatization is the adaptation to a new environment. It is possible to judge its efficiency only after the formation of a self-reproducing population with a stable integrated gene pool capable of long existence in a series of generations. Unfortunately, today there is too little evidence for phenomena of this kind. However, there are plenty of examples demonstrating the opposite. Ricker (1972) summarized a vast amount of data on transplantations, mainly of North American salmonid populations, and showed that return of the first generation to a "foreign" river is quite frequent, but to a much lesser extent than the return to the "native" river; in the following generations the return to the "foreign" river in an overwhelming majority of the cases decreases sharply or is not observed at all. Withler (1982) analyzed an extensive amount of published data on Pacific salmon transplantations and came to the conclusion that the activities aimed at producing new anadromous stocks within a natural area were quite inefficient.

All these and other facts are evidence of the unique and conservative character of local adaptations formed by selection through thousands of generations in a specific environment. When we withdraw a population from its "own" historically formed environment and transfer it into a new environment, the resources of genetic stability usually turn out to be insufficient.

It should be taken into account, however, that successful acclimatization, as a rule, also appears to be undesirable because of its ecological consequences: This may be accompanied by the extinction of local populations either on the basis of food competition or, which is apparently more frequent, due to dissemination of diseases to which the local forms have not developed immunity (for details see: Altukhov, 1990; Hindar et al., 1991).

Our long-term monitoring approach based on the previously discussed principles leads to the conclusion that human interference and exploitation of natural populations may lead to a redistribution of the intra- and interpopulation components of genic diversity. Even though this redistribution may take different directions depending on the mode of human interference, the accompanying effects are negative. Let us now consider the applicability of these principles to agricultural populations.

Genetic processes in agricultural populations

As examples of the genetic processes taking place in agricultural populations, we will consider monitoring data which have been obtained for 52 spring barley (*Hordeum vulgare* L.) varieties grown in East Siberia (A. Pomortsev and B. Kalabushkin, unpublished) and various hen (*Gallus gallus* L.) breeds (I. Moiseyeva and L. Bannikova, unpublished).

Electrophoretic analysis of seed storage proteins (hordeins) in barley varieties has shown that the diversity of different populations has changed significantly: If earlier a mixture of different biotypes (old varieties) prevailed, at present, strain varieties (new varieties) predominate. The levels of variation were found to be considerably higher in the old varieties compared to the new strain varieties: In 31% of the old varieties three or four alleles were found while among new varieties this proportion was equal to zero. In fact, in the predominant part of new varieties only one allele was observed ($\chi^2 = 9.22$; $p < 0.01$). It is therefore clear that for the past 60 years considerable changes have occurred in the genotypic composition of barley varieties cultivated on the territory of East Siberia. These changes have most probably been caused by the breeding practice, i.e., new varieties have been founded by only one or a few plants.

The same trend of loss of genetic diversity in time has also been observed during the monitoring of hen populations. Among factors involved in the loss of genetic variability in farmed poultry, we note a sharp reduction of the number of breeds used for commercial purposes. The commercial crosses include only four to seven breeds out of 603 listed in Somes' catalogue (Somes, 1985). As for Russia, approximately 30 out of 80 old breeds have not been preserved (or not found) by now. This is equivalent to a 37.5% reduction of the genetic resources in breed composition. An analysis of the dynamics of genetic variability in poultry breeding and its more precise quantitative estimation agrees well with the above-mentioned facts. This analysis was based on experimental data (our own and from literature) on biochemical polymorphisms of 38 hen populations of foreign (Mediterranean and Asiatic) and home (Russian) origin, including wild ancestors of domestic hens

(Red Jungle Fowl). The analysis was based on 16 loci controlling the formation of egg and blood proteins. Six loci were found to be polymorphic (Ov, G3, G2, Tf, Alb, Es-1) while 10 turned out to be monomorphic (Amy-3, Es-8, Pgm, Phi, To, Mdh, Ldh, Es-D, Hb-1, Hb-2). Each population was characterized by the allele frequencies of the six loci

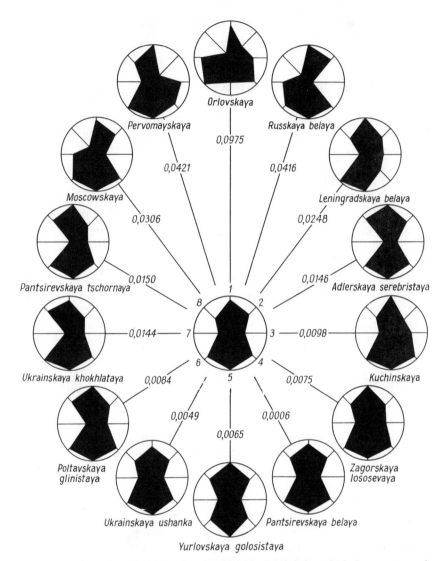

Fig. 2. Genetic profiles of home-produced hen breeds and their hypothetical ancestor population. The radii show the frequencies of the following alleles: 1 — Ov A; 2 — G 3A; 3 — G 3B; 4 — G 2B; 5 — Tf B; 6 — Alb B; 7 — Es-1A; 8 — Es-1B. In the middle of the circle the frequency is equal to zero, on the perimeter equal to 1. The lines connecting hen breeds with the ancestor population are estimates of genetic distances according to M. Nei. (Ov: ovalbumin; G: globulin; Tf: transferrin; Alb: albumin; Es: esterase.)

Table 4. Genetic diversity in hens for 16 loci encoding of blood and egg proteins and 18 loci of morphophysiological traits

Group of breeds	Number of breeds	Biochemical loci				Number of lost alleles responsible for morphophysiological characters (%) **
		Number of alleles per locus*	Percentage of polymorphic loci*	Average heterozygosity H ± (s.e.)	Number of lost alleles (%) **	
Red Jungle Fowl	1	1.44	31.25	0.091 (0.042)	9 (39.1)	19 (51.4)
Hypothetic ancestor population	37	1.62	37.50	0.090 (0.042)	7 (30.4)	0 (0)
Russian and Ukranian	14	1.44	37.50	0.085 (0.041)	9 (39.1)	9 (24.3)
Mediterranean	6	1.38	31.25	0.071 (0.035)	11 (47.8)	11 (29.7)
Asian	11	1.62	37.50	0.097 (0.045)	7 (30.4)	2 (5.4)
Commercial	6	1.38	31.25	0.093 (0.043)	9 (39.1)	13 (35.1)

* A locus was considered polymorphic if the frequency of the rarest allele exceeded 0.01.
** 23 alleles have been analyzed for biochemical loci and 37 alleles — for morphophysiological loci.

showing variation and an expected genetic structure of a hypothetical ancestral population was determined on the basis of averaging allele frequencies for 37 breeds.

The number of alleles *per* locus (Tab. 4) was, as a rule, lower in the groups of commercial and Mediterranean breeds, while relatively high estimates were obtained for wild hens, for the hypothetical ancestral population and for the group of Asian hens. The genetic profiles of several Russian commercial hen breeds are presented in Fig. 2. It is clear that some breeds have diverged substantially from the hypothetical ancestor population (e.g., Leningradskaya belaya, Orlovskaya, Moskovskaya, etc.) while others are very closely related to it (e.g., Yurlovskaya golosistaya, Pantsyrevskaya belaya, Ukrainskaya ushanka, etc.).

It is very interesting that the breeds (nine populations) that show the closest relationships to the ancestor population are characterized by the highest level of intrapopulation heterozygosity ($H = 0.213$) and a low interpopulation gene diversity ($G_{ST} = 0.0975$) as compared to the breeds (five populations) which are genetically most remote from the ancestor population ($H = 0.183$; $G_{ST} = 0.2311$).

From a statistical point of view this is perhaps not surprising. An important point, however, is that close genetic relationships to the ancestral population turns out to be linked with a less pronounced breed specialization: Almost all nine breeds in this group are used for both meat as well as egg production, whereas the five breeds which have diverged the most from the ancestor population are more specialized, either as egg-layers (e.g., Russkaya belaya, Moskovskaya) or meat fowl (e.g., Leningradskaya belaya). The Orlovskaya breed, which is genetically most remote from the ancestor population, has its origin from fighting fowl.

A considerable interpopulation genetic differentiation and a decreased level of heterozygosity characteristic for the second group of breeds clearly indicate that the process of selection in this case was accompanied by a loss of genic diversity, i.e., homozygotization of breeds. This is apparently the same unfavorable genetic process that has been mentioned above for barley and natural fish populations.

Summary and Conclusion

Genetic monitoring of natural populations of commercially valuable fish under anthropogenic influence revealed unfavorable genetic processes. This was caused by ignoring the historically formed subpopulation structure in the course of economical exploitation. In general, anthropogenic influence leads to the redistribution of the intra- and interpopulation components of genetic diversity. A decrease in the intrapopulation genic diversity was detected in some cases of artificial propagation. This process is maladaptive and may lead to an irreversible degradation or loss of variation even after the external influence factor has ceased. An increase of intrapopulation genic diversity was observed in the case of monitoring self-reproducing populations under pressure of selective fishing. This process is adaptive, but its final result is also the degradation of populations, since the adaptation cost turns out to be excessively high: It may involve the substitution of migrating (anadromous) highly productive populations by commercially useless residual forms. The evolutionary formed levels of genic diversity of populations and species are disturbed not only in the course of exploitation of natural populations, but also in the process of breeding and improvement of agricultural plants and animals. This has been demonstrated for domesticated organisms like barley and hens. It is therefore necessary to reevaluate the utilization of natural populations and the breeding strategies of domesticated animals and plants. Conservation genetics must play a key role in this process!

Acknowledgements
I would like to thank Volker Loeschcke, Jürgen Tomiuk, Michael M. Hansen, and two anonymous reviewers for valuable comments on the manuscript. Financial support from the Russian State Program "Frontiers in Genetics," the University of Aarhus, and the International Science Foundation is gratefully acknowledged.

References

Altukhov, Yu. P. (1974) *Populatsionnaya genetika ryb.* Moscow: Pischevaya Promyschlennost (Transl. from Russian by Fish. Mar. Serv., Canada, Transl. Ser. No 3548, 1975).

Altukhov, Yu. P. (1981) The stock concept from the viewpoint of population genetics. *Can. J. Fish. Aquat. Sci.* 38: 1523–1538.

Altukhov, Yu. P. (1990) *Population genetics: diversity and stability.* Harwood Academic Publ., London.

Altukhov, Yu. P. (1992) *Effects of fishing on the genetic resources of aquatic organisms.* FAO Expert Consultation, Rome, 9–14 November. Background paper, pp. 1–30.

Altukhov, Yu. P. and Blank, M. L. (1991) Computer modelling of genetic processes in structured populations. *Dokl. Rus. Acad. Nauk* 319: 1467–1472.

Altukhov, Yu. P. and Blank, M. L. (1992) Genetic dynamics of population systems with varying parameters of the structure and selection pressure. *Dokl. Rus. Acad. Nauk* 326: 1068–1072.

Altukhov, Yu. P. and Kalabushkin, B. A. (1974) Stable polymorphism in modern and ancient populations of molluscs, *Littorina squalida. Dokl. Akad. Nauk SSSR* 215: 1477–1480.

Altukhov, Yu. P. and Rychkov, Yu. G. (1970) Population systems and their structural components: genetic stability and variability. *Zh. Obshch. Biol.* 31: 507–526 (in Russian).

Altukhov, Yu. P. and Salmenkova, E. A. (1991) The genetic structure of salmon populations. *Aquaculture* 98: 11–40.

Altukhov, Yu. P. and Varnavskaya, N. V. (1983) Adaptive genetic structure and its connection with intrapopulational differentiation for sex, age and growth rate in sockeye salmon, *Oncorhynchus nerka* (Walb.). *Genetika* 19: 796–807.

Blank, M. L. and Altukhov, Yu. P. (1992) Genetic dynamics in populations with a complicated migration structure. *Dokl. Rus. Akad. Nauk* 324: 1309–1313.

Gall, G. A. E., Bartley, D. and Bentley, B. (1992) Geographic variation in population genetic structure of chinook salmon from California and Oregon. *Fishery Bulletin. U.S.* 90: 77–100.

Hindar, K., Ryman, N. and Utter, F. M. (1991) Genetic effects of cultured fish on natural fish populations. *Can J. Fish. Aquat. Sci.* 48: 945–957.

Kalabushkin, B. A. (1976) Genetic variability in modern and mid-holocoenic populations of *Littorina squalida, Zh. Obshch. Biol.* 3: 369–377 (in Russian).

Krogius, F. V. (1979) On the relationship between freshwater and marine life-spans of sockeye salmon from Lake Dal'neye. *Biol. Morya* 3: 24–29.

Nei M. (1975) *Molecular population genetics and evolution.* North-Holland Publ. Comp., Amsterdam.

Ricker, W. E. (1972) Hereditary and environmental factors affecting certain salmonid populations. *In:* Simon, R. C. (ed.), *The stock concept of Pacific salmon.* University of British Columbia Press, Vancouver, pp. 19–160.

Salmenkova, E. A. (1989) General results and aims of the population genetic studies of salmonids. *In:* Kirpichnikov, V. S. (ed.), *Genetics in aquaculture.* Nauka, Leningrad, pp. 7–29 (in Russian).

Somes, G. (1985) International registry of poultry genetic stocks. *Connect Exp. Stat. Bull.* 469: 1–95.

Soulé, M. E. (1987) *Viable populations for conservation.* Cambridge Univ. Press, Cambridge.

Wilson, E. O. (1988) *Biodiversity.* National Academic Press, Washington, D.C.

Withler, F. G. (1982) Transplanting Pacific salmon. *Canad. Techn. Rep. Fish. Aquat. Sci.* 1079. Dept. of Fisheries and Oceans, Vancouver, pp. 1–27.

Zhivotovsky, L. A. (1980) Measure of intrapopulation diversity. *Zh. Obshch. Biol.* 41: 828–836 (in Russian).

Part VI

Genetic resource conservation

Conservation Genetics
ed. by V. Loeschcke, J. Tomiuk & S. K. Jain
© 1994 Birkhäuser Verlag Basel/Switzerland

Introductory remarks

Genetic improvement of domesticated plants and animals depends on the availability of genetic resources in the form of landraces, selected breeds, or diverse collections of other materials. For plant species a large number of local, national, and international gene banks collect, evaluate, and maintain genetic resources for use by plant breeders. For domestic animals a number of regional gene banks were established recently for the preservation of endangered breeds, as increasing intensification of animal production, breed substitution, and crossbreeding are threatening genetic variation of domestic livestock. On a global scale, a Global Animal Genetic Resources Data Bank was started for the flow of information among regional centers and to allow a regular appraisal of the breeds (Alderson, 1990). The *ex situ* conservation approach has made impressive technological strides in the cryopreservation of tissues, seed, molecular probes, DNA and gene libraries in plants (Moore et al., 1992) or sperm, oocytes, embryos or isolated genes in animals (Bodo, 1990). For a select group of major crop plants such collections have become so large ($10^4 - 10^5$ accessions, or even more) that the concept of core collections to include a well-designed representative sample of 1000–2000 accessions appears valuable (Frankel and Brown, 1984). Collecting activities might still continue for the underutilized or minor crop species and for the wild relatives of major crop species, which may now find increased usage with the emerging biotechnological advances in gene mapping and gene transfer.

Another form of *ex situ* conservation had begun much earlier through the establishment of botanical gardens which served as the key sources of many economic plants to the colonial European empires (see Frankel and Soulé, 1981; Harlan, 1992). Even *prior* to the rediscovery of Mendelism, numerous industrial crops had been developed through research and continued supply of resource materials. Public education and recreation, of course, provided the *raison d'etre* for many of these gardens. Similar economic and educational objectives also supported the establishment of zoological gardens and specialized museums or parks (Western and Pearl, 1989). Oldfield (1989) provides an excellent historical and ethnobiological survey of human interests in the medicinal plant and animal resources which are often cited in the economic argument for conserving natural ecosystems.

During the past decade or two, the role of botanical gardens and zoos for *ex situ* conservation has become extremely important as more and more taxa have made the Red Databook lists. This situation calls for both short- and long-term recovery plans involving off-site propagation, breeding, and eventual reintroduction into the natural habitats (Heywood et al., 1990; Gipps, 1991). Despite their limited spatial capacities, zoos have the potential to maintain a variety of self-sustaining captive populations of a few selected rare and endangered species. Since the need for careful genetic management of small populations has been recognized by the international zoo community, cooperative breeding plans have come into existence (Ralls and Ballou, 1986; see also Templeton and Read, this volume). Restoration ecology is dependent on the availability of suitable stocks or gene pools to be used in such reintroduction efforts. The World Conservation Monitoring Centre report (1992) on global biodiversity noted that nearly three million accessions of plant materials are cultivated in a total of 1500 botanical gardens. Avishai (1985), for example, reviewed this role for the Mediteranean botanical gardens, and considered them to be very crucial for not only the agricultural species but also for the native flora. Several examples of targeted conservation of native flora and related projects include: (1) the Reunion Island botanical garden, founded in 1987, plans to conserve 60% of the island's rare and endangered taxa; (2) Kew Botanical Gardens has a program to salvage all the St. Helena flora; (3) the National Botanic Garden in Lucknow, India, has been renamed as the National Botanical Research Institute to signify its new role in conservation and genetic resource evaluation; (4) Rancho Santa Ana Botanical Garden in California has an outstanding molecular genetic research program to develop new biosystematic tools; (5) the Missouri Botanic Garden is the headquarters for the Center for Plant Conservation (Falk, 1990) and also a leader in tropical biodiversity issues. Such examples of major developments worldwide are too many to enumerate (Bramwell et al., 1989; Shan-An et al., 1990; Solbrig, 1991). Of course, many gardens do need more funding and personnel in order to redirect their efforts towards conservation. Population genetics plays an important role in all of these genetic resource programs through its focus on describing and monitoring genetic variation. The three chapters to follow discuss some of these significant conservation-oriented contributions.

References

Alderson, L. (1990) *Genetic conservation of domestic livestock*. CAB International, Wallingford.

Avishai, M. (1985) The role of Mediterranean botanic gardens in the maintenance of living conservation-oriented collections. *In:* Gomez-Pompa, C. (ed.), *Plant conservation in the Mediterranean area*. Dr. W. Junk, Dordrecht, pp. 221–236.

Bodo, I. (1990) Organizer: Workshop 2.4 – Conservation of genetic resources. *Proc. 4th Wld Cong. Geneg. Appl. Livestock Prod.* XIV: 421–472.

Bramwell, D., Hamann, O., Heywood, V. and Synge, H. (1989) *Botanic gardens and the world conservation strategy.* Academic Press, London.

Falk, D. A. (1990) Integrated strategies for conserving plant genetic diversity. *Ann. Missouri Bot. Garden* 77: 38–47.

Frankel, O. H. and Brown, A. H. D. (1984) Current plant genetic resources – a critical appraisal. *In:* Chapra, V. C., Sharma, R. P., Sawhney, M. and Joshi, B. C. (eds), *Genetics – New Frontiers* 4. India Book House, New Dehli, pp. 1–11.

Frankel, O. H. and Soulé, M. E. (1981) *Conservation and evolution.* Cambridge University Press, Cambridge.

Gipps, J. H. W. (1991) *Beyond captive breeding: Reintroducing endangered species through captive breeding.* Clarendon, Oxford.

Harlan, J. R. (1992) *Crops and man.* 2nd edn. Amer. Soc. Agronomy, Madison.

Heywood, C. A., Heywood, V. H. and Wyse Jackson, P. (1990) *International directory of botanic gardens.* V. BGCS and WWF (Koelty Scientific Books).

Moore, H. D. M., Holt, W. V. and Mace, G. M. (1992) *Biotechnology and the conservation of genetic diversity.* Chapman and Hall, London.

Oldfield M. L. (1989) *The value of conserving genetic resources.* Sinauer, Sunderland, MA.

Ralls, K. and Ballou, J. D. (1986) Proceedings of the workshop on genetic management of captive populations. *Zoo Biol.* 5: 81–238.

Shan-An, H., Heywood, V. H. and Ashton, P. S. (1990) *Proceedings of the international symposium on botanic gardens.* Jiangsu Sci. Tech. Publ. House, Nanjink.

Solbrig, O. T. (1991) *From genes to ecosystems: a research agenda for biodiversity.* IUBS, Cambridge.

Western, D. and Pearl, M. (1989) *Conservation for the twenty-first century.* Oxford University Press, Oxford.

World Conservation Monitoring Centre (1992) *Global biodiversity: status of the earth's living resources.* Chapman and Hall, London.

Conservation Genetics
ed. by V. Loeschcke, J. Tomiuk & S. K. Jain
© 1994 Birkhäuser Verlag Basel/Switzerland   .

Optimal sampling strategies for core collections of plant genetic resources

A. H. D. Brown[1] and D. J. Schoen[2]

[1]CSIRO, Division of Plant Industry, GPO Box 1600, Canberra ACT, Australia
[2]Department of Biology, McGill University, Montreal, Quebec, Canada

Summary. A core collection of crop germplasm aims to represent the genetic diversity in a single collection or in a crop species with minimum similarity between its entries. Core collections have a major role to play in conserving genetic resources and using them in plant improvement. Core selection can be based on stratified sampling from groups of related accessions. Elementary neutral theory indicates that the relative number from each group should be proportional to its level of polymorphism. This procedure has some biases when alleles are finite in number, or heterotic, or deleterious. However, in general, the weighting strategy is in practice robust to these departures from the assumptions underlying theory. Variation in divergence among populations is a factor that merits attention. In general, weighting in conservation should include both elements of richness and degree of divergence.

Introduction

Gene banks around the world hold collections of the genetic resources of crop plants both for long-term conservation and for ease of access by plant breeders and scientists (see Plucknett et al., 1987 for an overview). Remarkable progress has occurred in assembling and conserving these resources, particularly over the last two decades. Indeed, plant germplasm collections are now facing major problems of size and organization. Such collections have grown so large and numerous as to jeopardize the very purposes for which they exist, namely, both the conservation and the use of the genetic diversity they hold (Holden, 1984). These *ex situ* collections are caught in the nexus between two conflicting demands: the need to preserve as much variation as possible *versus* the need to limit their growth. On the one hand is the impulse to acquire or collect every available variant and to conserve it because, unless this is done, it might perish before its future utility has been tested. On the other hand is the limit on the numbers of accessions that can be maintained, described, and used in plant improvement.

This dilemma is a fundamental one and is met in many other areas of conservation biology. The conservationist is aware of the extent of biodiversity at the DNA level, at the species level and at the ecosystem level, and of its current high rate of extinction (e.g., Wilson, 1988). This awareness is the driving force for strenuous efforts at conservation. Yet, it is clearly impossible to preserve everything. Attempts at holistic preservation will fail because of the pressure of human needs and human greed. Thus, sampling is an inevitable part of biological conservation (Brown, 1992).

Recognizing that the size of a plant germplasm collection could hinder its use, Frankel (1984) proposed that a collection could be reduced to what he termed a *"core collection"*. With minimum similarity between its entries, the core collection aims to represent the genetic diversity of a large collection, or a crop or a wild species, or group of species. The remaining accessions of a collection, those not included in the core, would not necessarily be discarded. In many instances they would form a reserve collection.

Structured or stratified sampling

Since this proposal was made, the rationale and purposes of core collections of crop germplasm have been further developed, as have procedures for their construction (Frankel and Brown, 1984; Brown, 1989a,b, 1993). The logical approach to implementing such collections is one of structured or stratified random sampling. The target collection (or area, or species) is first divided into a number of groups – the groups being genetically, ecologically, or geographically distinct. A sample is then drawn from each group. The salient question is, given a fixed total size for a core collection, what is the relative sample size or weight to be given to each group?

Several sampling strategies are at hand to answer this question (Brown, 1989b). In the absence of detailed genetic data about the individuals within the groups, the weighting can be constant for all groups, or proportional to group size, or proportional to the logarithm of group size (referred to as the C-, P-, and L-strategies, respectively). When genetic data are available, more sophisticated strategies are possible (Schoen and Brown, 1993). One of these – the so-called H-strategy – follows directly from a strict maximization of the total number of selectively neutral alleles in the combined sample.

The strategy is based on the theoretical model of selectively neutral, infinite alleles (Kimura and Crow, 1964). Since the model has some unrealistic assumptions, this chapter examines the behavior of the H-strategy for some numerical examples in which the major assumptions do not apply. But first we outline the H-strategy.

Maximizing allelic richness in multi-population samples under the infinite neutral allele model

The number of selectively neutral alleles (K) in a sample of size S random gametes from an equilibrium population of size N at a locus with mutation rate u is approximately

$$K \approx \theta \log_e[(S + \theta)/\theta] + 0.6, \tag{1}$$

where $\theta = 4Nu > 0.1$ and $S > 10$. The error for K is less than 10% when $S > 4$ and $\theta > 1$.

Consider two geographically isolated populations (or groups of accessions), 1 and 2. Assume that the allelic diversity at a locus A in each population is in neutral equilibrium, indexed in the ith population by the parameter θ_i. Samples of size n_1 and n_2 unrelated gametes are now drawn from each population, where $S = n_1 + n_2$. The expected total number of distinct alleles at locus A in the composite sample from both populations is

$$K \approx \theta_1 \log_e[1 + (n_1/\theta_1)] + \theta_2 \log_e[1 + (n_2/\theta_2)] + \text{constant}. \tag{2}$$

The maximum number of alleles (K) for a fixed total size (S) follows from differential calculus and occurs when

$$n_1/n_2 = \theta_1/\theta_2. \tag{3}$$

In other words, the two groups should be sampled in proportion to the average of estimates of their θ_i for comparable loci. This result is readily extended to more than one locus and more than two populations (Schoen and Brown, 1993). Thus, the relative contribution of the ith population or a group should be in direct proportion to its diversity as measured by its value of θ_i.

The individual estimates of the θ_i can be obtained either a) directly from the observed number of alleles *per* locus (Ewens, 1972), or b) from a transformation of estimates of Nei's gene diversity (h)

$$\theta = h/(1 - h),$$

or c) using estimates of additive genetic variance of neutral characters (Clayton and Robertson, 1955), or total phenotypic variance if the environmental variance is assumed to be equal, or d) most crudely as proportional to population size, since this is directly related to θ in the model.

Since the sampling strategy will often be implemented using esitmates of gene diversity (h), it was called the H-strategy (Schoen and Brown, 1993). We now turn to the question of how well does this simple strategy perform, in comparison with the strategy that would maximize allelic richness in a number of situations that depart from the assumptions of the neutral model.

Heterotic model

Suppose that the distribution of allele frequencies at a locus is more "even" than would occur under the neutral model. In such allelic profiles, the variance of allele frequency among the alleles at a locus is less than that for alleles at a locus in neutral equilibrium. The most extreme case of such a distribution would be when all the alleles at a locus are equally frequent. This case represents a family of examples indexed by a single parameter k_i, the number of alleles at the A locus in the ith population. Thus, population 1 has k_1 alleles which all have frequency $(k_1)^{-1}$. The probability of sampling a particular allele at least once in a sample of n_1 is

$$1 - [(k_1 - 1)/k_1]^{n_1}.$$

From this follows the expected number of alleles in a composite sample of n_1 from population 1 and n_2 from population 2 as

$$K = k_1 - k_1[1 - 1/k_1]^{n_1} + k_2 - k_2[1 - 1/k_2]^{n_2}. \qquad (4)$$

The algebraic solution for the optimal proportion when total sample size is fixed is more complex than that for the neutral model. For our purposes a numerical solution will suffice. Tab. 1 gives the optimum proportion for the first population when it has two alleles and the second has more than two alleles. The value is compared with that obtained from using the H-strategy, where the value of θ_i for the population, computed from the gene diversity h_i ($=[1 - 1/k_i]$) is

$$\theta_i = k_i - 1.$$

These values indicate that in situations of complete evenness, the H-strategy gives a very good approximation to the actual optimum proportion that should be sampled from the less polymorphic population. In general, the H-strategy overvalues the more polymorphic (or diverse) population. The major departure from optimum occurs when the two populations differ greatly in polymorphism ($k_1 \ll k_2$) and when the total sample size (S) is small.

How well does the H-strategy perform for populations with equal allelic richness but different evenness and therefore varying gene diver-

Table 1. Optimum proportion of a total sample (S) that should be taken from a diallelic population 1 when the number of equally frequent alleles in a second population 2 is k_2, in comparison with the proportion computed from formula (3)

Allelic diversity		Optimal sample (n_1/S)			
k_1	k_2	S = 5	S = 50	S = 500	H-strategy (3)
2	3	0.39	0.37	0.37	0.33
2	4	0.33	0.30	0.29	0.25
2	5	0.29	0.25	0.24	0.20
2	10	0.20	0.14	0.13	0.10
2	20	0.15	0.08	0.07	0.05
2	100	0.11	0.02	0.02	0.01

Table 2. Effect of varying evenness (or gene diversity) of allelic frequencies in population 2 on the optimal sample proportion from population 1, when both populations have equal allelic richness (diallelic, triallelic)

Allele frequencies (p_2, q_2, r_2)	Optimal sample (n_1/S)			
	$S = 5$	$S = 50$	$S = 500$	H-strategy
Diallelic case – $p_1 = 0.75$, $q_1 = 0.25$, $k_1 = k_2 = 2$				
0.5, 0.5, 0	0.45	0.67	0.70	0.38
0.6, 0.4, 0	0.46	0.62	0.64	0.39
0.7, 0.3, 0	0.48	0.54	0.55	0.45
0.8, 0.2, 0	0.53	0.45	0.44	0.56
0.9, 0.1, 0	0.65	0.32	0.27	0.73
Triallelic case – $p_1 = 0.6$, $q_1 = 0.2$, $r_1 = 0.2$, $k_1 = k_2 = 3$				
0.4, 0.3, 0.3	0.45	0.59	0.61	0.40
0.5, 0.25, 0.25	0.47	0.55	0.56	0.43
0.7, 0.15, 0.15	0.56	0.44	0.43	0.59
0.8, 0.1, 0.1	0.64	0.37	0.33	0.71
0.9, 0.05, 0.05	0.75	0.30	0.20	0.85

sity? In the diallelic case suppose that the first population has allelic frequencies $p_1 = 0.75$, and $q_1 = 0.25$. Tab. 2 shows the optimal sampling weight for this population when the allele frequencies in the second population (p_2, q_2) range widely from those typical of heterotic models to values typical of deleterious mutation/selection balance. Again, the comparisons of optimal fractions with those from the H-strategy illustrate that the H-strategy overweights the more diverse population of the pair, and this causes discrepancies from the true optimal weighting. For example, the optimal proportion for population 1 is higher than the H-strategy would suggest when $p_2 = q_2 = 0.5$ with $h_2 > h_1$, i.e., population 2 is overrated. The same trend is apparent in the triallelic case (Tab. 2).

Deleterious mutant/balance models

At the other extreme to heterotic models are allele frequency profiles that are more "uneven" than those obeying the neutral allele model. In such cases the allele frequencies are overdispersed, often with one highly frequent allele and several very rare ones. This type of distribution can be modeled with a two-parameter family of examples.

Suppose population 1 has one "adapted" allele with frequency p_1 and k_1 "deleterious" alleles, all of which have frequency $(1 - p_1)/k_1$ or q_1/k_1. Likewise, population 2 has one adapted allele with frequency p_2, and k_2 "deleterious" alleles all with frequency $(1 - p_2)/k_2$ or q_2/k_2. If n_1 are sampled from population 1 and n_2 are sampled from population 2, the expected allelic recovery is

$$K = (k_1 + 1) - (q_1)^{n_1} - k_1[1 - q_1/k_1]^{n_1} + (k_2 + 1)$$
$$- (q_2)^{n_2} - k_2[1 - q_2/k_2]^{n_2}. \tag{5}$$

Assuming again that a fixed total sample of size S ($=n_1 + n_2$) is made, the questions are: what is the optimum proportion (n_1/S) from population 1 that would maximize K and how does this compare with the proportion derived from the H-strategy? Tab. 3 gives some answers to these questions arranged in three categories. It is immediately clear that in all three categories, the H-strategy is a good guide to the actual

Table 3. Examples of deleterious mutation-selection allelic profiles (with p_1 and p_2 the frequency of the favored allele in populations 1 and 2, and k_1 and k_2 the number of equally frequent rare deleterious alleles). The optimum proportion to sample from population 1 (n_1/S) is compared with that derived from the H-strategy

				Optimal n_1/S		H-strategy
k_1	k_2	p_1	p_2	S = 5	S = 50	(n_1/S)
(a) Equal allelic richness, unequal gene diversities						
1	1	0.8	0.9	0.63	0.36	0.68
		0.9	0.95	0.69	0.42	0.68
		0.8	0.95	0.73	0.29	0.82
2	2	0.8	0.9	0.68	0.42	0.69
		0.9	0.95	0.71	0.51	0.68
		0.8	0.95	0.77	0.41	0.83
10	10	0.8	0.9	0.72	0.79	0.70
		0.9	0.95	0.72	0.96	0.68
		0.8	0.95	0.79	0.97	0.84
(b) Allelic richness negatively related or unrelated to gene diversity						
1	2	0.8	0.8	0.45	0.33	0.48
		0.8	0.9	0.62	0.24	0.67
		0.9	0.95	0.68	0.31	0.67
		0.8	0.95	0.73	0.22	0.82
2	10	0.8	0.8	0.45	0.17	0.48
		0.8	0.9	0.68	0.22	0.69
		0.9	0.95	0.71	0.35	0.68
		0.8	0.95	0.77	0.31	0.83
1	10	0.8	0.8	0.40	0.09	0.46
		0.8	0.9	0.61	0.11	0.67
		0.9	0.95	0.68	0.18	0.67
		0.8	0.95	0.73	0.15	0.81
(c) Allelic richness positively related to gene diversity						
1	2	0.9	0.8	0.31	0.43	0.30
		0.95	0.9	0.29	0.35	0.32
		0.95	0.8	0.23	0.49	0.27
2	10	0.9	0.8	0.27	0.09	0.29
		0.95	0.9	0.28	0.04	0.31
		0.95	0.8	0.20	0.03	0.16
1	10	0.9	0.8	0.27	0.06	0.28
		0.95	0.9	0.28	0.04	0.30
		0.95	0.8	0.20	0.03	0.16

optimum proportion when the sample size is small (i.e., S = 5). In larger samples (S = 50), however, the H-strategy can be substantially misleading. Let us consider some of these examples.

Tab. 3a shows examples where the populations have the same allelic richness ($k_1 = k_2$), but where the gene diversity is greater in population 1 than in population 2 because $p_1 < p_2$. When there are few alleles to sample (k small) the H-strategy is misleading because it gives too much weight to the more diverse population. In such cases a limited sample from the more diverse population quickly captures the only alleles available and it is more efficient to devote more of the total effort toward getting the rarer variants in the second and less diverse population.

The second group of examples (Tab. 3b) has unequal allelic richness which is negatively correlated with or unrelated to gene diversity. Thus, a high diversity index (due mainly to lower p_1) is a poor guide to allelic richness. Therefore, the H-strategy is even more misleading that it is in the first set (Tab. 3a). Tab. 4 shows the effect of the H-strategy on the expected number of alleles recovered (S = 50). The values of most concern are obtained when gene diversity is negatively related to allelic richness, such as in the last case where the H-strategy is about 70% as efficient as the optimal strategy.

The final set (Tab. 3c) is examples where the differences in allelic richness are positively related to those in gene diversity (low k_1 goes with high p_1). As expected, the H-strategy performs much better in this set.

In general, the relationship between allelic richness and gene diversity is more likely to be positive. At the species level, Hamrick et al. (1979) found a correlation coefficient of 0.80 between the species mean gene diversity and mean allelic richness for a wide range of plant species. However, the general covariance pattern between populations within plant species has yet to be analyzed. For populations of wild barley, *Hordeum spontaneum*, the value was 0.86 (Nevo et al., 1979). Fig. 1 illustrates three examples of this relationship for allozyme data in the species, *Hordeum spontaneum*, *Zea mays* (Doebley et al., 1984) and *Sorghum bicolor* (Morden et al., 1990).

Table 4. Recovery of alleles (out of $k_1 + k_2 + 2$ alleles) for the H-strategy compared to that for the optimal strategy for six two-population cases of deleterious alleles from Table 3b when gene diversity is unrelated or negatively related to allelic richness (S = 50)

k_1	k_2	p_1	p_2	Expected number of alleles	
				Optimal	H-strategy
1	2	0.8	0.8	4.92	4.87
1	2	0.8	0.9	4.65	4.14
2	10	0.8	0.8	8.86	7.93
2	10	0.8	0.9	6.62	5.39
1	10	0.8	0.8	8.65	7.36
1	10	0.8	0.9	6.31	4.53

364

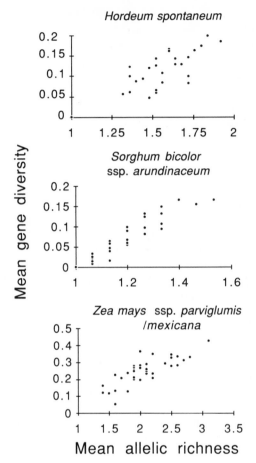

Fig. 1. The relationship between mean allelic richness and mean gene diversity for allozyme polymorphism in populations of wild relatives of barley, sorghum, and maize.

Overall, in cases of deleterious mutation/selection balance, the H-strategy overvalues loci with few alleles and undervalues loci with very rare alleles.

Variation in divergence among populations

A basic assumption for the H-strategy is that all populations are completely isolated and therefore that the alleles at a locus are unique to each population. However, partial isolation is a more prevalent and relevant condition in natural populations. Formerly (Brown, 1989b; Brown and Briggs, 1991), the principles derived from a model of complete isolation were extended to more general situations with the argument that loci could be conceived in two classes. The first class would contain loci that were fully differentiated among the populations that share none of the alleles at such loci. The H-strategy applies to this class. The second class is composed of completely undifferentiated loci with alleles shared among populations or groups. Alleles that are shared are much less sensitive to allocation strategies – any of the source populations can be used in any convenient proportion. Hence, when the two classes are combined, the H-strategy for differentiated loci dominates the sample allocation. In analogous fashion, the so-called "private" alleles that are present in only one population, are the basis of estimating migration between populations (Slatkin, 1985).

This argument however assumes, that in the set of partially differentiated populations, all are equally divergent from one another. This would plainly not be the case if the set comprises one group of several populations with high levels of migration between them, and an isolated population that is remote from the group. Another factor that would lead to unequal levels of divergence among pairs of populations is variation in the time since divergence from their common ancestor.

One way to study the effect on differing levels of divergence among populations is with a simple model of several $(r + 1)$ populations in which the first r populations are genetically "redundant" and have an identical allelic profile at a locus, whereas the last $(r + 1)^{th}$ population contains alleles that are not present in the other populations. The variation is assumed to be in equilibrium under the neutral infinite allele model.

Suppose that the value of theta in the first r redundant populations is θ_1 and that in the unique population is θ_2. Clearly in this situation there is extreme inequality in divergence among the populations. First, suppose that the redundancy is known to the collector. The optimum deployment of the total S sample would follow from the H-strategy. The number to sample from the unique population is

$$n_{r+1} = S\theta_2/(\theta_1 + \theta_2),$$

whereas the number to sample from a bulk of all the r redundant populations is

$$n_1 = S\theta_1/(\theta_1 + \theta_2).$$

If these sizes are used, the expected number of alleles is the optimal for a total sample of size S and is

$$K_0 \approx (\theta_1 + \theta_2) \log_e[1 + S/(\theta_1 + \theta_2)]. \tag{6}$$

However, if the redundancy (or variation in divergence) is not known in advance, sampling of $r + 1$ distinct populations would proceed in proportion to their theta values such that

$$n_{r+1} = S\theta_2/(r\theta_1 + \theta_2)$$

Table 5. Effect of unequal divergence among populations on allele recovery. The optimum proportion to sample from the unique population (n_{r+1}/S), and the expected number of alleles recovered, compared with actual proportion and recovery when the H-strategy is used and redundancy ignored

					Redundant sampling					
		Optimal sampling			$r = 2$			$r = 10$		
			K_0 (Eq. (6))			K (Eq. (7))			K (Eq. (7))	
θ_1	θ_2	n_{r+1}/S	$S = 5$	$S = 50$	n_{r+1}/S	$S = 5$	$S = 50$	n_{r+1}/S	$S = 5$	$S = 50$
1	1	0.50	2.5	6.5	0.33	2.5	6.4	0.09	2.1	5.6
1	2	0.67	2.9	8.6	0.50	2.9	8.5	0.17	2.3	7.0
1	10	0.91	4.1	18.8	0.83	4.1	18.7	0.50	3.5	15.8
2	1	0.33	2.9	8.6	0.20	2.9	8.5	0.05	2.6	7.6
2	2	0.50	3.2	10.4	0.33	3.2	10.2	0.09	2.8	8.7
2	10	0.83	4.2	19.7	0.71	4.1	19.4	0.33	3.5	15.6
10	1	0.09	4.1	18.8	0.05	4.1	18.7	0.01	4.1	18.2
10	2	0.17	4.2	19.7	0.09	4.2	19.5	0.02	4.1	18.6
10	10	0.50	4.5	25.1	0.33	4.4	24.5	0.09	4.2	20.9

and

$$n_i = S\theta_1/(r\theta_1 + \theta_2) \quad i = 1, \ldots, r.$$

The expected number of alleles from such a sample is

$$K \approx \theta_1 \log_e[1 + Sr/(r\theta_1 + \theta_2)] + \theta_2 \log_e[1 + S/(r\theta_1 + \theta_2)]. \quad (7)$$

Tab. 5 has some illustrative values of these formulae. It shows the optimal proportion of the sample that should be devoted to the genetically unique population (n_{r+1}/S), in comparison with the actual proportion when redundancy is overlooked. The latter include two levels of redundancy, low ($r = 2$) and high ($r = 10$). When the level of redundancy is low, and thus the variation in divergence is low, there is little effect of variation in divergence on allele recovery. In more extreme cases ($r = 10$), a marked effect emerges, particularly when the unique population ($k + 1$) is more polymorphic ($\theta_2 > \theta_1$).

Discussion

This study has focused on the question of how to maximize the number of distinct alleles retained in a stratified sample taken from a large structured collection. We have assumed that the collection can be divided readily into groups that are in some way genetically distinct. The division into groups or strata is usually based on geographic origin of the accessions in the collection. It might well include the clustering of like populations into groups based on marker genetic data.

From neutral theory, it then follows that the optimal sampling strategy is to take from each group a random sample, the size of which is in proportion to the level of within-group polymorphism for that group. It is necessary to add that every group is represented in the core by at least one gamete. This is because the distinctive assumption guarantees that the first gamete sampled from each new group contributes a new unique allele at some locus.

We then went on to test whether this so-called H-strategy can be a reliable guide to germplasm sampling when some of its assumptions do not hold. If the lessons from the two population (two group) cases are general, then a number of factors can render the strategy misleading. These situations are as follows:

(1) a strictly finite and limited number of alleles in any population, rather than an infinite number as the model assumes;
(2) alleles maintained by balancing selection leading to inflated estimates of polymorphism relative to the allelic richness;
(3) a negative relation between gene diversity and allelic richness, particularly for rare deleterious alleles;
(4) differences in divergence between populations, particularly when the more divergent groups are not the more polymorphic ones.

When any of these situations occurs a discrepancy arises between the truly optimal strategy for that situation and the sampling pattern derived from the H-strategy. Except for the case of equal evenness (Tab. 1), the discrepancy usually increases with increasing total sample size.

It would therefore appear that several factors could operate against the general reliability of the H-strategy. Yet, when we applied this strategy to allozyme data for collections of several wild species related to crops (Schoen and Brown, 1993), the strategy performed well. It improved allele recovery at target loci over those that used no genetic data. Two possible reasons for this result are (1) that the bias in group weighting arising from the breakdown of an assumption for one locus is balanced by an opposite bias from the deviation in some other way at another locus, i.e., that the H-strategy is an "average" strategy, and (2) that in general, the relationship between allelic richness and gene diversity tends on the whole to be positive rather than negative (as Fig. 1 illustrates).

Variation in divergence is likely to be the most important factor that could lead to serious biases. We formulated an alternative approach to using genetic data from marker loci, the so-called M or maximization strategy (Schoen and Brown, 1993). Unlike the H-strategy which uses the genetic data as a statistical sample to estimate the parameters of an underlying hypothetical model of breeding history, the M-strategy makes a deterministic use of the marker data. Through linear programming, it searches all possible samples for the actual combination of samples that maximizes the number of observed alleles at the marker loci. The only stochastic element in the M-strategy arises from the sampling of the marker genes to represent the whole genome. Thus, the M-strategy, which guarantees the maximizing of allelic richness for the marker loci, takes direct account of variation in divergence at those loci. If the pattern of differentiation and diversity among groups for the target loci parallels that of the marker loci, the M strategy should lead to high target allele retention in the core collection.

Both strategies assume all alleles are equivalent in their value for conservation. More complex value systems are conceivable. For example, alleles that are highly specialized, rare and localized in occurrence could be discounted in favor of more frequent or broadly adapted alleles (as discussed by Marshall and Brown, 1975; or Brown, 1989b). In part, the H- or M-strategies rely on the stratification of the total collection to achieve a representative sample of a species across all habitats, or separate areas. Therefore, a genetically rich area does not entirely displace samples from other less polymorphic areas.

Species richness, distinctiveness and community conservation

It is feasible to develop an analogue of our H-strategy at other levels of choice in conservation. Thus, where the number of species on isolated islands or in separate habitats is some function of island area (frequently exponential; MacArthur and Wilson, 1967), the question is – how much of each island, or type of habitat is needed to maximize the number of species conserved, given a fixed total acreage of reservation? Such a procedure again assumes that the species are of equal value, but it achieves some ecological coverage from the clustering procedure and the guarantee of one sample from each cluster or some portion of each island.

The use of species richness as a major criterion for evaluation has been criticized recently by several authors. Vane-Wright et al. (1991) introduced a weighting index for each species based on the branching order, or topology of the cladogram representing the hierarchic relationships among the species. The index of "taxic" diversity thus assigned preference based on cladistic distinctiveness. It is inversely proportional to the number of hierarchical groups to which a species belongs in the cladogram, i.e., the number of nodes crossed when tracing from that species back to the root of the cladogram.

Faith (1992) described an alternative approach that aims to measure phylogenetic diversity of a subset of taxa from a cladogram. It takes account of branch lengths as well as the topology of the minimum spanning cladogram and attempts to maximize the underlying "feature" diversity, where "features" are cladistically informative characters. It results in a bias in favor of the extremely contrasted taxa. Crozier's (1992) approach is similar, but is described in genetic terms; it takes account of genetic divergence along the branches of a cladogram. The species with the highest priority for conservation is that with the highest overall probability of having unique character states.

However, it is not clear whether such deterministic procedures that bias towards taxic, phylogenetic or genetic uniqueness result in an optimal sample for conservation. Should the Welwitschia of this world win overwhelming weight? Aside from the other practical factors involved in conservation choices, there are the random factors and future evolutionary pressures that will impinge on conserved communities. For these, some concept of richness, either allelic richness, or species richness is appropriate. Richness is understood simply as the number of different types and is not weighted by past evolutionary divergence according to hypotheses of ancestral state.

The question therefore is: should one maximize the diversity of the products of evolution, or attempt to conserve the potential for future diversification, the process of evolutionary change (Frankel and Soulé, 1981)? Our conclusion is that both criteria (distinctiveness and richness) must be factored into conservation schemes in a way that sustains the evolutionary process. A core sampling strategy that entails stratification into biologically sensible groups and sampling within them in proportion to diversity, therefore includes both divergence and richness in sampling. Genetic distances may well help the choice of what taxa to keep; genetic polymorphism will indicate how much to keep.

Acknowledgement
We thank Dr. J. Doebley for genetic data from Sorghum and Zea.

References

Brown, A. H. D. (1989a) The case for core collections. *In:* Brown, A. H. D., Frankel, O. H., Marshall, D. R. and Williams, J. T. (eds), *The use of plant genetic resources.* Cambridge University Press, Cambridge, pp. 136–156.

Brown, A. H. D. (1989b) Core collections: A practical approach to genetic resources management. *Genome* 31: 818–824.

Brown, A. H. D. (1992) Human impact on plant gene pools. *Oikos* 63: 109–118.

Brown, A. H. D. (1993) The core collection at the crossroads. *In:* Hodgkin, T., Brown, A. H. D., van Hintum, T. J. L. and Morales, E. A. V. (eds), *Core collections of plant genetic resources.* John Wiley and Sons Ltd., Chichester.

Brown, A. H. D. and Briggs, J. D. (1991) Sampling strategies for genetic variation in *ex situ* collections of endangered plant species. *In:* Falk, D. A. and Holsinger K. E. (eds), *Genetics and conservation of rare plants.* Oxford University Press, Oxford, pp. 99–119.

Clayton G. and Robertson, A. (1955) Mutation and quantitative variation. *Amer. Nat.* 89: 151–158.

Crozier, R. H. (1992) Genetic diversity and the agony of choice. *Biol. Conserv.* 61: 11–15.

Doebley, J. F., Goodman, M. M. and Stuber, C. W. (1984) Isoenzymatic variation in *Zea* (Gramineae). *Syst. Bot.* 9: 203–218.

Ewens, W. J. (1972) The sampling theory of selectively neutral alleles. *Theor. Pop. Biol.* 3: 87–112.

Faith, D. P. (1992) Conservation evaluation and phylogenetic diversity. *Biol. Conserv.* 61: 1–10.

Frankel, O. H. (1984) Genetic perspectives of germplasm conservation. *In:* Arber, W., Llimensee, K., Peacock, W. J. and Starlinger, P. (eds), *Genetic manipulation: Impact on man and society*. Cambridge University Press, Cambridge, pp. 161–170.

Frankel, O. H. and Brown, A. H. D. (1984) Plant genetic resources today: a critical appraisal. *In:* Holden, J. H. W. and Williams, J. T. (eds), *Crop genetic resources: Conservation and evaluation*. George Allen & Unwin Ltd., London, pp. 249–257.

Frankel, O. H. and Soulé, M. E. (1981) *Conservation and Evolution*. Cambridge University Press, Cambridge.

Hamrick, J. L., Linhart, Y. B. and Mitton, J. B. (1979) Relationships between life history characteristics and electrophoretically detectable genetic variation in plants. *Annu. Rev. Ecol. Syst.* 10: 173–200.

Holden, J. H. W. (1984) The second ten years. *In:* Holden, J. H. W. and Williams, J. T. (eds), *Crop genetic resources: Conservation and evaluation*. George Allen & Unwin Ltd., London, pp. 277–285.

Kimura, M. and Crow, J. F. (1964) The number of alleles that can be maintained in a finite population. *Genetics* 49: 725–738.

MacArthur, R. H. and Wilson, E. O. (1967) *The theory of island biogeography*. Princeton University Press, Princeton, N.J.

Marshall, D. R. and Brown, A. H. D. (1975) Optimum sampling strategies in genetic conservation. *In:* Frankel, O. H. and Hawkes, J. G. (eds), *Crop genetic resources for today and tomorrow*. Cambridge University Press, Cambridge, pp. 53–80.

Morden, C. W., Doebley, J. R. and Schertz, K. F. (1990) Allozyme variation among the spontaneous species of *Sorghum* section Sorghum (Poaceae). *Theor. Appl. Genet.* 80: 296–304.

Nevo, E., Zohary, D., Brown, A. H. D. and Haber, M. (1979) Genetic diversity and environmental asociations of wild barley, *Hordeum spontaneum*, in Israel. *Evolution* 33: 815–833.

Plucknett, D. L., Smith, N. J. H., Williams, J. T. and Murthi Anishetty, N. (1987) *Gene banks and the world's food*. Princeton University Press, Princeton, N.J.

Schoen, D. J. and Brown, A. H. D. (1993) Conservation of allelic richness in wild crop relatives is aided by assessment of genetic markers. *Proc. Natl. Acad. Sci. USA* 90: 10623–10627.

Slatkin, M. (1985) Rare alleles as indicators of gene flow. *Evolution* 39: 53–65.

Vane-Wright, R. I., Humphries, C. J. and Williams, P. H. (1991) What to protect? – Systematics and the agony of choice. *Biol. Conserv.* 55: 235–254.

Wilson, E. O. (1988) *Biodiversity*. National Academic Press, Washington DC.

Conservation Genetics
ed. by V. Loeschcke, J. Tomiuk & S. K. Jain

Conservation genetics and the role of botanical gardens

H. Hurka

Spezielle Botanik, University of Osnabrück, D-49069 Osnabrück, Germany

Summary. Botanical gardens once played a key role in plant taxonomic research. For the majority of gardens, this is no longer the case. In recent years, botanical gardens have turned to conservation as one of their major goals. A "Botanical Gardens Conservation Strategy" was launched by the International Union for Conservation of Nature and Natural Resources (IUCN) Botanical Gardens Conservation Secretariat with emphasis given to wild plants of economic use. It is stated by advocates of this policy that botanical gardens are *ex situ* centers *par excellence.* However, the role that botanical gardens can play in conservation of plant species may be questioned. Doubts include the problem of *in situ vs. ex situ* collections in general, and the present status of collections held by botanical gardens in particular. Normally, genetic variation within species, the concern of conservation genetics, is totally underrepresented in botanical gardens, a considerable percentage of plants is mislabeled, and proper documentation is weak. It is nevertheless argued that botanical gardens can play an active role in conservation efforts. First, they can influence broad public opinion and create proper education programs. Second, they shoud establish *ex situ* collections for local wild plants and could propagate and provide material for reintroduction programs. To achieve these goals botanical gardens have to base their activities on a sound basis, and have to collaborate with research institutions and conservationists.

Introduction

Overall discussions on *in situ vs. ex situ* conservation are covered extensively in the literature. Conservation techniques in plants include seed banks, tissue cultures and cryopreservation, cloning by cuttings, bulbs, corms and tubers, and maintaining whole plants. They are well described (Bramwell et al., 1987; Bermejo et al., 1990; Brown and Briggs, 1991; Eberhart et al., 1991), whereby much emphasis is placed on seed banks (Hawkes, 1987; Blixt, 1992). The problems imposed by orthodox and recalcitrant seeds is well known, although research on germination physiology and long-term storage conditions for wild plants is badly needed, especially for tropical species. Their seeds retain their viability only for short periods. They do not build up a persistent seed bank in nature, but instead seem to persist as seedlings for many years. The theory of sampling strategies is well developed and covers aspects of allelic richness, heterozygosity, genetic richness, minimum sample sizes, ecotypic variation, etc. (Marshall and Brown, 1975; Frankel and Soulé, 1981; Brown and Briggs, 1991). The problems of

372

using molecular markers in genetic conservation are also addressed (Hamrick et al., 1991), and the theory of core collections is important in handling large *ex situ* collections (Brown and Schoen, this volume). Sampling practicalities and guidelines have been developed which are based on the sampling theory (e.g., Center for Plant Conservation in Falk and Holsinger, 1991), although the problem is not settled yet.

The main role of *ex situ* collections in biological conservation is to supply propagating material for maintaining critical natural populations, or for reintroduction programs. Yet, it appears that the experience and evaluation of any success in this role is still limited. Much research and accumulation of practical experience is necessary.

All these facts are well known and will not be addressed in this chapter. Instead, facing the world's biodiversity crises, I will focus on the presumed and sometimes highly advocated role of botanical gardens in conservation efforts. As an acting director of a university botanical garden for more than 10 years, I feel experienced enough to have a fairly realistic view.

Biodiversity in crises

Statements that "biological diversity has reached a crisis state, and that human activities are destroying the natural world and its biota at an ever increasing rate" are meanwhile common place. We know that approximately 1.5 to 1.8 million plant and animal species have been named to date. The estimate of the total number of existing species is uncertain and varies between 5 and 30 million (Wilson, 1988). The current loss of species *per* year is difficult to assess, mainly because of three reasons: (a) The total number of existing species is not known; estimates vary by a factor of ten. (b) Species extinction rates depend on the sizes of the biogeographic areas, on the fragmentation of the areas, and on the sizes of the island fragments. These parameters vary considerably from country to country. (c) The total distribution patterns of the species already described are often not well known, so that we do not know which species have been wiped out by logging of parts of the tropical rain forests, to say nothing of the fate of undescribed species. Nevertheless, a scenario of the species extinction rates may be developed (Wilson, 1988), in which about 18 000 species *per* year will be lost by the destruction of tropical rain forests. Given 10 million species on the whole earth, a crude estimate of 1 in 1000 gives a loss of 10 000 species *per* year. Compared to the extinction rates as judged from changes of the fossil record, the mass extinction caused by humans exceeds the pre-hominid extinction by a magnitude of 1 to 10 000. Some scientists predict that if the present trend continues, some 25% of the world's species will be lost in the next 25 to 50 years. Deforestation, desertification, and destruction of wetlands are widespread phenomena; and even in economically important species, genetic diversity is being lost (McNeely, 1992). By combining data on the rate of destruction of tropical rain forests with the theory of island biogeography, Raven has estimated that no less than 60 000 plants, nearly 1 in 4 of the world total, could become extinct by the middle of the next century.

The case of tropical rain forests

Human population growth and poverty are important factors when conservation and sustainable utilization of living resources are considered, particularly in tropical countries. Both lead to pressure on the natural resources and further degradation, and prevent sound development and sustainable utilization.

By 1980, 40% of tropical rain forests had already disappeared (FAO/ UNEP, 1982). The annual deforestation was estimated to be about 75 000 km². An additional loss of 38 000 km² of other forest types, e.g., drier forests, monsoon forests, etc. has to be added. Almost all rain forests which are not protected as reserves or national parks will be seriously altered or destroyed in 30 years, and all rain forests of the world will be negatively affected in about 80 years. Presumably, only few areas in Amazonia and Central Africa will remain rather unaffected because of the remoteness of some parts of these areas.

Why shall we conserve biodiversity?

There is hardly any disagreement about the necessity of conserving the world's biodiversity. Why do we think that biodiversity is of much importance?

The reasons given span a wide field. On the one hand, we see pure economic arguments: a loss of the world's most fundamental capital stock – its genes, species, habitats, and ecosystems (McNeely, 1992). The loss of the living richness of the planet has serious implications for our common future as the World Commission on Environment and Development recognized in its 1987 report (WCED, 1987). Biological resources, when managed appropriately are renewable by the so-called sustainable development. Some colleagues argue that "the only certain way to save tropical biodiversity is to use it sustainably" (Janzen, 1992). The ecosystems themselves, apart from their species richness, serve important environmental aspects. These include maintaining hydrological cycles, regulating climates, contribution to soil formation, storing and cycling nutrients, absorbing and converting pollutants. They also provide sites for recreation, and serve the tourist industry (McNeely et al., 1990). Many scientific activities which concern the value of biodiversity have in common and now generally acknowledge the cost and benefit of biodiversity. Biodiversity is required to satisfy the vital needs of mankind. Nature is useful.

On the other hand, there are ethic concerns. Frankel was probably one of the first who introduced this aspect in the field of conservation genetics. He argued in 1974 that genetics has social responsibilities. One such is "to help in establishing an evolutionary ethic, as part of our social ethics, which will make acceptable and indeed inevitable for civilized man to regard the continuing existence of other species as an integral part of his own existence." However, to argue in the scientific and political world that biodiversity has a value in itself, is still not very popular and is considered by only a few advocates (e.g., Ehrenfeld, 1988; Naess, 1989). This idea, for instance, was not incorporated in the so-called Brundtland Report "Our Common Future" (WCED, 1987), but it seems that it is a major driving force in green movements. This idea is also part of our cultural inheritance.

How to conserve biodiversity?

Politicians and scientists now agree that a priority list of global centers for the preservation of biological diversity is necessary. Questions such as the size of the conserved area, whether the focal point of conservation biology should be the ecosystem or the population, or whether keystone species may be used to monitor whole ecosystems, and whether endangered species should be preserved, are now being addressed by biological research. In a recent article, Soulé and Mills (1992) point to the current controversy between species conservation which focuses on endangered and keystone species, and a habitat approach focusing on ecosystem processes and patch dynamics. A novel index to establish priorities for conservation areas was proposed by Vane-Wright et al. (1991) based on the information content of cladistic classifications and giving a measure for taxonomic distinctness. Many discussions appear academic, with no clear effect on stopping the destruction. It is therefore dependent on public opinion and public pressure to promote actions in favor of nature conservation. The major problems of conserving the world's biodiversity lie not in inadequate biological knowledge, but rather in social, economic, and political issues.

The simplest solution for conserving biodiversity is to save as many habitats as possible. It may be too late before a scientific recommendation can be given for an effective conservation program, but this does not deny the importance of conservation genetics and conservation biology. Their fields, however, are biotope and species management rather than conserving the primary biodiversity.

Ex situ collections and botanical gardens

Without any doubt, the world's primary biodiversity as a whole can only be conserved *in situ*. This, however, does not at all discredit *ex situ* collections, which intend to conserve genetic resources and – hopefully – a limited number of selected species. The reasons why a species should be conserved *ex situ* are subjective despite sophisticated arguments. The only arguments which are hardly debated are those which relate to the "usefulness" of the species. Consequently, the World Conservation Strategy (IUCN/UNEP/WWF, 1980) defines conservation as "the management of human use of the biosphere so that it may yield the greatest sustainable benefit to present generations while maintaining its potential to meet the needs and aspirations of future generations." Three main objectives of living resource conservation are identified by the World Conservation Strategy which have to be met to ensure that the harvest of living natural resources can be sustained. These general objectives of the World Conservation Strategy are (Hamann, 1987): (a) to maintain essential ecological processes and life-support systems, (b) to preserve genetic diversity, and (c) to ensure that the utilization of species and ecosystems is sustainable. The overall objective is to promote, to support, and to formulate efforts for conserving plant genetic resources and plants of actual or potential economic value.

Botanical gardens are considered to be "a vital link" within plant conservation programs. In 1984, the IUCN/WWF Plants Conservation Programme was launched. The aim of the program was:

(1) to draw botanical gardens closer into the conservation network;
(2) to co-ordinate them as the *ex situ* network for threatened plants;
(3) to promote the use of botanical gardens to educate the public about plants and their conservation.

In 1989, the IUCN Botanical Gardens Conservation Secretariat (BGCS) published "the Botanical Gardens Conservation Strategy" (WWF/IUCN/BGCS, 1989). The BGCS was charged with the task to "monitor and promote research into the conservation of useful plants, including germplasm of medicinal species, wild crop relatives, primitive land races of crops, and so-called minor crops which are not covered by other germplasm or organisations." The emphasis given to wild plants of economic use is new compared to the 1984 Plant Conservation

Program. The main emphasis in 1984 was the *ex situ* preservation of threatened plants (see also Simmons et al., 1976), but no forcible argument was given as to why threatened plants *per se* should be the target of conservation. How can one be enthusiastic about the role of botanical gardens in conserving germplasm of medicinal plants, wild crop relatives, primitive land races of crops, and so-called minor crops when they cannot even encompass within their declared role the local conservation of threatened species?

Preservation of the endangered native flora of the USA, is however, the aim of the Center for Plant Conservation, located at the Missouri Botanical Garden. The Center wants to develop an information system and databases on the conservation status and the biology of endangered species and the conservation technology applied. Falk and Holsinger (1991) proposed that "both the population dynamics and the genetics of rare species are key areas of research in biological conservation." A network was established between the National Collection of Endangered Plants of the United States – a living collection maintained at 21 botanical gardens – and a seed bank maintained within the National Germplasm System of the US Department of Agriculture. The emphasis is the integration of *in situ* and *ex situ* methods.

Such an organized network to save endangered species on a broad basis is an exception. There are only few initiatives by some gardens and other authorities. These initiatives are – if at all – only loosely coordinated, and they often miss sound scientific bases, despite all the encouragement by the Botanical Gardens Conservation Secretariat.

Let us consider some facts about botanical gardens and their collections. Heywood (1992) states that "botanical gardens are *ex situ* centers par excellence and hold enormously diverse, if numerically inadequate, collections. *Ex situ* is appropriate for the conservation of germplasm of wild species in determined circumstances ...." There are about 1500 botanical gardens in the world. The number of species cultivated in botanical gardens may exceed 40 000 (Heywood, 1992). It is the "largest assemblage of biodiversity outside nature." About half of the world's botanical gardens regularly offer seed lists (Index Seminum) that list those taxa for which seeds are available free of charge. An increasing number of botanical gardens now include seeds of wild origin. Indeed, some gardens have adopted the policy of offering only seeds collected in the wild.

Many botanical gardens house special collections of high scientific value. Often overlooked, but extremely important kinds of collections are those build up by experimental taxonomists or cytologists in universities or other laboratories associated with botanical gardens. These collections contain a wide range of wild material of the genera and species concerned. They are often not included in the botanical gardens' database of accession, despite being held in the gardens or associated institutions. Thus, they will be omitted from global surveys like those of the Botanical Gardens Conservation Secretariat. As Raven (1981) correctly stated: "Unfortunately, such collections are often dismantled or simply deteriorate after the specialists who built them up are no longer active at the respective institutions. Although they are often of very great value internationally, they may, if they are not actively utilized, come to be viewed as a drain upon limited resources of the institution where they are housed. Even when financial considerations are not limiting, it is difficult to provide for such collections the meticulous and sustained care that is essential for their survival without the attention of a specialist who is deeply concerned for them." In essence, this is not only true for those special collections, it describes the status of collections in many botanical gardens in general.

As institutions, botanical gardens are as diverse as the collections they hold. Once, botanical gardens played a key role in plant taxonomic research and in plant introductions from one continent to the other. Those days are mostly over, at least for the majority of the gardens. Today, many botanical gardens face an identity crisis. Transition from a former botanical garden to a public park has become the reality for some gardens, for others it is not yet decided. It is difficult for many botanical gardens to convince research institutions and funding bodies that useful functions are being performed. In recent years an increasing number of botanical gardens is looking to conservation as one of their major goals as outlined above.

It is relatively easy to debate about programs to save the plant species of the world. However, to play an active and successful role in conservation is another story. Present collections of botanical gardens are for the most part meaningless for conservation purposes, despite Heywood's appraisal of "*ex situ* centers *par excellence*." The problems include the following:

(a) Most species are represented by only very few individuals (far from meeting the so-called "basic rule of conservation genetics" by Frankel and Soulé, 1981).
(b) The original locality of the accessions and their subsequent handling is often not known.
(c) The bulk of collections is propagated for generations within gardens and inbreeding and hybridization is common.
(d) The exchange system among botanical gardens has led to extreme genetic depauperation.
(e) A high percentage of plants is mislabeled, either from erroneous determination or inadvertent misplacement of the correct label.

In summary, genetic variation within species, the concern of conservation genetics, is normally totally underrepresented in botanical gardens.

In addition, the danger of unintentional selection for floral and other traits is always present. Apart from this, the role that botanical gardens can play in conservation of plant species may be questioned in general (Ashton, 1987, 1988).

What then is or may be the role of botanical gardens in conservation efforts?

Two major aspects in conservation efforts appear to me as appropriate for botanical gardens. One is influencing broad public opinion, the other is playing an active part in conservation.

I mentioned the driving forces of green movements on public opinion. Botanical gardens could effectively support public awareness of the threat and loss of plants we are confronted with by displaying biodiversity and by telling the public about plants, about their usefulness, and their inherent value. They should also display and explain the function of ecosystems. Proper education programs would be highly supportive.

In playing an active part in conservation programs, botanical gardens should concentrate on the local flora. They should establish proper *ex situ* collections for local wild plants, and in doing so, not only concentrate on threatened and endangered species. They could propagate and provide the material for reintroduction programs. To achieve these goals, they have to base their activities on the well established theories and practical guidelines for *ex situ* collections. They have to improve their documentation systems. They have to provide the necessary facilities as, for instance, proper seed storage facilities. They have to collaborate effectively with herbaria and other research institutions. Finally, they should seek the help of local conservationists.

For many gardens, it is not primarily financial shortage or lack of personnel, it is often indolence, lack of proper knowledge, and unawareness of the real problems. To quote Heywood (1992) once again: "Conservation ... requires more than words if it is to be fully implemented. Now is the time for action." We have heard those words for many years, and from many authorities. Will botanical gardens finally wake up to reality?

Acknowledgements
I thank Dr. A. H. D. Brown, CSIRO, Canberra, Australia, for many discussions on the subjects concerned, and for comments on the present manuscript.

References

Ashton, P. S. (1987) Biological considerations in *in situ* vs *ex situ* plant conservation. *In:* Bramwell, D., Hamann, O., Heywood, V. and Synge, H. (eds), *Botanic gardens and the world conservation strategy.* Acad. Press, London, pp. 117–130.

Ashton, P. S. (1988) Conservation of biological diversity in botanical gardens. *In:* Wilson, E. O. (ed.), *Biodiversity*. Nat. Acad. Press, Washington, DC., pp. 269–278.

Bermejo, J. E. H., Clemente, M. and Heywood, V. (1990) *Conservation techniques in botanic gardens*. Koeltz Scient. Books, Koenigstein, Germany.

Blixt, S. (1992) Gene banks for plant conservation. *In:* Sandlund, O. T., Hindar, K. and Brown, A. H. D. (eds), *Conservation of biodiversity for sustainable development*. Scandinavian Univ. Press, Oslo, pp. 204–213.

Bramwell, D., Hamann, O., Heywood, V. and Synge, H. (1987) *Botanic gardens and the world conservation strategy*. Acad. Press, London.

Brown, A. H. D. and Briggs, J. D. (1991) Sampling strategies for genetic variation in *ex situ* collections of endangered plant species. *In:* Falk, D. A. and Holsinger, K. E. (eds), *Genetics and conservation of rare plants*. Oxford Univ. Press, New York, pp. 99–119.

Eberhart, S. A., Roos, E. E. and Towill, L. E. (1991) Strategies for long-term management of germplasm collections. *In:* Falk, D. A. and Holsinger, K. E. (eds), *Genetics and conservation of rare plants*. Oxford Univ. Press, New York, pp. 135–148.

Ehrenfeld, D. (1988) Why put a value on biodiversity? *In:* Wilson, E. O. (ed.), *Biodiversity*. Nat. Acad. Press, Washington DC., pp. 212–216.

Falk, D. A. and Holsinger, K. E. (1991) *Genetics and conservation of rare plants*. Oxford Univ. Press, New York.

FAO and UNEP (1982) *Tropical forest resources*. Food and Agriculture Organization, and United Nations Environment Programme, Rome, Nairobi.

Frankel, O. H. (1974) Genetic conservation: our evolutionary responsibility. *Genetics* 78: 53–65.

Frankel, O. H. and Soulé, M. E. (1981) *Conservation and evolution*. Cambridge Univ. Press, Cambridge.

Hamann, O. (1987) The IUCN/WWF plants conservation programme in action. *In:* Bramwell, D., Hamann, O., Heywood, V. and Synge, H. (eds), *Botanic gardens and the world conservation strategy*. Acad. Press, London, pp. 31–43.

Hamrick, J. L., Godt, M. J. W., Murawski, D. A. and Loveless, M. D. (1991) Correlations between species traits and allozyme diversity: implications for conservation biology. *In:* Falk, D. A. and Holsinger, K. E. (eds), *Genetics and conservation of rare plants*. Oxford Univ. Press, New York, pp. 75–86.

Hawkes, J. G. (1987) A strategy for seed banking in botanic gardens. *In:* Bramwell, D., Hamann, O., Heywood, V. and Synge, H. (eds), *Botanic gardens and the world conservation strategy*. Acad. Press, London, pp. 131–149.

Heywood, V. H. (1992) Conservation of germplasm of wild plant species. *In:* Sandlund, O. T., Hindar, K. and Brown, A. H. D. (eds), *Conservation of biodiversity for sustainable development*. Scandinavian Univ. Press, Oslo, pp. 189–203.

IUCN and WWF (1984) The IUCN/WWF plants conservation programme 1984–1985. International Union for Conservation of Nature and Natural Resources, Gland, Switzerland.

IUCN, UNEP and WWF (1980) World conservation strategy. International Union for Conservation of Natural Resources, Gland, Switzerland.

Janzen, D. H. (1992) A south-north perspective on science in management, use, and economic development of biodiversity. *In:* Sandlund, O. T., Hindar, K. and Brown, A. H. D. (eds), *Conservation of biodiversity for sustainable development*. Scandinavian Univ. Press, Oslo, pp. 27–52.

Marshall, D. R. and Brown, A. H. D. (1975) Optimum sampling strategies in genetic conservation. *In:* Frankel, O. H. and Hawkes, J. G. (eds), *Crop genetic resources for today and tomorrow*. Cambridge Univ. Press, Cambridge, pp. 53–80.

McNeely, J. A. (1992) The biodiversity crises: challenges for research and management. *In:* Sandlund, O. T., Hindar, K. and Brown, A. H. D. (eds), *Conservation of biodiversity for sustainable development*. Scandinavian Univ. Press, Oslo, pp. 15–26.

McNeely, J. A., Miller, K.-R., Reid, W., Mittermeier, R. and Werner, T. (1990) *Conserving the world's biological diversity*. International Union for Conservation of Nature and Natural Resources. World Resource Institute, World Bank, World Wide Fund for Nature US, and Conservation International, Washington DC.

Naess, A. (1989) *Ecology, community and lifestyle*. Cambridge Univ. Press, Cambridge.

Raven, P. H. (1981) Research in botanical gardens. *Bot. Jahrb.* 102: 53–72.

Simmons, J. B., Beyer, R. I., Brandham, P. E., Lucas, G. and Parry, V. T. H. (1976) *Conservation of threatened plants*. Plenum Press, New York.

Soulé, M. E. and Mills, L. S. (1992) Conservation genetics and conservation biology: a troubled marriage. *In*: Sandlund, O. T., Hindar, K. and Brown, A. H. D. (eds), *Convervation of biodiversity for sustainable development*. Scandinavian Univ. Press, Oslo, pp. 55–69.

Vane-Wright, R. I., Humphries, C. J. and Williams, P. H. (1991) What to protect? – Systematics and the agony of choice. *Biol. Conserv*. 55: 235–254.

WCED (1987) *Our common future. World Commission on Environment and Development*. Oxford Univ. Press, Oxford.

Wilson, E. O. (1988) The current study of biological diversity. *In:* Wilson, E. O. (ed.), *Biodiversity*. Nat. Acad. Press, Washington, DC, pp. 3–18.

WWF, IUCN and BGCS (1989) The botanic gardens conservation strategy. World Wide Fund for Nature, International Union for Conservation of Nature and Natural Resources, Botanic Gardens Conservation Secretariat, Gland and Richmond.

Conservation Genetics
ed. by V. Loeschcke, J. Tomiuk & S. K. Jain
© 1994 Birkhäuser Verlag Basel/Switzerland

Animal breeding and conservation genetics

J. S. F. Barker

Department of Animal Science, University of New England, Armidale NSW 2351, Australia

Summary. Conservation genetics in an animal breeding context relates both to questions of preservation of rare and endangered breeds or populations, and to utilization with planned genetic change to improve viability, productivity, and efficiency of production. In the developed world, preservation is the primary issue, and various organizations exist which are committed to the preservation of rare and endangered breeds. In the developing world, breeds as such often are not defined or recognized, but many local populations exist that are adapted to and integrated into existing production systems. The genotypes of at least some of these populations could well also be crucial for future production systems, but many are threatened, primarily by crossbreeding with breeds introduced from the developed world. However, not all can be conserved, and priorities will have to be set for preservation, for development (breeding programs) and for evaluation for future programs. Some priorities will be set for pragmatic reasons, but the primary rational reason must be that a breed is in some way genetically unique, and makes a substantial contribution to the genetic diversity of the species. Thus, measures of genetic distance are essential to quantify the degree of genetic differentiation among populations, but such measures must be based on a large sample of loci. Although this has been emphasized many times, it still seems not to be adequately appreciated, and the effect of using a limited sample of loci is illustrated with an example from swamp buffalo populations. Comparative estimates of distances based on electrophoretic variation and direct DNA variation (both mitochondrial and genomic) are needed as a basis for future work on conservation of the global domestic animal diversity. Finally, studies of feral populations and wild relatives of domestic animals will provide a link between natural populations and domestic animal populations, and bring together these two areas, which to now have been largely separate.

Introduction

Interest and work in genetic conservation in the context of animal breeding predates the recognition of conservation genetics as a discipline within genetics, and the two have been largely separate, but concurrent, in their recent development. Plant conservation, in relation to plant breeding, has an even longer history. The distinction between conservation genetics and conservation in animal and plant breeding is that the former has dealt with natural populations, largely at the species level, and the latter with populations (usually defined breeds and varieties) *within* the domesticated species.

In this chapter, I will review past efforts in conservation in relation to animal breeding, discuss the present situation, and compare the needs and approaches with those of conservation of natural populations.

The earlier recognition of genetic conservation in plant and animal breeding should not be surprising, given the importance to mankind of our domesticated species. But this importance, in fact, our dependence on their products of food and fiber, their use as draft animals etc., has meant continuing efforts by breeders to improve productivity through both between breed and within breed selection, thus exacerbating loss of genetic diversity. Concern about reduction in genetic diversity was first expressed in terms of loss of breeds and varieties, or loss of animal genetic resources. I believe this concern was first publicly expressed at a meeting of the then existing Standing Advisory Committee on Agriculture, in Copenhagen, in 1946, which led to FAO (the Food and Agriculture Organization of the United Nations) being given the role of cataloging, maintaining, and using animal genetic resources. Initial efforts were devoted to publications describing types and breeds of cattle in various parts of the world, and to meetings on animal breeding under tropical and sub-tropical conditions, and related topics. Some attention was given to the conservation and use of animal genetic resources at these meetings, but it was not until 1966 that FAO convened the first study group on the evaluation, utilization, and conservation of animal genetic resources (FAO, 1967).

In the 25 years since then, FAO has continued to take a lead position, sometimes together with UNEP (United Nations Environment Program), and the topic of animal genetic resources has been considered at various symposia, workshops, and conferences. In the scientific community, the conservation of animal genetic resources was reviewed and discussed at all four World Congresses on Genetics Applied to Livestock Production (Mason, 1974; Lauvergne, 1982; Barker, 1986; Bodo, 1990). While there has been a great deal of talk, there is a clear contrast in action between countries of the developed world and the less developed countries. In the former, various non-government national organizations have effective conservation programs, e.g., the Rare Breed Survival Trust in the UK, and the American Minor Breed Conservancy in the USA. In 1991, a beginning was made for the internationalization of these activities, in the formation of Rare Breeds International, with founder membership in 30 countries. Alderson (1990) gives an overview of the philosophy and methods of these conservation programs. In other countries, such as Hungary, China, and some South American countries, government agencies have had major responsibility for conservation programs, many of which have been very effective. For the rest of the world, the main result of the extensive discussions and study has been increased awareness, and an evolving recognition of what needs to be done.

This contrast in apparent action between the developed and the developing world parallels another contrast between two aspects of animal genetic resource conservation, *viz.* the maintenance of breeds and populations that are in danger of extinction, and the continued and necessary improvements in productivity through animal breeding. Failure to recognize these as two distinct aspects has been compounded by varying usage of terminology, particularly of the terms conservation and preservation (e.g., see Barker, 1986). The following definitions have been accepted by various FAO Expert Consultations (FAO, 1984, 1992) to ensure consistency and to facilitate future activity:

(i) *Animal genetic resources* – All species, breeds, strains, and populations of livestock and poultry having economic, scientific, and cultural interest to mankind for agriculture now or likely to have such interest in the future.

(ii) *Conservation* – The management of human use of the biosphere so that it may yield the greatest sustainable benefit to present generations while maintaining its potential to meet the needs and aspirations of future generations. Thus conservation is positive, embracing preservation, maintenance, sustainable utilization, restoration and enhancement of the natural environment.

(iii) *Preservation* – That aspect of Conservation by which a sample of an animal genetic resource population is designated to an isolated process of maintenance, by providing an environment free of the human forces which might bring about genetic change. The process may be *in situ*, whereby the sample consists of live animals in a natural environment, or it may be *ex situ*, whereby the sample is placed, for example, in cryogenic storage.

(iv) *Conservation by management* – That aspect of Conservation by which a sample, or the whole of an animal population is subjected to planned genetic change with the aim of sustaining, utilizing, restoring or enhancing the quality and/or quantity of the animal genetic resource and its products of food, fibre or draught animal power.

(v) *Threatened* (*species, breed, strain or population*) – A term used to describe an animal genetic resource population which is subject to some force of change, affecting the likelihood of it continuing indefinitely, either to exist, or to retain sufficient numbers to preserve the genetic characteristics which distinguish it from other populations. Threatened is a generic term embracing more precise descriptions such as Endangered or Vulnerable.

However, there are still problems with these definitions. While the English language distinguishes preservation and conservation, in some other languages there is only one word to cover both aspects. FAO currently is developing an integrated global program for the management of animal genetic resources. Clearly, the terminology must be such as to allow unambiguous communication and understanding, as the global management must encompass all research, development, education, training, and communication activities associated with conservation.

In the global consideration of management of animal genetic resources, the fundamental distinction is not between those breeds that are endangered and those that are not, but between those which are perceived to have no or little current utility and those which do have current utility or could have in the immediate future. For each of these categories, the necessary actions are then preservation or utilization. In current FAO activities, Hammond (personal communication) uses passive and active conservation to describe these two aspects.

Animal breeding – What is the problem?

The aim of animal breeding is to improve productivity, product quality, and the efficiency of production by either or both of selection within breeds (or strains) or utilization of differences among breeds (or strains) through crossbreeding, grading-up to a superior breed by repeated backcrossing, or formation of a synthetic population. Thus, future improvement of livestock to meet human needs is dependent on genetic variation – both the variation within breeds, and the variation between breeds, strains, and populations. Genetic variation is the basic material of the animal breeder – we use it to mold our animal populations to our needs, and loss of variation will restrict the options available to meet unpredictable future requirements. While loss of variation within breeds or populations is continually countered by the introduction of new variation through mutation (Franklin, 1981; Hill and Keightley, 1988), the genetic variation present as differences among breeds, strains or populations cannot be readily regenerated. Each breed or strain is the product of the random processes of genetic drift (due to finite population size), as well as separate adaptation and evolution, perhaps over many centuries, with differing selection pressures imposed by climate, endemic parasites and diseases, available nutrition and criteria imposed by man. Each is thus likely to represent a unique combination of genes.

How then does animal breeding impinge on the dichotomy between the developed and the developing world? In the former, animal breeding has been and is primarily driven by market forces – by competition among breeders and among breeds. Some hundreds of defined breeds have been produced in the developed world over the past 200 years, and those that are already extinct, or are presently rare and endangered, presumably have been tested by market forces and found wanting. The scale of this potential loss of variation is not trivial. For example, Maijala et al. (1984) found that about one-third of the 737 recognized livestock breeds in Europe were in danger of extinction. At the other end of the spectrum, there is extreme concentration of use on just one breed, as exemplified by dairy cattle, where the Holstein–Friesian breed probably produces about 80% of all milk in the developed world. However, market forces consider only the present and near future; breeds that are currently rare or endangered may be of value in the more distant future. Thus, in the developed world, preservation, as insurance, is the primary issue.

The picture in the developing world is rather different in a number of ways. Firstly, breeds as such may not be clearly defined and recognized, but there exist a multitude of strains or geographically separated populations. Local populations may have different names, but without apparent differences in phenotype; a change in phenotype may occur without change in name, or all populations may have just one name and be phenotypically similar. For example, Hardjosubroto and Astuti (1980) stated that there are four breeds of indigenous duck in Indonesia, *viz.* Alabio, Tegal, Bali, and Manila (Muscovy). Yet in Java, local populations (all ostensibly Tegal) often are identified by the name of the town or area where they are found, and are considered different from other such local populations. Strictly, Tegal ducks are those found around the town of Tegal on the north coast of Java, and Hetzel (1982) noted that at least some of these local populations are morphologically distinct. As another example, the swamp buffalo of southeast Asia comprise a number of geographically separate populations, but all are phenotypically similar, and distinct breeds are not recognized. What should be the approach to conservation when there are many strains or populations, about which there is little information – either for productive performance or for genetic differences among them?

Secondly, while some breeds and strains clearly are endangered, others are not numerically small but are still potentially under threat. These latter breeds are being used for production, and generally are considered to be well-adapted to the prevailing climatic – husbandry – economic conditions. However, relatively little research has been done with them, and their production performance and adaptability is generally poorly known. Nevertheless, it is clear that average production levels (of milk yield, growth rate or whatever) of these populations in *their environment* is less than that of the breeds being used in the environment of the developed world. While basic quantitative genetic theory states that the mean phenotype of a population for some trait is a measure of its mean genotype, comparing mean phenotypes of two populations as a measure of their comparative genetic merit is valid only if all individuals of the two populations have been raised and measured in the same environment at the same time. Regardless of this, high production of breeds in developed countries, as compared with that of indigenous strains in less-developed countries has led to unrealistic expectations of the potential for rapid improvement of productivity in the less-developed countries through importation. In some cases, where the aim was to replace an indigenous strain by an imported breed, the program was a dismal failure – the imported breed simply lacked adaptation, and failed to breed or to survive. In other cases, notably with pigs and poultry, where management and husbandry technology was imported with the improved breed, some successful and persistent production systems have developed. More commonly, the aim has been to produce crossbred animals, by importing males only or semen to be used with indigenous females. The F_1 crossbreds generally are superior in performance to the indigenous strains (FAO, 1987; Bondoc et al., 1989), so that such crossbreeding programs have been successfully developed in some cases. But most have failed, because of inadequate infrastructure to maintain production of F_1 animals, or to develop a breeding program beyond the F_1, or for various other reasons. As a result, indigenous strains have been subject at least to genetic dilution, and some may well have been lost. In the less-developed world, therefore, there is a need for preservation of some strains, but the primary issue is utilization with increased emphasis on development and genetic improvement of the indigenous strains.

There is a pressing need to increase livestock yields in developing countries, where levels of animal protein intake are low on average. This increase in yield may be achieved by selection in some indigenous populations, by crossbreeding between indigenous strains, or by crossbreeding indigenous strains with exotic breeds where that is appropriate. Only the first of these strategies will, in itself, ensure continued maintenance of each indigenous population. However, there is also increasing recognition that the existing indigenous strains and populations are likely to carry valuable genes – genes controlling specific physiological, behavioral, and parasite or disease resistance traits. Thus, the genotypes of some of these strains could well be crucial to future sustainable animal production systems, at least in the stressful environments where they have evolved by natural selection, and undoubtedly many will contribute also to developed country production systems. Within the framework of conservation in the developing countries, there is then a need to ensure that unique indigenous strains are not lost.

Conservation issues in animal breeding

This realization of the potential value of indigenous breeds – both for immediate use and for future animal production systems, was voiced most strongly during discussions at the FAO/UNEP Technical Consultation on Animal Genetic Resources Conservation and Management (FAO, 1981). This consultation represents a real milestone in the consideration of animal genetic resources, in defining what needed to be done to facilitate development of a global program. Since then, methodologies for such a program have been researched (reviewed by Hodges, 1990a), including procedures for description and characterization of breeds and the development of data banks for this information (FAO, 1986a,b,c), and for the establishment of genebanks as repositories for frozen animal genetic material (Hodges, 1990b,c).

Nevertheless, it remains to convert these activities into an effective global program of action, where the key issues are:

(i) identification, characterization, and enumeration of all breeds and strains of livestock used in animal agriculture, i.e., documentation of existing resources;
(ii) specification of those breeds and strains at risk of extinction, with appropriate preservation measures initiated for those perceived to have little or no current utility, i.e., preventing loss of resources;
(iii) development of breeding programs to enhance the productivity of those indigenous breeds perceived to have major utility, either now or in the immediate future, i.e., utilization;
(iv) designing and implementing a comprehensive information network;
(v) coordinating all activities.

It is ironic that the major concern is loss of diversity, when the existing diversity is so great that its true magnitude is not known. There are probably in excess of 4000 breeds and strains of livestock in the world. Obviously, the existing diversity is so great that it will not be possible to preserve every breed or strain that might be in danger of extinction, nor to set up development programs for all breeds.

Clearly, priorities will have to be set, both in choosing breeds for preservation and for development programs.

Setting priorities for preservation and development

Here again, it is the enormous diversity of breeds and strains in the developing world that is of major concern. In these countries, financial and other resources are very limited, and the option for preserving rare strains and breeds as live animal populations is far less possible than in developed countries. Thus, cryopreservation as semen or embryos is likely to be preferred, and the costs of such preservation measures are small relative to potential future benefits (Smith, 1984).

But even so, it is inevitable that choices will have to be made: which one (or ones) of some set of breeds should be preserved, which should be let go? Preservation is for the future, it is insurance to cater to possible future needs. We cannot know what those needs will be, nor can we know what genotypes or alleles in existing breeds might best meet those needs. Thus, the only rational criterion for setting preservation priorities is the probability that the breed or strain in question is genetically unique, as first suggested by Miller (1977). An unequivocal definition of uniqueness is not possible; the pragmatic definition must be "sufficiently different". Thus, for some set of breeds or strains of a particular species that are endangered, priorities for preservation will be based on estimates of genetic relationships among them, with those that are most different from the others being the primary candidates for preservation.

In setting priorities for breed development programs, the criterion is not just one of genetic uniqueness. Here, the aim is immediate improvement of productivity, and the primary criteria to be considered (FAO, 1992) are:

(i) the breed possesses one or more highly desirable attributes in terms of productivity and/or adaptation;
(ii) the breed is endangered, or is not efficiently utilized;
(iii) the breed should be one whose improvement could have the potential to influence large populations, either of the same breed in one or more countries, or other very similar breed types.

However, where a genetic improvement program is to be initiated, there is still the question of identifying which particular population (or populations) should be used. Ideally, the adapted population (or populations) that is superior for the economic traits of interest should be chosen as the base for the breeding program. Such information will rarely be available, so that knowledge of the genetic relationships among the populations again would be helpful. If all populations were closely related, it would not matter which population was chosen, as all would provide a similar genetic base. If the populations were not all closely related, rational choice would be more difficult. One might be chosen on some more subjective grounds (e.g., availability of infrastructure, facilities and expertise), but the breeding program should then include introduction of males (or semen) from other populations, both to broaden the genetic base for selection and genetic improvement, and with appropriate planning and design, to gain information on the comparative genetic merit of the different populations.

As noted previously, many of the livestock populations in the developing world are not in danger of extinction; they are not numerically small and they are being used in existing production systems. At the same time as some are included in development programs, there is a need to plan for future programs in others. Studies on the comparative productivity and adaptability of the available breeds and geographical populations (evaluation studies to determine comparative genetic merit) could be done to allow rational choice of populations for future programs. But given the wealth of strains and populations that exist within each species, it will not be possible to evaluate all of them. The solution to this dilemma (Barker, 1980) is to determine the genetic relationships among the breeds and populations, so that they may be grouped into sets that are genetically similar, and then to include in evaluation studies one representative from each set.

Clearly, knowledge of the genetic relationships among breeds, strains, and populations within each livestock species is and will be of major significance in the conservation of animal genetic resources, and in the development of breeding programs for increased productivity.

Estimation of genetic relationships among strains, populations or breeds

Ideally, we would wish to determine which populations are genetically unique, or at least to quantify the magnitude of the genetic differences among the set of populations being considered.

Baker and Manwell (1984) suggested that, while not ideal, simple comparison of breeds for frequencies of alleles at single loci could be useful. For 84 pair-wise comparisons among eight breeds and four loci, they found 74 to be significant, and concluded that these breeds were genetically distinct. This approach could be of some use, for example, if data were available for only a few loci, and decisions had to be made on priorities for preservation, but it is not in itself sufficient, and may be misleading. There are a number of problems. Firstly, each locus in each population should be tested for deviations from Hardy-Weinberg equilibrium. If all populations are in Hardy-Weinberg proportions, allele frequencies would be tested, while if there are deviations from Hardy-Weinberg proportions in some or all populations, genotype frequencies should be tested. Secondly, even with only a small number of populations and a few loci, the interpretation of many pair-wise tests becomes very difficult: how different are the breeds or populations from one another? Instead of all possible pair-wise comparisons, each locus could be tested separately, say, using the contingency chi-squared statistic. This reduces the number of statistical tests, but the problem of interpretation remains. Some loci may indicate significant differentiation, while some may not, and there is still no objective quantification of the degree of differentiation among the populations.

These two problems of reducing the data to simplify interpretation and quantifying the degree of differentiation may be overcome by use of a measure of genetic distance, and by constructing from all the pair-wise distances a single diagram that best represents all relationships among the populations, i.e., a phylogeny.

Genetic distance measures have, of course, been widely used in evolutionary genetic studies, to describe the genetic structure of populations of a species, or to determine evolutionary relationships among species, and will be of value in the conservation genetics of natural populations (Crozier and Kusmierski, this volume). If two populations are, for geographic or ecological reasons, genetically isolated, they will tend to accumulate different alleles, due to selection, mutation or genetic drift. Thus, if allele frequency data are available for only a few loci, the estimated genetic distances and the phylogeny will not be reliable. Lewontin (1974) suggested using 50 or more loci to average out effects of selection or drift. In particular, use of a large number of loci, rather than a large number of individuals per locus, will reduce the variance of the genetic distance estimate (Nei and Roychoudhury, 1974; Nei, 1978). However, Archie et al. (1989) noted that the number of individuals sampled per population should not be too small, say at least 20.

A number of genetic distance measures have been proposed, and the methodologies have been extensively reviewed (e.g., Wright, 1978; Nei, 1987; Chakraborty and Rao, 1991), but Nei's standard genetic distance (Nei, 1972, 1973) has been most widely used in studies of population differentiation and in evolutionary biology.

Genetic distance studies of livestock breeds date at least from the early 1970s (e.g., Zetner et al., 1972; Kidd, 1974), while the possibility of using genetic variation in blood groups and in blood and milk proteins to measure relationships among breeds was advanced even earlier (reviewed by Rendel, 1967). A preliminary literature survey has revealed about 100 reports of genetic distance studies in livestock. The rationale for these studies has not been conservation, but interest in breed origins and evolution, and phylogenetic relationships. Many of these studies refer to breeds and populations in Europe, although some include currently rare breeds or ones restricted in their distribution, and others include indigenous strains and populations in Asia. An overall review of these studies may then provide a useful baseline for future work specifically directed towards questions concerning the conservation of animal genetic resources.

Such a review is yet to be done, and there are likely to be some problems in interpretation of the results. First, many of the studies, particularly earlier ones, primarily report blood group allele frequencies, while later ones more commonly use electrophoretic blood protein frequencies. Human studies have shown distances based on electrophoretic systems, HLA antigens and mt DNA sequences to be highly correlated, while the correlations of each of these with blood group derived distances were lower (Ryman et al., 1983; McLellan et al., 1984; Nei, 1985). The reason for this is not clear, but may relate to the blood group loci often involving complex dominance relationships among alleles, so that allele frequency estimation is more subject to error. Secondly, the work has been done in many different laboratories, and there may well be heterogeneity in laboratory techniques, or even differences in allelic designations. Thirdly, only a few loci have been analyzed in a number of studies, and the number of individuals sampled per population also is often small. The need to use a large number of loci was emphasised previously, and as this has been well-known for some time, there should be no place for future studies based on very few loci in addressing questions of relationship among populations in genetic conservation.

In a study of swamp buffalo populations in southeast Asia that is still in progress (Barker et al., 1991; Mukherjee et al., 1991), we set out to develop electrophoretic techniques for 70 loci. To date, methods have been developed for 58 loci (Tan et al., 1992), but so far, only 39 of these have been assayed for all individuals in the 17 populations sampled. Of these 39, 14 are polymorphic in one or more populations. Additional data may well change the estimated relationships among the populations, but these data may be used to illustrate the effect of using only a few loci. As it is only the comparison between results using different numbers of loci, and not the absolute magnitude of the estimated distances that is of interest, only polymorphic loci have been used. Using the program BIOSYS-1 (Swofford and Selander, 1989), genetic distances among populations were estimated using Nei's (1972) standard distance, and dendrograms constructed using the unweighted pair group method with arithmetic means (Sneath and Sokal, 1973). This was done including all 14 polymorphic loci (Fig. 1), and for a subset of seven of these loci chosen at random (Fig. 2). There are clearly very marked differences between the two dendrograms and the correlation between the two sets of estimated pair-wise genetic distances is only 0.39.

With 14 loci used, populations that are closer geographically generally tend to cluster together, and the major separation into two groups is biologically reasonable as the larger group of 12 are all swamp buffalo, and the smaller group of five are river buffalo populations (Murrah breed) and three populations from Sri Lanka where karyotype

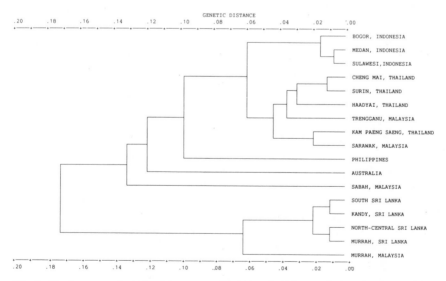

Fig. 1. Dendrogram of relationships among 17 swamp buffalo populations based on Nei's standard genetic distances derived from allele frequencies at 14 polymorphic loci.

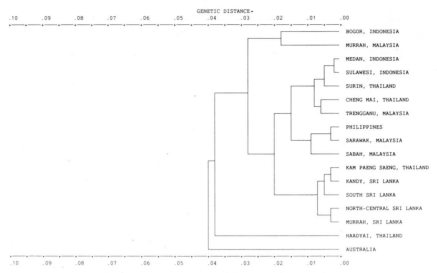

Fig. 2. Dendrogram of relationships among 17 swamp buffalo populations based on Nei's standard genetic distances derived from allele frequencies at seven polymorphic loci.

analysis has shown them to be river buffalo. On the other hand, the estimated genetic distances based on only seven of these loci are much smaller, the discrimination among populations is much weaker, and geographically close populations do not cluster together to the same extent.

Protein electrophoresis is inexpensive and easy, but is limited by the relatively low level of variability, with about two-thirds of loci investigated expected to be monomorphic in all populations and low levels of heterozygosity for at least a proportion of the polymorphic loci. Thus, direct analysis of DNA is being more widely suggested and used for distance studies. Either mt DNA or genomic DNA may be used, and analyses may be of RFLPs, micro- or minisatellite loci or even direct sequencing. These procedures require more sophisticated laboratory equipment, and are more expensive than electrophoresis, which limits their utilization in developing countries. In addition, while high levels of variation may be detected by direct DNA analysis, there is need for caution in measuring genetic distance using non-coding regions, because of genomic turnover mechanisms (Amos and Hoelzel, 1992). Further, although protein and mt DNA variation have been shown to produce highly correlated distances among human races (as noted above), the same may not be true for livestock populations. In modern animal breeding, gene flow between populations is primarily due to migration of male animals, and the same is likely to have been true in earlier times, or for movement of animals among local populations of endemic livestock. Distances estimated from mt DNA variation thus may be greater than those from protein loci or genomic DNA. These questions of possibly different distances and the resulting dendrograms are about to be studied by my colleagues and I, by obtaining data on mt DNA and genomic variation in the buffalo samples for which we have electrophoretic results.

Wild relatives, feral populations and conservation genetics

This paper started by noting that conservation genetics of natural populations and conservation of animal genetic resources have been largely concurrent, but separate, in their development. Clearly, both have the common goal of providing the basis for rational decisions concerning the conservation of biodiversity, and the same techniques and analytical methods are being used to identify unique populations.

For natural populations, the endpoint in application of genetic knowledge is the conservation of a species, or of a unique population of a species – for its own sake and *in toto*. Animal genetic resources conservation also is driven by the need to conserve biodiversity. However, there is the additional pragmatic reason that a conserved population may be of value in future production systems, or even that a particular gene may be of value in the future. In this context, feral populations and wild relatives of our domestic livestock also offer potentially important opportunities for the introduction of new genes (Woodford, 1992; van Vuren and Hedrick, 1989). Many of these wild species already are classified as vulnerable or endangered (IUCN, 1990). Examples include the kouprey, *Novibos sauveli*, of Thailand, Laos, Vietnam and Cambodia, and the gaur, *Bos gaurus*, of India (relatives of cattle), and species related to domestic buffalo – the tamaraw, *Bubalus mindorensis*, from Mindoro in the Philippines, and two species of *Anoa*, *A. depressicornis* and *A. quarlesi*, from Indonesia. The conservation genetics of these species and other wild relatives of domestic livestock would link together studies of natural populations and of domestic animal populations – to the advantage of both.

References

Alderson, L. (1990) *Genetic conservation of domestic livestock.* CAB International, Wallingford.

Amos, B. and Hoelzel, A. R. (1992) Applications of molecular genetic techniques to the conservation of small populations. *Biol. Conserv.* 61: 133–144.

Archie, J. W., Simon, C. and Martin, A. (1989) Small sample size does decrease the stability of dendrograms calculated from allozyme-frequency data. *Evolution* 43: 678–683.

Baker, C. M. A. and Manwell, C. (1984) Indigenous breeds: Anachronism or asset? *Proc. 2nd Wld Cong. Sheep and Beef Cattle Breeding*, Vol II, paper P1.

394

Barker, J. S. F. (1980) Animal genetic resources in Asia and Oceania – The perspective. *Proc. SABRAO Workshop on Animal Genetic Resources in Asia and Oceania*. Tropical Agriculture Research Center, Tsukuba, Japan, pp. 13–19.

Barker, J. S. F. (1986) Coordinator: Round Table H – Preservation and management of genetic resources. *Proc. 3rd Wld Cong. Genet. Appl. Livestock Prod.* XII: 469–500.

Barker, J. S. F., Mukherjee, T. K., Hilmi, M., Tan, S. G. and Selvaraj, O. S. (1991) Genetic identification of strains and genotypes of swamp buffalo and of goats in southeast Asia: rationale for their study. *ACIAR Proceedings* 34: 13–15.

Bodo, I. (1990) Organiser: Workshop 2.4 – Conservation of genetic resources. *Proc. 4th Wld Cong. Genet. Appl. Livestock Prod.* XIV: 421–472.

Bondoc, O. L., Smith, C. and Gibson, J. P. (1989) A review of breeding strategies for genetic improvement of dairy cattle in developing countries. *Anim. Breed. Abst.* 57: 819–829.

Chakraborty, R. and Rao, C. R. (1991) Measurement of genetic variation for evolutionary studies. *In:* Rao, C. R. and Chakraborty, R. (eds), *Handbook of statistics*, Vol. 8. Elsevier Science Publishers, Amsterdam, pp. 271–316.

FAO (1967) Report of the FAO Study Group on the Evaluation, Utilization and Conservation of Animal Genetic Resources, FAO, Rome.

FAO (1981) Animal genetic resources conservation and management. Proceedings of the FAO/UNEP Technical Consultation. FAO Animal Production and Health Paper 24. FAO, Rome.

FAO (1984) Animal genetic resources conservation by management, data banks and training. FAO Animal Production and Health Paper 44/1. FAO, Rome.

FAO (1986a) Animal genetic resources data banks. 1: Computer systems study for regional data banks. FAO Animal Production and Health Paper 59/1. FAO, Rome.

FAO (1986b) Animal genetic resources data banks. 2: Descriptor lists for cattle, buffalo, pigs, sheep and goats. FAO Animal Production and Health Paper 59/2. FAO, Rome.

FAO (1986c) Animal genetic resources data banks. 3: Descriptor lists for poultry. FAO Animal Production and Health Paper 59/3. FAO, Rome.

FAO (1987) Crossbreeding *Bos indicus* and *Bos taurus* for milk production in the tropics. FAO Animal Production and Health Paper 68. FAO, Rome.

FAO (1992) The management of global animal genetic resources. FAO Animal Production and Health Paper 104. FAO, Rome.

Franklin, I. R. (1981) Population size and the genetic improvement of animals. *In:* Barker, J. S. F., Hammond, K. and McClintock, A. E. (eds), *Future developments in the genetic improvement of animals*. Academic Press Australia, Sydney, pp. 181–196.

Hardjosubroto, W. and Astuti, M. (1980) Animal genetic resources in Indonesia. *Proc. SABRAO Workshop on Animal Genetic Resources in Asia and Oceania*, Tropical Agriculture Research Center, Tsukuba, Japan, pp. 189–204.

Hetzel, D. J. S. (1982) Evaluation of native strains of ducks in the SABRAO region. *Proc. 2nd SABRAO Workshop on Animal Genetic Resources in Asia and Oceania*, SABRAO, Kuala Lumpur, pp. 135–153.

Hill, W. G. and Keightley, P. D. (1988) Interrelations of mutation, population size, artificial and natural selection. *In:* Weir, B. S., Eisen, E. J., Goodman, M. M. and Namkoong, G. (eds), *Proc. Second Internat. Conf. Quant. Genet.* Sinauer, Sunderland, MA, pp. 57–70.

Hodges, J. (1990a) Animal genetic resources. A decade of progress, 1980–1990. FAO Animal Production and Health Paper. FAO, Rome.

Hodges, J. (1990b) Manual on establishment and operation of animal gene banks. FAO Animal Production and Health Paper. FAO, Rome.

Hodges, J. (1990c) Review of regional animal gene banks and recommendations from Hannover Workshop on associated topics raised by the Tenth Committee on Agriculture. *FAO Animal Production and Health Paper* 80: 51–57.

IUCN (1990) IUCN Red List of Threatened Animals. IUCN, Gland, Switzerland.

Kidd, K. K. (1974) Biochemical polymorphisms, breed relationships, and germ plasm resources in domestic cattle. *Proc. 1st Wld Cong. Genet. Appl. Livestock Prod.* 1: 321–328.

Lauvergne, J. J. (1982) Moderator: Round Table B – Breeds Conservation. *Proc. 2nd Wld Cong. Genet. Appl. Livestock Prod.* 6: 71–135.

Lewontin, R. C. (1974) *The genetic basis of evolutionary change*. Columbia University Press, New York.

Maijala, K., Cherekaev, A. V., Devillard, J-M., Reklewski, Z., Rognoni, G., Simon, D. L. and Steane, D. E. (1984) Conservation of animal genetic resources in Europe. Final Report of an E.A.A.P. Working Party. *Livest. Prod. Sci.* 11: 3–22.

Mason, I. L. (1974) Moderator: Round Table A – The conservation of animal genetic resources. *Proc. 1st Wld Cong. Genet. Appl. Livestock Prod.* 2: 13–93.

McLellan, T., Jorde, L. B. and Skolnick, M. H. (1984) Genetic distances between the Utah Mormons and related populations. *Amer. J. Human Genet.* 36: 836–857.

Miller, R. H. (1977) The need for and potential application of germ plasm preservation in cattle. *J. Hered.* 68: 365–374.

Mukherjee, T. K., Barker, J. S. F., Tan, S. G., Selvaraj, O. S., Panandam, J. M., Yushayati, Y. and Sreetharan (1991) Genetic relationships among populations of swamp buffalo in southeast Asia. *ACIAR Proceedings* 34: 34–40.

Nei, M. (1972) Genetic distance between populations. *Am. Nat.* 106: 283–292.

Nei, M. (1973) The theory and estimation of genetic distance. *In:* Morton, N. E. (ed.), *Genetic structure of populations.* University of Hawaii Press, Honolulu, pp. 45–54.

Nei, M. (1978) Estimation of average heterozygosity and genetic distance from a small number of individuals. *Genetics* 89: 583–590.

Nei, M. (1985) Human evolution at the molecular level. *In:* Ohta, T. and Aoki, K. (eds), *Population genetics and molecular evolution.* Japan Scientific Societies Press, Tokyo, pp. 41–64.

Nei, M. (1987) *Molecular evolutionary genetics.* Columbia University Press, New York.

Nei, M. and Roychoudhury, A. K. (1974) Sampling variances of heterozygosity and genetic distance. *Genetics* 76: 379–390.

Rendel, J. (1967) Studies of blood groups and protein variants as a means of revealing similarities and differences between animal populations. *Anim. Breed. Abst.* 35: 371–383.

Ryman, N., Chakraborty, R. and Nei, M. (1983) Differences in the relative distribution of human gene diversity between electrophoretic and red and white cell antigen loci. *Hum. Hered.* 33: 93–102.

Smith, C. (1984) Genetic aspects of conservation in farm livestock. *Livest. Prod. Sci.* 11: 37–48.

Sneath, P. H. A. and Sokal, R. R. (1973) *Numerical taxonomy.* Freeman, San Francisco.

Swofford, D. L. and Selander, R. B. (1989) BIOSYS-1: A computer program for the analysis of allelic variation in population genetics and biochemical systematics (Release 1.7). Illinois Natural History Survey, Champaign, Illinois.

Tan, S. G., Selvaraj, O. S., Mukherjee, T. K. and Barker, J. S. F. (1992) Methodology for the electrophoretic study of southeast Asian buffaloes. *Buffalo J.* 2: 103–107.

van Vuren, D. and Hendrick, P. W. (1989) Genetic conservation in feral populations of livestock. *Conser. Biol.* 3: 312–317.

Woodford, M. H. (1992) The wild relatives of domestic animals. *FAO Animal Production and Health Paper* 104: 227–233.

Wright, S. (1978) *Evolution and the genetics of populations.* Vol. 4. *Variability within and among natural populations.* University of Chicago Press, Chicago.

Zetner, K., Rohrbacher, H., Schleger, W. and Pirchner, F. (1972) Relationships between Austrian cattle breeds as inferred from blood group and serum protein frequencies. *XIIth Europ. Conf. Anim. Blood Groups Biochem. Polymorph.*, pp. 131–135.

Scenarios

Conservation Genetics
ed. by V. Loeschcke, J. Tomiuk & S. K. Jain
© 1994 Birkhäuser Verlag Basel/Switzerland

Introductory remarks

We present a few scenarios as outlines of some important topics not covered in the preceding chapters in order to illustrate significant ecological and population genetic options and developments in conservation biology. They show interconnections, some wishful thinking, but mostly feasible and necessary ideas in practice (what we do know, can know, and should know for useful applications). In considering these and others, we soon realize also that population genetics is not always the centerpiece in many conservation programs, but nevertheless, more of it should be thoughtfully learned and taught.

the great stability of the females' hierarchy, even across a number of generations.

Kawamoto et al. (1984) estimate the migration rates to be about 41% per generation in *Macaca fascicularis*, which was attributed to the female's persistence at her natal place and the large proportion of males emigrating to another troop (see also, de Jong et al., this volume). Evidence for intergroup migration of only one sex exists for many primate species. Usually, migration is linked to males which are typically in a minority at reproductive ages.

Contrarily, the common chimpanzee is described as living in communities rather than in groups, because males and anoestrus females range separately within prescribed areas. Males, however, are the resident sex and stay in the area of their birth while females normally emigrate to another area and community. In the most studied populations of gorillas, the majority of females apparently also leaves the group of their birth (Harcourt and Stewart, 1983). The relationship between the kind of social hierarchy of groups within a species and its territorial behavior has to be considered when genetic breeding programs propose the exchange of individuals from one to another group because of the close correlation to the aggressiveness of the species. High acceptance of individuals from other groups, for example, seems to be present in the loose, temporary groupings of chimpanzees. Gorillas also are tolerant of exchanges of individuals from group to group, whereas the adult males of, for example, *Papio* species are perhaps the other extreme; they punish straying adult males very severely, but juvenile males are often accepted into strange groups (Hall, 1968).

A similar behavior was described in a semi-free living group of barbary macaques, *Macaca sylvana*, which was studied for the reproductive success of three adult males. A matrilinear group was established with two adult and two subadult unfamiliar males. This caused several problems in the beginning when the two subadult males were seriously injured and therefore removed from the group. Later, a juvenile male frequently supported one of the males, which was the beginning of a coalition between them. The rank reversals also occurring in natural populations were preceded by severe fights. The genetic studies disclosed the reproductive success of all three males and demonstrated that, once established, the adult female hierarchy is remarkably stable and the reproductive success of females depends on their degree of dominance (Witt et al., 1981; Schmitt et al., 1981).

The evolution of social complexity is assumed to be favored by inbreeding, but limited by the genetic disadvantage of inbreeding (Wilson, 1975) which is largely a function of the rate of migration by individuals between social units or troops. In captive groups, the relationship among individuals and therewith the level of inbreeding is of interest for designing breeding programs which minimize the risk of

404

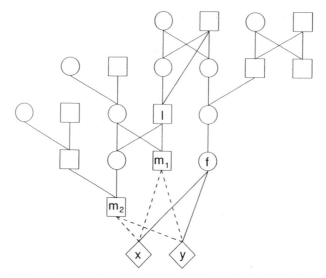

Fig. 1. Pedigree of lemurs, *Varecia variegata variegata*, living in the zoological garden of Saarbrücken (Germany). Females are indicated by circles and males by squares. The paternity for the individuals, x and y, is unknown. The mother of the individuals, x and y, is marked with f and the potential fathers with m_1 and m_2 (from Loeschcke et al., 1992).

inbreeding effects. A breeding group of lemurs, *Varecia variegata varie-gata*, living in the zoological garden of Saarbrücken (Germany) has been analyzed to demonstrate the use of genetics (Fig. 1; see also Loeschcke et al., 1992).

Based on behavioral studies on the relationship of individuals, the inbreeding coefficients of each individual can be calculated assuming unrelated ancestors. In the breeding group, there are female (f) and two adult males, (m_1) and (m_2). The paternity for the progeny (x) and (y) is unknown. The inbreeding coefficients for the progeny (x and y) of female (f) are of interest and separately calculated for each of the potential fathers (m_1 and m_2). In the case of male m_1 the inbreeding coefficient is 0.031, and for male m_2, it is 0.016. For both individuals, the inbreeding coefficient is fairly low. However, genetic analyses can only provide the exact knowledge on the relationship of the individuals, which then serves as the basis for a reliable assessment of the pedigree and of the resulting inbreeding coefficients. And the behavioral observations often do not give reliable indicators of paternity in social groups (Smith, 1982). The study on a group of squirrel monkeys, for example, demonstrates the greater accuracy of genetic markers (VandeBerg et al., 1990) as the paternity in 5 of 74 offspring had to be reassessed following genetic analysis. Furthermore, two mother-infant relationships had to be corrected because of the exchange of progeny among females. There-fore, genetic analyses support the design of breeding programs which minimize, for example, by the exclusion of individuals from breeding groups, the increase of inbreeding level.

When breeding plans are designed to minimize the loss of genetic diversity, the social organization evolved in natural populations must also be maintained in breeding groups for minimizing aggressions which can decrease the reproductive success in breeding groups (see Ceska et al., 1992). For reintroduction programs, however, this approach might be still too naive for some primate species in which non-inherited behavior is of high importance for individual survival in the wild, e.g., the experience during the period of maternal care and the learning of managing situations in the wild from group members. Young gorillas, for example, appear to learn largely what to eat and what not to eat by observing their mother and other group members (Schaller, 1963). Even traditional behavior is observed in some primate species: A group of Japanese monkeys learned to wash potatoes and wheat in the sea, and this behavior spread from one to other individuals (Frisch, 1968). In chimpanzees, the use of tools differs among local populations since ecological differences in the habitats may influence the choice of the objects (Jay, 1968).

Summarizing, we have to apply both the genetic and ethological knowledge on natural populations of primates for their successful conservation. The minimum requirements are: (1) The reserves must have a certain size and structure which is not only adequate for the maintenance of genetic variability, but also for the maintenance of the social organization. (2) The social structure of breeding groups should be close to that evolved in natural populations. The selection of individuals for an exchange between different breeding groups must take into account the migration behavior of the species in the wild.

References

Ceska, V., Hoffmann, H.-U. and Winkelsträter, K.-H. (1992) *Lemuren im Zoo – Aktuelle Forschungsergebnisse, Artenschutz, Perspektiven*. Paul Parey, Berlin-Hamburg.

Frisch, J. E. (1968) Individual behavior and intertroop variability in Japanese macaques. *In:* Jay, P. C. (ed.), *Primates – Studies in adaptation and variability*. Holt, Rinehart and Winston, New York, pp. 243–252.

Ganzhorn, J. U. and Kappeler, P. M. (1993) Lemuren Madagaskars. Tests zur Evolution von Primatengemeinschaften. *Naturwissenschaften* 80: 195–208.

Hall, K. R. L. (1968) Social organization of the old-world monkeys and apes. *In:* Jay, P. C. (ed.), *Primates – Studies in adaptation and variability*. Holt, Rinehart and Winston, New York, pp. 7–31.

Harcourt, A. H. and Stewart, K. J. (1983) Interactions, relationships and social structure: the great apes. *In:* Hinde, R. (ed.), *Primate social relationships*. Sinauer, Sunderland, MA, pp. 307–314.

Jay, P. C. (1968) *Primates – Studies in adaptation and variability*. Holt, Rinehart and Winston, New York.

Kawamoto, Y. and Ischak, TB. M. (1981) Genetic differentiation of the Indonesian crab-eating macaque (*Macaca fascicularis*): I. Preliminary report on blood protein polymorphism. *Primates* 22: 237–252.

406

Kawamoto, Y., Nozawa, K. and Ischak, TB. M. (1981) Genetic variability and differentiation of local populations in the Indonesian crab-eating macaque (*Macaca fascicularis*). *Kyoto Univ. Overseas Report Stud. Indon. Macaque* 1: 15–39.

Kawamoto, Y., Ischak, TB. M. and Supriatna, J. (1984) Genetic variations within and between troops of the crab-eating macaque (*Macaca fascicularis*) on Sumatra, Java, Bali, Lombok and Sumbawa, Indonesia. *Primates* 25: 131–159.

Loeschcke, V., Tomiuk, J. and Siegismund, H. R. (1992) Möglichkeiten der Erkennung und Erhaltung von genetischer Variabilität in Zuchtgruppen. *In:* Ceska, V., Hoffman, H.-U. and Winkelsträter, K.-H. (eds), *Lemuren im Zoo – Aktuelle Forschungsergebnisse, Artenschutz, Perspektiven.* Paul Parey, Berlin-Hamburg, pp. 263–271.

Nozawa, K., Shotake, T., Ohkura, Y. and Tanabe, Y. (1977) Genetic variations within and between species of Asian macaques. *Japan. J. Genetics* 52: 15–30.

Nozawa, K., Shotake, T., Kawamoto, Y. and Tanabe, Y. (1982) Population genetics of Japanese monkeys: II. Blood protein polymorphisms and population structure. *Primates* 23: 252–271.

Read, A. F. and Harvey, P. H. (1986) Genetic management in zoos. *Nature* 322: 408–410.

Schaller, G. B. (1963) *The mountain gorilla. Ecology and behavior.* The University of Chicago Press, Chicago.

Schmitt, J., Ritter, H., Schmidt, C. and Witt, R. (1981) Genetic markers in primates: pedigree patterns of a breeding group of barbary macaques (*Macaca sylvana* Linnaeus, 1758). *Folia Primatol.* 36: 191–200.

Shotake, T. (1981) Population genetical study of natural hybridization between *Papio anubis* and *P. hamadryas. Primates* 22: 285–308.

Smith, D. G. (1982) Use of genetic markers in the colony management of nonhuman primates: a review. *Lab. Anim. Science* 32: 540–545.

Templeton, A. R. (1986) Coadaptation and outbreeding depression. *In:* Soulé, M. E. (ed.), *Conservation biology. The science of scarcity and diversity.* Sinauer, Sunderland, MA, pp. 105–116.

VandeBerg, J. L., Aivaliotis, M. J., Williams, L. E. and Abee, C. R. (1990) Biomechanical genetic markers of squirrel monkeys and their use for pedigree validation. *Biochem. Genet.* 28: 41–56.

van Schaik, C. P. and Kappeler, P. M. (1994) Life history, activity period and lemur social system. *In:* Kappeler, P. M. and Ganzhorn, J. U. (eds), *Lemur social systems and their ecological basis.* Plenum Press, New York (in press).

Wilson, E. O. (1975) *Sociobiology: The new synthesis.* Harvard University Press, Cambridge, MA.

Witt, R., Schmidt, C. and Schmitt, J. (1981) Social rank and Darwinian fitness in a multimale group of barbary macaques (*Macaca sylvana* Linnaeus, 1758). *Folia Primatol.* 36: 201–211.

Conservation Genetics
ed. by V. Loeschcke, J. Tomiuk & S. K. Jain
© 1994 Birkhäuser Verlag Basel/Switzerland

B: Heavy metal tolerance, plant evolution and restoration ecology

S. K. Jain

Department of Agronomy and Range Sciences, University of California, Davis, CA 95616, USA

Two major *in situ* conservation activities are: Setting aside nature reserves (various kinds of protected areas representative of remnant natural communities), and restoration/reconstruction of functioning plant and animal communities on degraded or polluted/distributed lands. An essential by-product of industrial civilization, as noted by A. D. Bradshaw (a pioneer in restoration ecology with a strong emphasis on evolutionary ecology), is the toxic metal waste left on mined lands. Bradshaw and his colleagues have studied genetic variation in metal tolerance in many plant species and have applied this knowledge in the successful choices of species and genotypes for revegetating derelict areas.

Several important phases of this work are outlined here. First, as reviewed in a book by Bradshaw and Chadwick (1980), by using laboratory assay procedures, basic physiological adaptations and limitations are determined for many plant species sampled from the mined and adjacent areas. Some species show very specific and narrow adaptive range; others might even have broad- and cross-tolerance involving several metal elements (Schat and Ten Bookem, 1992). Second, one also finds clinal variation near the mine boundaries which provides clues to the selective and gene flow patterns (Bradshaw, 1984). Third, experimental assays provide useful comparative information on the relative amounts of genetic variation for metal tolerance (Tab. 1). Bradshaw (1992) discussed some likely patterns in the past evolution and the genetic constraints for future adaptive changes such that only certain species or populations might be more likely to succeed in restoration programs. And finally, one must soon recognize the importance of understanding certain key ecosystem features – in this case, the potential need for legumes and certain nurse species as sources of mineral nutrition and shade, the need to provide for good soil conditions, and the factors for speeding up the successional stages. Evolutionary genetics of metal tolerance clearly helped here in conjunction with the other disciplines of biology, agronomy, and soil-water-mineral relations.

Table 1. Percentage of copper-tolerant individuals found in normal populations of various grass species, in relation to the presence of the species on copper-polluted waste and whether the plants collected were tolerant of copper.[a] (Bradshaw 1984)

Species	Percentage occurrence of tolerant individuals	Presence of species on mines		Tolerance of collected adult plants
		On waste	Margins	
Holcus lanatus	0.16	+	+	+
Agrostis tenuis	0.13	+	+	+
Festuca ovina	0.07	−	+	−
Dactylis glomerata	0.05	+	+	+
Deschampsia flexuosa	0.03	+	+	+
Anthoxanthum odaratum	0.02	−	+	−
Festuca rubra	0.01	+	+	+
Lolium perenne	0.005	−	+	−
Poa pratensis	0.0	−	+	−
Poa trivialis	0.0	−	+	−
Phleum pratense	0.0	−	+	+
Cynosurus cristatus	0.0	−	+	−
Alopecurus pratensis	0.0	−	+	−
Bromus sp.	0.0	−	+	−
Arrhenatherum elatius	0.0	−	+	−

[a]Unpublished data of C. Ingram.

Above all, this work over the past three decades has elegantly supported the dictum of adaptive learning in applied biology in which each experiment is also a learning step toward better applications (Slobodkin, 1988).

In a recent volume (Kareiva et al., 1993), several contributors refer to the potential evolutionary responses of plants and animals under increased heat stress, pollution, shifting latitudinal species range, changing biotic interactions, etc. Genetic variation in stress tolerance (Bradshaw and Hardwick, 1989) and the desired rates of evolutionary changes are basic to any predictions here. Referring to metal tolerance and its complex genetic basis, Geber and Dawson (1993) note that air pollution tolerance evolves rapidly (few traits, major genes), whereas responses to climatic shifts may involve many traits but ubiquitous quantitative genetic responses might be expected here, too. How rapid are these changes is not clear as different authors seem to be talking about different time scales, e.g., a few years or decades, to centuries. Another major concern in these writings is the role of marginal populations which may or may not have much genetic variation; however, migration and hybridization as well as stress-induced novel variation would often furnish it. And in many cases, finally, several authors claim that phenotypic plasticity would buy time as new genetic variation becomes available. Many of these are highly conjectural suggestions (much of this requires rigorous research and synthesis), and some even appear a bit farfetched. In the past, climate has acted as a powerful selective agent to give rise to wind-pollination and polyploidy in plants; to cite this from a geological scale for current climatic changes is a leap of faith. Ecologists have begun to take interest in genome evolution and the correlations between genome size and certain ecological traits (longevity, cold tolerance), but extrapolation from this to predict optimistic plant survival at large is premature. Many such deductions must be tempered with a mix of the restoration ecology work (including the highly recommended transplant experiments) as illustrated above.

Templeton (1991) has produced a success story for the re-establishment of collared lizards in the Ozark glades. Many forestry projects routinely select appropriate stocks for reintroduction projects. Both laboratory-based preselection and selection under natural environments play an important role. Buckeley (1989) cited numerous examples of ecosystem restoration in the United Kingdom using selected species mixes along with some attention paid to the genetic choices. However, it is evident from surveying hundreds of projects in progress (e.g., Berger, 1990; National Research Council, 1992; reports in the journal *Restoration and Management Notes*, and abstracts in *Biological Conservation*) that most of them only casually refer to genetics or adaptation and simply require decisions in the choices of species such that population genetics or evolution are entirely out of the picture.

Several reviews have reported on a tally of successes or failures in reintroductions of propagated or artificially bred species into natural areas. Seldom do we learn from them anything about the genetic variation or lack of it. We may ask if these fail due to a lack of certain basic colonizing features. Perhaps we should consider the knowledge on colonizing species, and not treat them all as undesirable exotic invaders. Conservation of many plant communities in remnant habitat fragments could provide useful information. For example, certain readjustments in the pollination and dispersal requirements, at least initially, might make a vital difference in a restoration project. Questions are frequently raised about the metapopulation dynamics of populations at the margin, adapting to stress and loss of variation: Is gene flow capable of retaining some cohesiveness? How far do genes move? Is this likely to evolve also under disturbance and habitat losses? Population genetics offers powerful methodology to deal with these questions and is indeed attracting many ecologists. Genetic basis and evolution of stress tolerance within short periods of time is a complicated issue; laboratory experiments might only weakly simulate the natural world. Relative importance of different kinds of mutations in different parts of the entire genetic machinery is also a topic deserving careful scrutiny before making too many optimistic predictions (see Berry et al. 1993, for several useful reviews). Restoration projects offer many opportunities for population geneticists to get involved.

References

Berger, J. J. (1990) *Environmental restoration*. Island Press, Washington.

Berry, R. J., Crawford, T. J. and Hewitt, G. M. (1993) *Genes in ecology*. Blackwell, Oxford.

Bradshaw, A. D. (1984) The importance of evolutionary ideas in ecology and vice versa. *In:* Shorrocks, B. (ed.) *Evolutionary ecology*. Blackwell, Oxford, pp. 1–25.

Bradshaw, A. D. (1992) The biology of land restoration. *In:* Jain, S. K. and Botsford, L. W. (eds), *Applied population biology*. Kluwer, Dordrecht, pp. 25–44.

Bradshaw, A. D. and Chadwick, M. J. (1980) *The restoration of land*. Blackwell, Oxford.

Bradshaw, A. D. and Hardwick, K. (1989) Evolution and stress-genotypic components. *Biol. J. Linn. Soc.* 37: 137–155.

Buckeley, G. P. (1989) *Biological habitat reconstruction*. Bellhaven Press, London.

Geber, M. A. and Dawson, T. E. (1993) Evolutionary response of plants to global change. *In:* Kareiva, P. M., Kingsolver, J. G. and Huey, R. B. (eds), *Biotic interactions and global change*. Sinauer, Sunderland.

Kareiva, P. M., Kingsolver, J. G. and Huey, R. B. (1993) *Biotic interactions and global change*. Sinauer, Sunderland.

National Research Council, U.S. (1992) *Restoration of aquatic systems*. National Academy Press, Washington, D.C.

Schat, H. and Ten Bookem, W. M. (1992) Metal-specificity of metal tolerance syndromes in higher plants. *In:* Baker, A. J., Proctor, J. and Reeves, R. D. (eds), *The vegetation of ultrafamic (Serpentine) soils*. Intercept, Andover, pp. 337–352.

Slobodkin, L. B. (1988) Intellectural problems of applied ecology. *BioScience* 38: 337–342.

Templeton, A. R. (1991) Genetics and conservation biology. *In:* Seitz, A. and Loeschcke, V. (eds), *Species conservation: A population-biological approach*. Birkhäuser, Basel, pp. 15–30.

Conservation Genetics
ed. by V. Loeschcke, J. Tomiuk & S. K. Jain
© 1994 Birkhäuser Verlag Basel/Switzerland

C: Genetic conservation and plant agriculture

S. K. Jain

Department of Agronomy and Range Sciences, University of California, Davis, CA 95616, USA

The *ex situ* programs for crop plants are well-known in terms of gene banks and international institutes which have thousands of accessions and offer very useful genetic resources for breeders. Population geneticists have contributed in many ways, including: (1) sampling theory for designing germplasm collecting strategies, (2) analyzing genetic variation and evolution in landraces, (3) evidence for specialized adaptations in some cases (e.g., host-pathogen coevolution), and (4) an overview of the diversity and systematics of crop-related wild and weedy flora. Several polemic issues and new challenges have come forth. Since wild and weedy relatives have been underrepresented in most collections, as are numerous endemic or semidomesticated species, new collecting priorities aim to emphasize these resources. We might try to learn more about the ecogeographical patterns of variation, beyond simply mapping the origins or collecting routes, and plant breeders might try wider use of them, too. Many current changes in agriculture with lower inputs and greater sustainability demand such biological knowledge. Breeding for disease and pest resistance as well as integrated control projects need more population genetic insights into the nature of coevolutionary systems. The ideas about core collection may not be so novel (see Harlan, 1992), practical, or even necessary in the light of new needs, so much that is unknown, and every so many serendipitous findings of useful germplasm materials. Rice is a well-studied crop for these issues and would serve well as an example.

Knowledge of rice genus (*Oryza*) has given us some important evolutionary insights (Swaminathan, 1984). For example, we continue to discover the role of incipient speciation mechanisms as well as genome divergence in giving rise to many species and races. Genetics of these reproductive barriers and rice breeders' attempts to circumvent them have required numerous interpopulation variation studies. Comparison of annual and perennial weedy/wild accessions also provided a r-K dichotomy of life histories. The continuing search for genetic sources of resistance to diseases and pests has shown that their frequencies vary worldwide such that some gene sources were predictable from the

Table 1. Main features of morphological groups defined in Asia (Ahmadi et al., 1991)

Character	Temperate japonica		Tropical japonica		Indica		Atypical
	Asia	Madagascar (G2)	Asia	Madagascar (G4)	Asia	Madagascar (G5A/G5B)	Madagascar (G6)
Length of 1st leaf under flag leaf (cm)	Short	30.5 (24.6–36.2)	Long	46.3 (39.4–51.9)	Long	32.8 (23.0–40.1)	39.1 (29.8–57.7)
Width of 1st leaf under flag leaf (cm)	Narrow	10.3 (8.7–12.3)	Wide	16.0 (10.7–18.1)	Narrow	10.6 (8.0–13.8)	12.4 (9.7–17.3)
Tillering	Intermediate	14.8 (9.5–18.1)	Low	8.7 (5.5–16.6)	High	14.6 (10.9–23.2)	11.3 (8.2–19.7)
Plant height (cm)	Short	85 (74–101)	Tall	120 (97–137)	Tall	111 (73–135)	122 (104–145)
Culm diameter (mm)	Intermediate	4.3 (3.6–4.8)	Thick	5.3 (3.8–6.5)	Intermediate	4.3 (3.3–5.6)	5.0 (3.3–6.0)
Panicle length (cm)	Short	16.8 (15.2–19.5)	Long	24.4 (18.8–29.3)	Intermediate	20.7 (17.2–23.2)	22.1 (17.7–26.6)
Shattering (%)	Little	10.6 (4.7–24.6)	Intermediate	7.9 (1.7–17.9)	Much	17.4 (4.5–32.8)	12.5 (5.1–32.0)
Panicle secondary branches	Few	25.5 (22.1–30.5)	Many	44.0 (28.0–66.4)	Intermediate	31.5 (21.5–39.5)	35.9 (23.3–50.4)
Grain length (L, mm)	Short	8.5 (7.9–9.3)	Intermediate	9.4 (7.1–11.3)	Long	10.0 (7.8–11.8)	9.6 (7.6–12.1)
Grain width (W, mm)	Wide	3.5 (2.8–4.1)	Wide	3.6 (3.1–4.2)	Thin	2.9 (2.5–3.4)	3.3 (3.0–3.8)
Grain shape (L/W)	Bold	2.4 (2.1–2.7)	Big	2.6 (2.1–3.6)	Fine	3.4 (2.7–4.2)	2.8 (2.7–3.8)
100-grain weight (g)	Heavy	2.7 (2.0–4.9)	Heavy	3.0 (2.2–4.4)	Light	2.7 (2.2–3.5)	3.0 (2.0–3.8)
Phenol reaction	Negative	–	Negative	–	Positive	+	+/–

Varietal Group

[a]Figures in parentheses are ranges.

host-pest coevolution, but some were rare and unrelated to the pest or pathogen species or racial distributions (Chang, 1989). Therefore, rather large collections are routinely scored for such rare resources. New taxa have also been recently described (*Oryza punctata*); one taxon is known to have become extinct (*O. perennis* var. *formosana* in Taiwan; Kiang et al., 1979); accordingly, several *in situ* reserves in Nepal and Sri Lanka are in development stages (Vaughan and Chang, 1992). One often finds, as expected, cold tolerant landraces from high latitudes or high altitudes, but variety Silewah from Sumatra at 1200 m represents a surprise. Much of the sampling in the past has not necessarily followed any strategic plan as, in reality, many collections are made for various goals, programs, and institutions/researchers.

Thus, in our scenario, a genetic conservation program in rice continues to grow in size and utility, as well as scientific endeavor. I will discuss three recent developments in order to make this point. Ahmadi et al. (1991) recently studied rice accessions from Madagascar along with the well-known *indica* and *japonica* races. Finding morphological and allozyme results more or less concordant, they argued for the Madagascar populations to have inter-racial hybrid origin such that some coadapted gene associations of the parental races are still retained. Novel variation in Madagascar would have potential value in the breeding programs. Tab. 1 shows some key traits used in this study.

Table 2. Intrapopulational variation found in three wild and two cultivated rice populations in response to infection by **BB** pathogen race T7133 (Hamamatsu et al., 1991)

Population[a]	Plants (no.) showing given lesion length (cm) on leaves												Mean (cm)	Standard deviation (cm)	H[b]
	0	2	4	6	8	10	12	14	16	18	20	22			
Wild rice															
(Thailand)	7	3	4	1	4	6	2		1				5.5	4.01	0.193
NE4	8	16	5										2.0	1.17	0.360
CP20	7	19	4	6	5	2	1			1			4.1	3.57	0.390
NE88															
Land race															
(China)															
Ch 54					1			4	6	4	11	10	19.8	4.00	0.074
Ch 55			1	2	1	9	11	8	3				11.9	3.40	0.095
Control				(5	2)[c]		2	3	2	1			14.4	1.78	
Norin 8													0.8	0.62	
Java 4	6	3											1.8	1.89	
WA3	3	4	1												

[a]NE4 = annual, CP20 and NE88 = perennial, Ch 54 = lowland, Ch 55 = upland. [b]H = average genetic diversity estimated from isozyme variation. Cited from Barbier (1989) for wild rices and from Morishima (1989) for land races. [c]Tested in the experimental field; others were in a greenhouse.

Table 3. Relative contribution of resistance genes of rice accessions to antibiosis against BPH (Bharathi and Chillah, 1991)

Accession	Gene conferring resistance	Nymphal survival (%)	Total nymphal duration (d)	Eggs laid (no.)	Egg Hatch-ability (%)	Brachyp-terous adults developed (%)	Female insects developed (%)	Longevity (d)		Volume of honeydew excreted (μg)	Population buildup (no.)	Growth index	Resistance index
								Female	Male				
ARC6650	*Bph-3*	32.5 b	19.5 f	126.5 b	52.0 a	45.0 ab	42.5 a	7.0 a	6.4 ab	6.0 bc	48.3 cd	1.66 b	2580
PTB33	*Bph-3*	20.0 ab	23.8 a	84.0 b	55.0 ab	37.5 a	37.5 a	5.9 a	4.9 a	2.3 a	26.0 abc	0.84 ab	31339[a]
ARC10550	*bph-5*	15.0 a	23.2 b	86.0 a	51.0 a	40.0 a	40.0 a	5.6 a	5.7 ab	2.5 a	23.0 a	0.65 a	58767
ASD11	New dominant gene	32.5 b	21.8 cd	†17.0 b	66.0 bc	42.5 ab	45.0 a	6.8 a	4.5 a	4.3 ab	39.0 abc	1.50 ab	3729
Manoharsali	New dominant gene	30.0 b	20.2 f	124.0 b	79.5 ab	47.5 ab	52.5 a	7.1 a	6.7 ab	6.0 ab	82.8 de	1.49 ab	1483
T7	New dominant gene	30.0 b	19.3 f	134.0 b	55.0 ab	52.5 ab	47.5 a	6.8 a	6.3 ab	5.0 ab	66.3 de	1.54 ab	1915
V.P. Samba	New dominant gene	22.5 ab	22.4 bc	121.0 ab	68.0 ab	37.5 bcd	45.0 a	6.4 a	6.3 ab	4.8 ab	28.5 ab	0.99 ab	11487
V. Cheera	New dominant gene	32.5 b	20.9 de	124.0 b	67.0 bcd	47.5 ab	47.5 a	6.5 a	6.1 ab	5.3 ab	40.8 bcd	1.55 ab	3223

Column group header spanning the parameter columns: Level of antibiosis in each biological parameter[a]

[a]Exhibited by 60-d-old plants of the resistant accessions. In a column, means followed by the same letter are not significantly different at the 5% level by Duncan's Multiple Range Test.

In dealing with resistance to bacterial blight, Hamamatsu et al. (1991) described variation within and between populations of wild and cultivated rice. Tab. 2 gives a summary for just a single race pathogen out of five (annual populations have more between-lines variation and perennials have more within-line heterozygosis) for disease resistance scores. Taking all of their data, one finds evidence for both race-specific (oligogenic) and race-nonspecific (polygenic?) resistance. Likewise, for the rapidly coevolving pest, brown planthopper, Bharathi and Chelliah (1991) found several major genes for resistance which show contributions through numerous insect life history components (Tab. 3). In addition, they too find evidence for significant residual variation to be controlled by minor genes. Plant breeders have recognized this for a long time, but only in a few instances have they begun to pay attention to such population genetic findings.

Rice cultivation in many tropical countries has also required selection for acid-soil tolerant varieties. Several promising strains are available from Africa, Asia, and Brazil. Rao et al. (1993) reviewed the large amount of interdisciplinary research on this topic. Yield trials using N, P, K, Ca, and Mg for added minerals in various proportions have shown, on average, that Al-tolerant lines can yield up to 92% of the non-problem soil treatment, but Al-sensitive lines yield only about 48%. Also, one needs to develop certain tolerant strains of forage grasses and legumes to provide rotations in cropping schemes. Molecular biology and basic physiological researchers are in progress, but already conventional crossing, selection, and testing have resulted in notable advances. Conservation of numerous forage species collections and their evaluation were also valuable.

We present this brief sketch of rice genetic conservation with the expectations of greater interest in such studies by conservation biologists at large. In making economic arguments for why-to-conserve, crop and livestock or medicinal sources are cited, but in a global scheme of nature protection (e.g., WCMC, 1992). Frequently, a call for *in situ* reserves designed for crop genes is made, and yet population genetics, evolution or agricultural sciences are largely underemphasized, whereas species diversity or regional/global ecosystem issues dominate most such global biodiversity agendas. A better integration of genetic resource conservation issues with the nature protection projects will be greatly helped by the use of population genetic and evolutionary knowledge of agricultural species.

References

Ahmadi, N., Glaszman, J. C. and Rabary, E. (1991) Traditional highland races originating from intersubspecific recombination in Madagascar. *In: Rice genetics II*, Intern. Rice Research Inst., Manila, pp. 67–79.

Bharathi, M. and Chelliah, S. (1991) Genetics of rice resistance to brown planthopper *Nilaparvata lugens* (Stal) and relative contribution to resistance mechanisms. *In: Rice genetics II*, Intern. Rice Research Inst., Manila, pp. 255–261.

Chang, T. T. (1989) The management of rice genetic resources. *Genome* 31: 825–831.

Hamamatsu, C., Sato, Y-I. and Morishima, H. (1991) Variation in resistance to bacterial blight between and within rice populations. *In: Rice genetics II*, Intern. Rice Research Inst., Manila, pp. 744–745.

Harlan, J. R. (1992) *Crops and man*. Second Edition. American Soc. Agron., Madison.

Kiang, Y. T., Antonovics, J. and Su, L. (1979) The extinction of wild rice (*Oryza perennis* var. *formosana*) in Taiwan. *J. Asian Ecol.* 1: 1–19.

Rao, I. M., Zeigler, R. S., Vera, R. and Sarkarung, S. (1993) Selection and breeding for acid-soil tolerance in crops. *BioScience* 43: 454–458.

Swaminathan, M. S. (1984) Rice. *Sci. Amer.* 250: 81–94.

Vaughan, D. A. and Chang, T. T. (1992) *In situ* conservation of rice genetic resources. *Econ. Bot.* 46: 368–383.

World Conservation Monitoring Centre (1992) *Global biodiversity: status of the earth's living resources*. Chapman and Hall, London.

Conservation Genetics
ed. by V. Loeschcke, J. Tomiuk & S. K. Jain
© 1994 Birkhäuser Verlag Basel/Switzerland

D: Fragmented plant populations and their lost interactions

J. M. Olesen[1] and S. K. Jain[2]

[1]*Department of Ecology and Genetics, University of Aarhus, Ny Munkegade, Building 540, DK-8000 Aarhus C, Denmark*
[2]*Department of Agronomy and Range Sciences, University of California, Davis, CA 95616, USA*

There may be hundreds of species of orchids hanging on to fencerows as living dead, yet they do not form a viable population when restoration occurs because the pollinators disappeared with the forest. Isolated trees left standing in tropical pastures often have rotting piles of fruit beneath them, mute testimony to the extinct vertebrates that will not reappear even if the pasture trees are giving back terrain on which to grow (Janzen and Martin, 1982).

Plants interact with coteries of other organisms for different services. They interact with other plants, with animals, fungi, and microorganisms. Gene dispersers (i.e., pollinators, and fruit and seed dispersers) comprise an important group which must also be viewed in terms of numerous facets of reproductive ecology (i.e., seed set, mate choice, species persistence). These organisms vary in their dispersal ability and patterns of life history variation due to the attributes of both the plant and the animals, and due to the impact of human-induced habitat disturbances.

Plants and pollinators: A life and death issue

Mutualistic plant-animal interactions are of special interest in conservation biology. Numerous authors have predicted a "cascade of linked extinctions" due to losses of habitats and species in the species-rich tropical communities (Terborgh, 1986; Cox et al., 1991; Whitmore, 1991). Clearly, one expects to find here ample opportunities to study genetic and demographic components of minimum viable population size (MVP). Failure to set seed due to a lack of pollinators, for example, would trigger changes in genetic structure, mating system, and selection forces.

Examples of specialized plant-pollinator coevolution are not uncommon in the tropical forest literature. However, very little seems to be known specifically about the *population dynamics of the species involved under changing (deteriorating) conditions*. However, recent studies show that many plant species in disturbed habitats and with reduced population sizes show lowered seed set. Hand-pollination in some cases gave higher seed set which means pollinator activity was limited; in others, the plant itself may be poorer in reproductive performance (lower ovule number, poorer seed quality), or even the habitat might have deteriorated. It is not at all clear how much of this loss of reproductive rate is due to inbreeding depression (does a feedback drop between higher inbreeding and declining population size constitute an extinction vortex?) (see also Gabriel and Bürger, this volume). What sort of experiment would make possible to analyze such interactive (feedback) system? Are there examples of plant extinctions due to declining pollinator availability? Probably yes.

We briefly review plant-pollinator/disperser interactions in terms of plant population size (patch size) and interpatch isolation. Tab. 1 lists a series of steps toward the commonly postulated scenario of coordinated extinctions of mutualists; this brief review shows empirical support for some of the steps, but in no case has the entire extinction process been studied. Pollinator-limited seed set reduction is often the primary argument in developing any extinction scenario (see the excellent concise review by Ratchke and Jules, 1993; several chapters in Real, 1983, and Bock and Linhart, 1989). On the other hand, several recent books on plant-animal interactions have not dealt with these issues (Abrahamson, 1989; Price et al., 1991).

Table 1. Fragmentation/insularization processes and changed biotic interactions (invasions, etc.)

Smaller populations of both plant and animal species – Many coordinated extinctions of mutualists are likely due to the following changes:

1. N_e/N ratios even more skewed → genetic drift & loss of genetic variation.
2. Increased inbreeding → loss of heterozygosity. Inbreeding depression → lower seed output and poorer offspring survival.
3. Greater isolation of remnants (patches) → reduced gene flow and loss of genetic variation.
4. Pollinators less abundant → mutualists suffer in reproduction and persistence. Reduced seed output (pollinator-limited) and reduced gene flow.
5. Seed dispersers less abundant → mutualists suffer in dispersal-dependent colonization and again, due to reduced gene flow.
6. Thus, mutualistic interactions affect plants and animals together, in synergism with their genetic and demographic causes of extinction (PVA, lower recolonization, etc.). Interdependence and degree of specialization may vary.
7. Loss of keystone species → large number of coordinated extinctions (how common are these events?)
8. Some animals may switch to new host plants, or plants switch to new pollinators or adjust in part from animal-mediated allogamy to autogamy/anemophily.
9. Evolution to adjust toward increased inbreeding.
10. Also, nongenetic responses during some stress period to substitute for the loss of genetic adaptability.
11. Recovery of community structure with some species replacements (and even losses).

How sensitive are pollination interactions to reduced patch size?

In general, pollination interactions and seed set are expected to react negatively on a reduced plant group size until a threshold level is reached, beyond which interactions disappear and seed set is only governed by the degree of selfing (the so-called Allee effect). In general, all *Banksias* are dependent on pollinators (honey-eaters and honey possums) for seed set. Monitoring all known populations of the rare Western Australian *Banksia goodii* showed that many individual plants produced fewer seed-bearing cones due to fewer visits from pollinators in small populations than in large ones; four of the smallest populations did not produce any seed at all (Lamont and Barker, 1988; Lamont et al., 1993). *Argyroxiphium sandwicense* from Hawaii is monocarpic, self-incompatible, and has recently experienced decreasing population sizes; seed set is also very low due to reduced pollinator fauna in small populations (Huenneke, 1991). Fruit set of small populations (5–40 individuals) of *Acourtia runcinata* was hypothesized to be pollinator limited (Cabrera and Dieringer, 1992). However, different types of pollinators may react differently. Vogel and Westerkamp (1991) reported that 20 *Lysimachia vulgaris* plants studied in the Mainz Botanical Garden seem to be a sufficient number to sustain a population of the oligolectic pollinator *Macropis labiata*, a specialist bee on *Lysimachia*. Likewise, Petanidou et al. (1991) stress the importance of a threshold population size of *Gentiana pneumonanthe* and *G. cruciata* pollinated by the generalist *Bombus pascuorum*. In *G. pneumonanthe* seed set is dramatically reduced in populations with less than 30 plants. Both species are self-compatible, but the level of spontaneous selfing is relatively low. Much more work is needed on how sensitive are flower foragers with different food plant ranges to such habitat fragmentation and losses.

N_e is also influenced by the diversity of breeding systems found, for example, in the tropical rainforests (Bawa and Ashton, 1991). Bawa and Krugman (1991), in a thorough review of tropical tree genetics, concluded that we have a large variety of preliminary data on mating systems, genetic variation patterns, and population size changes. Most conservation decisions are, therefore, based on some highly generalized verbal models.

The geography of remnant habitats, in particular, for central versus marginal stands, may also influence the effect of fragmentation on plant-pollinator interactions. A review of how plants survive rather precariously near the periphery of their distributions is given by Olesen (1992). Studies that intentionally compare the pollination of plants in patches of varying size are rare, especially in relation to extinction risks. Linhart and Feinsinger (1980) report on plant-hummingbird interactions on various sized islands. Jennersten (1988) compared a "mainland" site (traditional farmland) and a fragmented site consisting of eight "habitat islands" (range: 0.1–0.5 ha, modern farmland) within an agricultural region. The islands had a lower abundance of flowering individuals of *Dianthus deltoides* and had fewer flower-visiting insects. Hand-pollination increased seed set up to 4.1 times in the fragmented site, but no differences were found at the mainland site. Fewer plant species in smaller groups on "islands" create temporal windows of resource availability for many pollinators and thus may cause a reduction in insect diversity. Jennersten et al. (1992) found another appropriate example of lowered seed set in small populations of *Senecio integrifolius*.

In a detailed study of plant group size and their increased isolation, Ouborg (1993) studied two plant species, *Salvia pratensis* and *Scabiosa columbaria*, threatened in the Netherlands. In addition, he analyzed a data set of species composition of Dutch grasslands (also see Bjilsma et al., this volume). Different members of the same pollinator guild may react differently to a change in plant group size. Sih and Baltus (1987) examined the effect of group size of *Nepeta cataria* on pollinator visitation rate (i.e., number of visits *per* flower *per* time) and pollinator limitation. *N. cataria* was pollinated by *Apis*, *Bombus*, and Halictidae. Visitation rate increased with larger group size for *Apis* and *Bombus* and decreased for Halictidae. Fruit set was lower in small groups for open pollinated flowers, but fruit set in hand-pollinated flowers was not affected by group size. Thus, pollen seemed to be the only resource that became in short supply when the size of the plant group decreased.

Plants with different flower type may demonstrate a large variation in sensitivity to pollinator limitation. *Echium vulgare* is generally visited by a guild of generalists and specialists so that the range of group sizes present may determine the outcome of pollinator visit analyses, and one might expect that each species may have its own optimum group size regarding visitation rate (Sowig, 1989). Jennersten et al. (1992) discussed that visitation rate of *Viscaria vulgaris* by long-tongued bumblebees is independent of patch size. Thus, long-tongued *Bombus* spp. experience the negative effects of fragmentation later than short-tongued ones. However, pre-dispersal seed predation by a moth (Geometridae: *Perizoma hydrata*) and a weevil (Curculionidae: *Sibinia viscariae*), and pathogen-attack by a sterilizing fungal smut (*Ustilago violacea*) is positively correlated to group size.

How does increased distance between patches affect pollination?

Most plant species have a diverse coterie of pollinators and seed dispersers 1, 2, . . . , *j*, ranked according to their decreasing ability to disperse genes. Increasing fragmentation should delimit the plant species into isolated local host plant groups with reduced between-patch genetic and ecological communication. Gene dispersal caused by species $j, j-1, \ldots$ should then stop in the mentioned order as fragmentation progresses. The first interactions to become extinct may accelerate the disappearance of later interactions.

Pollination mechanisms play an important role in determining the degree of gene flow between the different levels of plant groups. Most pollinator movements are short and they determine whether local plant groups are demographically independent, whereas the few extensive dispersal distances are important for the colonization of new patches and determine whether patches are connected on a larger space and time-scale. Between-patch gene flow has a more or less constant species-specific level (Hamrick and Loveless, 1989). Beyond a certain distance gene flow is not reduced substantially. Some interacting animals may even mediate a directional gene flow, i.e., they may fly directly over longer distances between pollen and seed donor and recipient (especially large bees, butterflies, bats, ants, fruit garden-tending birds).

Oostermeijer et al. (1992) discuss the impact of increasing metapopulation fragmentation in *Gentiana pneumonanthe* on important life-history parameters. The frequently observed alternative fixations and the size of the inbreeding coefficient indicate that there is very little gene flow between local plant groups. This may have been caused by an increased fragmentation.

How does increased distance between patches affect dispersal?

However, only few pollination studies are between-patch orientated. Part of the variation in pollinator limitation in *Nepeta cataria* is explained by the level of interpatch isolation (Sih and Baltus, 1987). Isolated patches received fewer visits from *Apis* and *Bombus* species than expected (see also, the example of rose clover given by Jain, this volume).

Increased distance between California vernal pools due to human activities reduces plant migration (Zedler and Black, 1992). Ten species of vernal pool plants were germinated from rabbit pellets found close to newly excavated artificial pools. Thus, rabbits foraging on seeds may enhance interpool seed dispersal rate or at least conserve *status quo*. In general, vernal pool plants do not seem to have any traits adapted to long-distance dispersal and the authors doubt if any of these plants are specialized to rabbit-mediated seed dispersal. In the vernal pool genus *Limnanthes* (see Jain, this volume), small remnant populations of the highly outbred taxon *L. douglasii* var. *rosea* declined rapidly due to fewer pollinator visits and highly reduced seed output. In contrast, several equally fragmented stands of *L. alba* seemed to survive through switch to partial selfing.

Geranium purpureum has probably evolved as a shade-tolerant forest herb, but is now, at a Spanish locality, found scattered under individual *Juniperus* bushes, i.e., at sites used by wild or domestic ungulates (Herrera, 1991). *G. purpureum* is able to disperse around single stands of *Juniperus* but not to cross between different *Juniperus* individuals. The seeds are probably adapted to epizoochory. However, today a few seeds are dispersed between *Juniperus* by being ingested by domestic verte-brate herbivores.

How do broken interactions affect the genetic and ecological structure of plant populations at different hierarchical levels?

All the classic genetics of populations becoming reduced in size is expected to accompany habitat fragmentation and a reduced gene flow. However, no careful experiments under natural conditions seem to have investigated if fragmentation in fact leads to genetic depletion. The total genetic variation maintained within a species can be partitioned hierar-chically in a way that is a function of the spatial structuring of the species' plant groups.

Dianthus deltoides changes from being protandrous to homogamous and, thus, towards more selfing in smaller patches. In the study by Jennersten (1988) the seeds of *D. deltoides* from the "island" site have less variability in germination time than seeds from the "mainland" site. Jennersten (1988) suggests that the "island" population may be more homogeneous genetically than the "mainland" population.

Ipomopsis aggregata is semelparous, but becomes more iteroparous when fruit set drops below 30–40%. A reduced pollinator abundance of *Ipomopsis aggregata* causes a lower fruit set and the plant then allocates surplus resources to production of rosettes and to iteroparity. Thus, life-history traits may be very flexible or plastic and may act as adaptive responses to changes in pollinator abundance. How genetically deter-mined this switching is, is unknown. Herrera (1991) has demonstrated that fruit characters of many plants are very conservative to changes in the fruiting environment. Studies of the genetic effects of fragmentation must be designed carefully in order to minimize the uncertainty caused by historical factors. Cox et al. (1991) discussed Samoa and Guam ecosystems and the role of flying foxes as "strong interactors" with many plant species. Relative reproductive success of plant species should be studied in Guam (where flying fox populations are virtually extinct) in comparison with Samoa where significant populations of flying fox are still extant.

Mating systems and dispersal patterns in tropical trees

Bawa and Ashton (1991) and Bawa and Krugman (1991) have thoroughly covered the conservation aspects of pollination and seed dispersal. They note emphatically that very little empirical information is available for making any demographic predictions, although most extinction scenarios seem quite likely (Whitmore and Slatyer, 1992). Similar observations are found in Gentry's (1990) work on four neotropical forests.

A common extinction vortex

Previously, we have addressed the consequences of smaller patch size becoming increasingly isolated. Increasing fragmentation of natural landscapes may lead to loss of gene dispersers that again causes extinction of plant species and so on. A kind of community or interaction extinction vortex is expected to take place. Such a vortex may also be started by other human disturbances, such as introduction of pests, exotics, and domestics. A few populations of *Atropa baetica* persist in southeast Spain. However, its endangered status in the region is caused by an intense herbivore pressure that precludes the dispersal of seeds (Herrera, 1986). This again may affect other pollinators and other plant species.

An example of the initiation of an extinction vortex with potentially catastrophic consequences is found in the S. African fynbos. Bond and Slingsby (1984) describe how an introduced pest, the Argentine ant (*Iridomyrmex humilis*), outcompetes native ant species. The latter disperse elaiosome-containing seeds of many native plant species. The Argentine ant does not forage for seeds and, thus, many plant species now rely on passive dispersal only. Increased fragmentation of fynbos makes it easier for the Argentine ants to invade from nearby agricultural land. The spatial distribution of plants become much more aggregated and between-patch pollen flow more difficult. Thus, the introduction of a pest may create a snowball effect throughout the whole fynbos bionome, i.e., a kind of floral or faunal collapse.

Pollinators may in their larval stage be dependent upon other plant species, sometimes even upon those from other habitats. Such cross-habitat dependency makes a system more vulnerable to human habitat destruction. In Madagascar, *Cynorkis uniflora* is among several very specialized orchids dependent upon a few *Sphingidae* spp. for its pollination (Nilsson et al., 1992). This orchid grows on mountain sides, but is dependent on nearby forest areas where the larval host plants grow. The rapid ongoing destruction of forests on Madagascar will destroy these orchid pollination systems. In such ecosystems, and in many others, the loss of a few pollinator or seed disperser species may lead to a cascade of linked extinctions. Tropical islands often have a depauperate pollinator and seed disperser fauna (Cox et al., 1991). Thus, island plants are very vulnerable to extinction due to loss of pollinators and seed dispersers. The Hawaiian *Freycinetia arborea* was pollinated by now extinct birds but saved by an introduction of a new pollinator. On the other hand, most animal-dispersed plants are dispersed by a range of animal species. So, when one animal species becomes extinct the remaining ones continue to disperse the plant. This is not necessarily the case with keystone species such as several members of *Ficus*, *Piper* and Lauraceae. Their fruits may be a limiting resource to many animals. Cox et al. (1991) review the dependency of island floras to a few key pollinators and seed dispersers, e.g., flying foxes. As stated earlier, the island of Guam has lost its flying fox populations, while Samoa has significant populations left. Several plant species on Guam were observed to have reduced fruit set and dispersal. Tropical euglossine bees forage over areas of many square kilometers (Janzen, 1974). However, destruction of tropical forests has eliminated many of the more specialized large bees. Each bee species visits many different plant species, mainly orchids. When some of the host plants disappear, the bees disappear, and then the rest of the food plant species. However, several authors have challenged sweeping assumptions about the generalist-specialist definitions (Schemske, 1984; Howe and Westley, 1986). Thus, loss of a keystone species can cause a cascade of extinctions or start an extinction vortex among both animals and other plants. A well known experiment on forest ecosystem dynamics in Amazonia is in progress (Bierregaard et al., 1992).

Even a cursory survey of recent pollination biology literature shows the growing interest in such conservation issues as pollinator specialization, seed set loss under habitat changes, or genetic causes of population failures to persist. We cite only a few studies here. Renner and Feil (1993) reviewed dioecy in tropical flora and focused on the unsettled issue of specialized pollinators. Karoly (1992) showed that pollinator availability can limit the reproductive rates even in a facultative autogamous species (*Lupinus nanus*). Two recent studies on self-incompatible perennials have shown that under unfavorable environments low genet numbers and therefore, much fewer self-incompatibility alleles could result in lower seed set or switch to vegetative reproduction (Aspinwall and Christian, 1992; Weis and Hermanutz, 1993). Overall, even this brief review of plant-animal interactions in the context of conservation biology clearly shows that (1) our intuitive or piecemeal arguments need much more empirical support, and (2) genetic and ecological consequences of habitat losses can be concurrently very well studied here (Bawa and Ashton, 1991). Only then would we expect new developments in our recommended actions for saving the ecosystems and their major biotic features.

Acknowledgements
We thank Drs. R. Thorp and P. McGuire for helpful discussions, and Marilee Schmidt for her cordial help in manuscript preparation.

References

Aspinwall, N. and Christian, N. (1992) Pollination biology, seed production, and population structure in queen-of-the-prairie *Filipendula rubra* (Rosaceae) at Botkin fen, Missouri. *Amer. J. Bot.* 79: 48–494.

Abrahamson, W. (1989) *Plant-animal interactions.* McGraw-Hill, New York.

Bawa, K. S. and Ashton, P. S. (1991) Conservation of rare trees in tropical rain forests: A genetic perspective. *In:* Falk, D. A. and Holsinger, K. E. (eds), *Genetics and conservation of rare plants.* Oxford Univ. Press, New York, pp. 62–71.

Bawa, K. S. and Krugman, S. L. (1991) Reproductive biology and genetics of tropical trees in relation to conservation and management. *In:* Gomez-Pompa, A., Whitmore, T. C. and Hadley, M. (eds), *Rainforest regeneration and management.* Parthenon, UNESCO, Paris, pp. 119–136.

Bierregaard, R. O., Lovejoy, T. E., Kapos, V., Santos, A. A. and Hutchings, R. W. (1992) The biological dynamics of rainforests. *BioScience* 42: 859–866.

Bock, J. H. and Linhart, Y. B. (1989) *The evolutionary ecology of plants.* Westview, Boulder.

Bond, W. and Slingsby, P. (1984) Collapse of an ant-plant mutalism: The Argentine ant (*Iridomyrmex humilis*) and myrmecochorous Proteaceae. *Ecology* 65: 1031–1037.

Cabrera, L. and Dieringer, G. (1992) Reproductive biology of a population of *Acourtia runcianata* (Asteraceae: Mutisieae) at the northeastern limit of its range. *Am. Midl. Nat.* 128: 83–88.

Cox, P. A., Elmqvist, T., Pierson, E. D. and Rainey, W. E. (1991) Flying foxes as strong interactors in South Pacific island ecosystems, a conservation hypothesis. *Cons. Biol.* 5: 1–7.

Gentry, A. H. (1990) *Four neotropical forests.* Yale Univ. Press, New Haven.

Hamrick, J. L. and Loveless, M. D. (1989) The genetic structure of tropical tree populations: Associations with reproductive biology. *In:* Bock, J. H. and Linhart, Y. B. (eds), *The evolutionary ecology of plants.* Westview, Boulder, pp. 129–146.

Herrera, C. M. (1986) Distribución, ecología y conservación de *Atropa baetica* Willk. (Solanaceae) en la Sierra de Cazorla. *Anales del Jardin Botanico de Madrid* 43: 387–398.

Herrera, J. (1991) Herbivory, seed dispersal and the distribution of a ruderal plant living in a natural habitat. *Oikos* 62: 209–215.

Howe, H. F. and Westley, L. C. (1986) Ecology of pollination and seed dispersal. *In:* Crawley, M. J. (ed.), *Plant ecology.* Blackwell, Oxford, pp. 185–206.

Huenneke, L. F. (1991) Ecological implications of genetic variation in plant populations. *In:* Falk, D. A. and Holsinger, K. E. (eds), *Genetics and Conservation of Rare Plants.* Oxford Univ. Press, New York, pp. 31–44.

Janzen, D. H. (1974) The deflowering of Central America. *Natural History* 83: 48–53.

Janzen, D. H. and Martin, P. S. (1982) Neotropical anachronisms: the fruits the gomphotheres ate. *Science* 215: 19–27.

Jennersten, O. (1988) Pollination in *Dianthus deltoides* (Caryophyllaceae): effects of habitat fragmentation on visitation and seed set. *Cons. Biol.* 2: 359–366.

Jennersten, O., Loman, J., Møller, A. P., Robertson, J. and Widén, B. (1992) Conservation biology in agricultural habitat islands. *In:* Hansson, L. (ed.), *Ecological principles of nature conservation.* Elsevier, London, pp. 399–424.

Karoly, K. (1992) Pollinator limitation in the facultatively autogamous annual, *Lupinus nanus* (Leguminosae). *Amer. J. Bot.* 79: 49–56.

Lamont, B. B., Klinkhamer, P. G. L. and Witkowski, E. T. F. (1993) Population fragmentation may reduce fertility to zero in *Banksia goodii* – a demonstration of the Allee effect. *Oecol.* 94: 446–450.

426

Lamont, B. B. and Barker, M. (1988) Seed bank dynamics of a serotinuos fire-sensitive *Banksia* species. *Austr. J. Bot.* 36: 193–203.

Linhart, Y. B. and Feinsinger, P. (1980) Plant-hummingbird interactions: effects of island size and degree of specialization on pollination. *J. Ecol.* 68: 745–760.

Nilsson, L. A., Rabakonandrianina, E., Razananaivo, R. and Randriamanindry, J.-J. (1992) Long pollinia on eyes: hawk-moth pollination of *Cynorkis uniflora* Lindley (Orchidaceae) in Madagascar. *Bot. J. Linn. Soc.* 109: 145–160.

Olesen, J. M. (1992) How do plants reproduce on their margin? *In:* Thanos, C. A. (ed.), *Plant-animal interactions in Mediterranean type ecosystems.* Univ. Athens, Athens, pp. 217–222.

Oostermeijer, J. G. B., den Nijs, J. C. M., Raijmann, L. E. L. and Menken, S. B. J. (1992) Population biology and management of the marsh gentian (*Gentiana pneumonanthe* L.), a rare species in The Netherlands. *Bot. J. Linn. Soc.* 108: 117–130.

Ouborg, N. J. (1993) *On the relative contribution of genetic erosion to the chance of population extinction.* Ph.D. Thesis, State Univ., Utrecht.

Petanidou, T., den Nijs, H. C. M. and Ellis-Adam, A. C. (1991) Comparative pollination ecology of two rare Dutch *Gentiana* species, in relation to population size. *Acta Hort.* 288: 308–312.

Price, P. W., Lewinsohn, T. M., Fernandes, G. W. and Benson, W. W. (1991) *Plant-animal interactions: Evolutionary ecology in tropical and temperate regions.* Wiley-Interscience, New York.

Ratchke, B. J. and Jules, E. S. (1993) Habitat fragmentation and plant-pollinator interactions. *Curr. Science* 65: 273–276.

Real, L. (1983) *Pollination biology.* Academic, London.

Renner, S. S. and Feil, J. P. (1993) Pollinators of tropical dioecious angiosperms. *Amer. J. Bot.* 80: 1100–1107.

Schemske, D. W. (1984) Limits to specialization and coevolution in plant-animal mutualisms. *In:* Nitecki, M. H. (ed.), *Coevolution.* Univ. Chicago Press, Chicago, pp. 67–110.

Sih, A. and Baltus, M.-S. (1987) Island size, pollinator behavior, and pollinator limitation in catnip. *Ecology* 68: 1679–1690.

Sowig, P. (1989) Effects of flowering plant's patch size on species composition of pollinator communities, foraging strategies, and resource partitioning in bumblebees (Hymenoptera: Apidae). *Oecol.* 78: 550–558.

Terborgh, J. (1986) Keystone plant resources in the tropical forest. *In:* Soulé, M. E. (ed.), *Conservation biology: The science of scarcity and diversity.* Sinauer, Sunderland, MA, pp. 330–344.

Vogel, S. and Westerkamp, C. (1991) Pollination: An integrating factor of biocenoses. *In:* Seitz, A. and Loeschcke, V. (eds), *Species conservation: a population biological approach.* Birkhäuser, Basel, pp. 159–170.

Weis, I. M. and Hermanutz, L. A. (1993) Pollination dynamics of arctic dwarf birch (*Betula glandulosa*; Betulaceae) and its role in the loss of seed production. *Amer. J. Bot.* 80: 1021–1027.

Whitmore, T. C. (1991) Tropical rainforest dynamics and its implications for management. *In:* Gomez-Pompa, A., Whitmore, T. C. and Hadley, M. (eds), *Rainforest regeneration and management.* Parthenon, UNESCO, Paris, pp. 67–90.

Whitmore, T. C. and Slatyer, J. A. (1992) *Tropical deforestation and species extinction.* Chapman and Hall, London.

Zedler, P. H. and Black, C. (1992) Seed dispersal by a generalized herbivore: rabbits as dispersal vectors in a semiarid California vernal pool landscape. *Amer. Midl. Nat.* 128: 1–10.

Conservation Genetics
ed. by V. Loeschcke, J. Tomiuk & S. K. Jain
© 1994 Birkhäuser Verlag Basel/Switzerland

E: Host-pathogen coevolution under *in situ* conservation

S. K. Jain

Department of Agronomy and Range Sciences, University of California, Davis, CA 95616, USA

In this final scenario, we briefly present the key argument for establishing genetic reserves: to allow the dynamic coevolutionary process for crop resistance to evolve under changing pathogen virulence. Population genetic studies play an important role in assessing variation patterns in the host *vis-à-vis* the pathogen, and in attempting to predict the dynamic properties of such agroecosystems. Genetic reserves have been proposed for the *in situ* conservation of many threatened wild or weedy relatives of crop plants, as well as certain landraces with a model of endemic balance in which useful sources of resistance genes evolve where the disease or pest is prevalent. In his classic book, Burdon (1987) provided cogent arguments for studying genetic variation and evolution in many such natural situations. Clearly, conservation genetics will be enriched in both theory and useful new applications. Also note that several researchers have suggested the role of coevolutionary forces in the evolution of sex and in the maintenance of polymorphisms (e.g., Anderson and May, 1982; Hamilton, 1980).

Genetic polymorphisms that affect disease resistance and pathogen specificity are widespread in plants. The so-called gene-for-gene model has led geneticists to define the race-specific resistance genes (vertical resistance, VR) in the host, and the matching virulence genes in the pathogen. Numerous recent studies require a modified version of this model and, in fact, finding polygenic, race-nonspecific resistance (horizontal resistance, HR), as noted in Scenario C on rice genetic resources, further complicates the genetic analyses of variation in natural populations.

Using simple models, however, several interesting population genetic features have been investigated. For example, a balancing or stabilizing selection defines how the evolution of a "super-race" (which could breakdown all resistance sources) might be kept in check in any particular population by the cost of unneeded virulence. The role of mutation and recombination in the origin of new pathogen races needs more research. McDonald et al. (1993) emphasized the role of molecular markers in such studies. Barrett (1987) and McDonald et al. (1989) have provided useful reviews of the theoretical and experimental issues.

The fine-scaled distribution of numerous virulence genes within a single field of barley with powdery mildew (pathogen *Erysiphe graminis* f. sp. *hordei*) was described with the use of RAPD markers. Use of isozyme markers similarly helped in tracking the long-distance dispersal of the cereal rust pathogen *Puccinia graminis*. In a long-term study, McDonald et al. (1989) described the barley composite crosses which are dynamically evolving populations, initially synthesized from very wide genetic base (e.g., intercrosses among many landraces), and showed continued genetic shifts avoiding an epidemic breakdown. Using cereal rusts as model organisms, Leonard (1987) studied the variable role of sexual *versus* asexual reproduction in the pathogen species along with its rapid population turnover. His experimental work also provided some estimates of selection coefficients for pathogenicity in different host backgrounds which suggested that selection in natural pathogen populations generally proceeds slower than in the breeders' nurseries.

Using gene-for-gene models and variation along this theme, equilibrium and relative stability have been explored and, in fact, breeders have been advised to consider a "dirty" crop approach in which a less than perfectly clean field (due to an imperfect matching by the VR system) is recommended. Anderson and May (1982) observed that epidemiological factors have often been left out in these studies, as also have the interesting features of frequency-dependent selection. We point out only briefly here that although maintenance of variation is a very likely outcome of such coevolutionary dynamics, stability properties are not quite certain yet.

In a model for tracking population sizes and gene frequencies, Frank (1991) found that ecological and demographic factors (e.g., birth and death rates, b and s, of the pathogen) might be even more critical than the cost of virulence, selection coefficient, or other genetic parameters. An important variable, the average host range of pathogens (δ), introduces significant asymmetry as the host resistance is more narrowly race-specific. An epidemic index ($\delta b - s$) defines the conditions for instability so that the frequency of epidemic is controlled mostly by the difference between the pathogen birth and death rates; furthermore, increasing the migration-mutation rates for pathogen would raise the epidemic risks profoundly, as also observed in the empirical studies. In general, levels of genetic variation would tend to decrease as the potential for epidemics rises. Frank (1991) recognized that such models do not include the catastrophes, or rare chance events of migration to a suitable patch for attacking a host population. Clearly, both theory and experimental work need to go much further.

What about the planning of *in situ* reserves in the meantime? Among many startups, we mention the reserve at Ammiad, Israel, established primarily for the wild wheat genetic resources. Long-term research by

Dinoor and coworkers (Dinoor and Eshed, 1987; also see Horovitz and Feldman, 1991, for a commentary on this project) has tracked the dynamics of several cereal pathogen systems. However, although several examples of coevolutionary outcomes are found in them, at the Ammiad reserve, environmental factors accounted for the host largely escaping the disease such that resistance and pathogenicity variation were not correlated. Such reserves will undoubtedly offer ample opportunities for further population genetic and evolutionary research on this issue.

References

Anderson, R. M. and May, R. M. (1982) Coevolution of hosts and parasites. *Parasitology* 85: 411–426.

Barrett, J. A. (1987) The dynamics of genes in populations. *In:* Wolfe, M. S. and Caten, C. E. (eds), *Populations of plant pathogens: their dynamics and genetics*. Blackwell, Oxford, pp. 39–53.

Burdon, J. J. (1987) *Diseases and plant population biology*. Cambridge Univ. Press, Cambridge.

Dinoor, A. and Eshed, N. (1987) The analysis of host and pathogen populations in natural ecosystems. *In:* Wolfe, M. S. and Caten, C. E. (eds), *Populations of plant pathogens: their dynamics and genetics*. Blackwell, Oxford, pp. 75–88.

Frank, S. A. (1991) Ecological and genetic models of host-pathogen coevolution. *Heredity* 67: 73–83.

Hamilton, W. D. (1980) Sex versus non-sex versus parasite. *Oikos* 35: 282–290.

Horovitz, A. and Feldman, M. (1991) Evaluation of the wild wheat study at Ammiad. *Israel J. Bot.* 40: 501–508.

Leonard, K. J. (1987) The host population as a selective factor. *In:* Wolfe, M. S. and Caten, C. E. (eds), *Population of plant pathogens: their dynamics and genetics*. Blackwell, Oxford, pp. 163–179.

McDonald, B. A. and McDermott, J. M. (1993) Population genetics of plant pathogenic fungi. *BioScience* 43: 311–319.

McDonald, B. A., McDermott, J. M., Goodwin, S. B. and Allard, R. W. (1989) The population biology of host-pathogen interactions. *Annu. Rev. Phytopath.* 27: 77–94.

Conservation Genetics
ed. by V. Loeschcke, J. Tomiuk & S. K. Jain
© 1994 Birkhäuser Verlag Basel/Switzerland

Concluding remarks

S. K. Jain[1] and J. Tomiuk[2]

[1]*Department of Agronomy and Range Sciences, University of California, Davis, CA 95616, USA*
[2]*Section of Clinical Genetics, University of Tübingen, Wilhelmstrasse 27, D-72074 Tübingen, Germany*

This book was largely aimed at defining and assessing the role of population genetics in conservation biology. The topics include the widely recognized problems of genetic variation losses under random drift, inbreeding, reduced migration rates, and increased selective pressures under stress or widely fluctuating environments, and they are mainly addressed against an ecological background. The necessity to additionally consider social behavior, metapopulation structure, asexuality, and certain population dynamic aspects for the conservation of some species is also treated briefly. The use of molecular variation in systematics and some strategic decisions about genetic variance maximization in the conserved population or community units are analysed. Finally, the case studies and scenarios illustrate the application of genetic information in conservation practices. The important role geneticists share with other biologists became clearer in our minds with the contributions to this book. The need to focus our research on genetic issues has been pointed out by several authors, e.g., the need for better designed experiments to evaluate inbreeding depression and its consequences in relation to changes in population size. Our knowledge in this field has rapidly advanced during the last decade (for an excellent treatment of inbreeding and outbreeding topics, see also Thornhill, 1993).

Based on theoretical models, breeding programs were first designed in order to minimize the increase of the level of inbreeding when the number of individals was limited. In present conservation efforts, the species-specific attributes are additionally considered to improve the long-term success of conservation. The monitoring of natural populations is necessary for discovering the long-term consequences of ecological factors on the population dynamics. Such studies and the knowledge about the genetic role of adaptive processes during the evolution of species are fundamental for any predictions on future population dynamics. Case studies of natural populations which can ascribe the extinction avoidance to the role of genetic variation are therefore necessary.

In his influential book *Viable populations for conservation*, Soulé (1987) asked whether bottlenecks lead to extinction and how well the various estimates of minimum viable population (MVP) from genetic and demographic analyses match (and if not, what numbers do we select). Most applications of population genetics in any nature conservation project (e.g., species survival plans, advising on the design and size of reserves) would have to incorporate demographic and environmental variables. Some notable exceptions to this include *ex situ* conservation and captive breeding projects in which the genetic principles of founding new populations, artificial selection, and controlled migration (hybridization) are the primary tools.

In the following paragraphs, we refer to some recent developments and certain specific research needs in conservation genetics.

Description of genetic variation with protein electrophoresis has been further improved by DNA sequencing techniques. DNA sequence data provide both general and gene-specific assays of variation; one can infer new ancestor-derivative relationships among species, genera and even higher units of organization; common descent is certainly more directly assayed with these rich details of genic and genomic identities. Population structure and gene flow patterns can now be described using a larger and ever increasing variety of genetic "markers." Obviously, the analysis of DNA sequences gives the ultimate information on genetic differences among species. Closely related species can be better differentiated at the DNA level than with techniques which use only morphological characters or protein electrophoresis. Species complexes (subspecies, races) which live sympatrically and are highly adapted to specific ecological niches must be known before reserves are created for their conservation or before breeding programs are initiated.

Aside from the statistical methodology, and independent from the applied techniques for describing genetic polymorphism, several questions arise in this area: what objective criteria exist to decide (1) which species to protect on a short-term scale because they are highly endangered, (2) which species to maintain in the long-term because they are important for the functioning of whole ecosystems or as potential future resources in, for example, agriculture or medicine, (3) if biological species and species as a unit to define genetic relationships (and distinctness) are now replaced by *genetic-distance*-based evolutionary branches (lineages), should we conserve species groups, species, or subspecies? How do we weigh various ecological, phenetic or ecosystem-oriented attributes along with the genetic variables?

fer. The conceptual and theoretical bases of many "rules of thumb" were understood well by the experts, but these required strong caveats in their application. Now, we can update our position on such guidelines by summarizing the results of a World Wildlife Fund workshop (Lacy, 1988). The summary recommendations were as follows: (1) One should not expect a single N_e or MVP value to apply to all populations or species; genetic and demographic processes must be understood. (2) Past geographic and historical information and measures of evolutionary divergence among populations, races, etc. should help us determine the demographic unit(s) for conservation action. (3) Variation analyses should incorporate distributional, morphological, molecular, ecological, and behavioral data. "No technique short of complete sequencing of the DNA of each organism (not now feasible) can reveal all genetic information variation" (Lacy, 1988), but meanwhile, one should not assume that any population is so devoid of variation that it is of little conservation value or will necessarily become extinct! (4) For genetic monitoring many markers should now be used for estimating relative fitnesses and to establish the adaptive role of genetic shifts, if any. (5) For reintroductions and new colony founders, more variable and highly heterozygous sources should be used.

It is evident that our guidelines have "evolved" considerably in these past 10 years. Moreover, long-term research along with many ongoing applications in parallel (a process of adaptive learning, so characteristic of applied fields involving decisions under uncertainty) are needed to overcome what Soulé (in Frankel and Soulé, 1981) once labelled as scientists' paralysis leading to inaction in a crisis. We hope that after the next 10 years, we shall be able to report on more refined and validated genetic guidelines derived from the actual conservation experiences.

Finally, we may glean from the conservation biology literature how our role is perceived by our colleagues. Among a very long list of recent volumes, a few are worth mentioning here. A book on remnant habitats (Saunders et al., 1987) dealt with the extinction risks for various plant and animal communities in depth, but only a few chapters casually referred to the potential importance of genetic factors. Many tropical forest conservation books have only a short treatment of mating systems and genetic variation patterns, whereas community dynamics and human impact tend to be the central topics. In dealing with the climatic change and deforestation, a recent book (Kareiva et al., 1993) has several incisive reviews of genetic variation in physiological traits and of fragmentation theory, but the role of population genetic research gets rapidly lost in the much broader conservation agenda. Western and Pearl's book (1989, *Conservation for the 21st Century*) has treated the topics of population genetics largely in relation to the *ex situ* needs (with Vrijenhoek's chapter as a notable exception). Several others essentially include a standard treatment of the old MVP rules, importance of adaptive variation, inbreeding scenarios, etc. – which almost read as clichés and are confining.

The point to be made is that we need to crystallize the essentials of population genetics and systematics to be directly used in conservation practice and to establish an increased role by collaboration in numerous, ongoing nature protection and restoration projects. The World Conservation Monitoring Centre at Cambridge (1992), U.K., has published, for example, a remarkable global biodiversity report in which the listing of nature reserves and the roles of numerous regional and national programs are detailed; MVP and the genetic management, variation assays, inbreeding levels, etc. are not included in the big picture, but the underlying need for additional population studies is implicit in most programs. Besides the agricultural species research in which genetics is clearly cited for its contributions, we should accept the challenge for becoming involved in many of the tasks of planning and implementing nature reserves.

References

Berry, R. J., Crawford, T. J. and Hewitt, G. M. (1992) *Genes in ecology.* Blackwell, London.

Bryant, E. H., Meffert, L. M. and McCommas, S. A. (1990) Fitness rebound in serially bottlenecked populations of the house fly. *Am. Nat.* 136: 542–549.

Frankel, O. H. and Soulé, M. E. (1981) *Conservation and evolution.* Cambridge University Press, Cambridge.

Goodnight, C. J. (1988) Epistasis and the effect of founder effects on the additive genetic variance. *Evolution* 42: 441–454.

Kareiva, P. M., Kingsolver, J. S. and Huey, R. B. (1993) *Biotic interactions and global change.* Sinauer, Sunderland, MA.

Lacy, R. C. (1988) A report on population genetics in conservation. *BioScience* 2: 245–247.

Saunders, D. A., Arnold, G. W., Burbidge, A. A. and Hopkins, A. J. M. (1987) *Nature conservation: the role of remnants of nature vegetation.* Surrey Beatty, CSIRO.

Seitz, A. and Loeschcke, V. (1991) *Species conservation: a population-biological approach.* Birkhäuser, Basel.

Soulé, M. E. (1987) *Viable populations for conservation.* Cambridge University Press, Cambridge.

Thornhill, N. W. (1993) *The natural history of inbreeding and outbreeding.* University of Chicago Press, Chicago.

Western, D. and Pearl, M. (1989) *Conservation for the 21st century.* Oxford University Press, Oxford.

World Conservation Monitoring Centre (1992) *Global biodiversity: states of the living earth.* Chapman and Hall, London.

Subject index

(The page number refers to the exact page in which the keyword first occurs)

The Polymerase Chain Reaction

Edited by

K.B. Mullis, La Jolla, CA, USA
F. Ferré, Immune Response Corp., Carlsbad, CA, USA
R. Gibbs, Baylor College of Medicine, Houston, TX, USA

1994. 432 pages. 125 illustrations. Hardcover. ISBN 3-7643-3607-2
Also available in softcover: ISBN 3-7643-3750-8

This is the first comprehensive handbook on polymerase chain reactions (PCR). Edited by the inventor of PCR and Nobel Prize winner in Chemistry for 1993, Kary Mullis, and two prominent experts in the field, it provides the most up to date methodological protocols from the world's leading laboratories, as well as exciting new techniques and enhanced applications not yet available in book form. Nearly forty chapters will inform the novice and experienced PCR user on how to optimize their results. Special sections include the latest on QPCR, non-isotopic detection, genetic analysis and much more!

The applications chapters are quite unique, with the foremost researchers providing not only protocols, but descriptions of how PCR has revolutionized their particular field. Future enhancements of PCR as well as new potential uses are discussed. Readers will learn how PCR has changed the face of diagnostic testing, cancer research, genetics, forensics, plant biology, DNA sequencing, gene therapy, and much more! Nearly forty chapters have been extensively reviewed and checked for accuracy and breadth of subject matter.

Special sections include the latest QPCR, non-isotopic detection, genetic analysis, and PCR and the World of Business, which includes a fascinating behind-the-scenes look at the legal battles between biotech giants Cetus and DuPont, as well as insights into the origins of PCR and the history of nucleic acid research. Every researcher working with PCR will want to own a copy of this book.

Birkhäuser Verlag • Basel • Boston • Berlin

Gene Therapeutics
Methods and Applications
of Direct Gene Transfer

Edited by
J.A. Wolff, University of Wisconsin Medical School, WI, USA

1994. 417 pages. Hardcover
ISBN 3-7643-3650-1

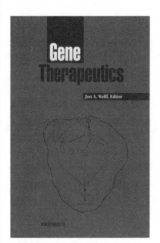

Edited by one of the premier researchers in the field, this is the first book to comprehensively cover the new methods of direct in vivo gene therapy with over 20 chapters by leading practioners of gene therapeutics. Although all areas of gene therapy are reviewed, of particular interest is its focus on direct in vivo gene therapy, which offers the promise of broad pharmaceutical effects.

The book is divided into three sections: scientific background, methods and mechanisms, and applications. Among the new methodologies for gene delivery are naked DNA, particle bombardment with „gene guns", electrotransfection, and liposomes. These methods are explored with applications for cardiovascular disease, brain tumors, immunization, arthritis, cystic fibrosis, Duchenne muscular dystrophy, and more.

Gene Therapeutics is essential reading for all pharmaceutical and biotechnology researchers, as well as clinicians interested in these exciting new technologies for the treatment of human disease.

Birkhäuser Verlag • Basel • Boston • Berlin

Molecular Ecology and Evolution: Approaches and Applications

Edited by

B. Schierwater, B. Streit, Zool. Institut d. J.W. Goethe-Univ., Frankfurt, Germany
G.P. Wagner, Dept. of Biology, Yale Univ., New Haven, USA
R. DeSalle, American Museum of Natural History, New York, NY, USA

1994. Approx. 400 pages. Hardcover. ISBN 3-7643-2942-4 (EXS 69)
Also available in Softcover: ISBN 3-7643-2943-2

The past 25 years have witnessed a revolution in the way ecologists and evolutionary biologists approach their disciplines. Modern molecular techniques are now reshaping the spectrum of questions that can be addressed while studying the mechanisms and consequences of the ecology and evolution of living organisms.

Molecular Ecology and Evolution: Approaches and Applications describes, from a molecular perspective, several methodological and technical approaches used in the fields of ecology, evolution, population biology, molecular systematics, conservation, and development. Modern techniques are introduced, and older, more classic ones refined. The advantages, limitations, and potentials of each are discussed in detail, and thereby illustrate the widening range of cross-field research and applications which this modern technology is stimulating.

The book will serve as an important textbook for graduate and advanced undergraduate students, and as a key reference work for researchers.

Birkhäuser Verlag • Basel • Boston • Berlin